MECHANISMS OF POLYMER
DEGRADATION AND STABILISATION

MECHANISMS OF POLYMER DEGRADATION AND STABILISATION

Edited by

GERALD SCOTT

Professor of Polymer Science,
Aston University,
Birmingham, UK

ELSEVIER APPLIED SCIENCE
LONDON and NEW YORK

ELSEVIER SCIENCE PUBLISHERS LTD
Crown House, Linton Road, Barking, Essex IG11 8JU, England

Sole Distributor in the USA and Canada
ELSEVIER SCIENCE PUBLISHING CO., INC.
655 Avenue of the Americas, New York, NY 10010, USA

WITH 61 TABLES AND 83 ILLUSTRATIONS

© 1990 ELSEVIER SCIENCE PUBLISHERS LTD

British Library Cataloguing in Publication Data

Mechanisms of polymer degradation and stabilisation.
1. Polymers. Degradation & Stabilisation
I. Scott, Gerald, 1927–
668.9

ISBN 1-85166-505-6

Library of Congress Cataloging-in-Publication Data

Mechanisms of polymer degradation and stabilisation/edited by Gerald Scott.
  p. cm.
Includes bibliographical references.
ISBN 1-85166-505-6
1. Polymers—Deterioration. 2. Antioxidants. I. Scott, Gerald, 1927–
QD381.9.D47M43 1990
620.1'92'0422—dc20   90-3226
                                                        CIP

No responsibility is assumed by the Publisher for any injury and/or damage to persons or property as a matter of products liability, negligence or otherwise, or from any use or operation of any methods, products, instructions or ideas contained in the material herein.

Special regulations for readers in the USA

This publication has been registered with the Copyright Clearance Centre Inc. (CCC), Salem, Massachusetts. Information can be obtained from the CCC about conditions under which photocopies of parts of this publication may be made in the USA. All other copyright questions, including photocopying outside of the USA, should be referred to the publisher.

All rights reserved. No part of this publication may be reproduced, stored in a retrieval system, or transmitted in any form or by any means, electronic, mechanical, photocopying, recording, or otherwise, without the prior written permission of the publisher.

Printed by The Universities Press (Belfast) Ltd.

Preface

The purpose of this publication is two-fold. In the first place it is intended to review progress in the development of practical stabilising systems for a wide range of polymers and applications. A complementary and ultimately more important objective is to accommodate these practical developments within the framework of antioxidant theory, since there can be little question that further major advances in the practice of stabilisation technology will only be possible on a firm mechanistic foundation.

With the continual increase in the number of commercial antioxidants and stabilisers, often functioning by mechanisms not even considered ten years ago, there is a need for a general theory which will allow the potential user to predict the performance of a particular antioxidant structure under specific practical conditions. Any such predictive tool must involve a simplified kinetic approach to inhibited oxidation and, in Chapter 1, Denisov outlines a possible mechanistic approach with the potential to predict the most useful antioxidant to use and the limits of its usefulness.

In Chapter 2, Schwetlick reviews the current state of knowledge on the antioxidant mechanisms of the phosphite esters with particular emphasis on their catalytic peroxidolytic activity. Dithiophosphate derivatives show a similar behaviour but for quite different reasons and, in Chapter 3, Al-Malaika reviews information available from analytical studies, particularly using ^{31}P-NMR spectroscopy, to elucidate the complex chemistry that leads to the formation of the antioxidant-active agents.

With the increasing use of high-energy irradiation to sterilise polymers for use in hygienic applications, it is increasingly necessary to find effective radioprotective antioxidants for such polymers as polypropylene. In Chapter 4, Carlsson and Chmela review the nature of radiation damage in polypropylene and report recent advances in the development of protective agents under these conditions.

The stabilisation of polyphenylene oxide (PPO) and its blends has proved to be an unexpectedly difficult problem under conditions of UV irradiation, due to the fact that the primary mechanism appears to involve photolysis of PPO rather than photo-oxidation. Recent progress in understanding this process is reviewed by Pickett, who also outlines the most effective approaches to its stabilisation.

In Chapter 6, Gugumus challenges the generally accepted view that hydroperoxide photolysis is the primary cause of the initiation of photo-oxidation in polyethylene. The molecular photo-elimination mechanisms proposed satisfy the chemistry, but it remains to be seen whether these processes and the more traditional free-radical mechanisms are mutually exclusive.

Chromatographic techniques are widely used in the identification of antioxidants and stabilisers in polymers and, in Chapter 7, Munteanu reviews in considerable detail the application and utility of the various high-performance chromatographic procedures now available.

Gerald Scott

Contents

Preface   v

List of Contributors   ix

1. A Theoretical Approach to the Optimisation of Antioxidant Action   1
   E. T. DENISOV

2. Mechanisms of Antioxidant Action of Phosphite and Phosphonite Esters   23
   K. SCHWETLICK

3. Antioxidant Mechanisms of Derivatives of Dithiophosphoric Acid   61
   S. AL-MALAIKA

4. Polymers and High-Energy Irradiation: Degradation and Stabilization   109
   D. J. CARLSSON & S. CHMELA

5. Photodegradation and Stabilization of PPO® Resin Blends . .   135
   J. E. PICKETT

6. Photo-oxidation and Stabilization of Polyethylene 169
   F. GUGUMUS

7. Analysis of Antioxidants and Light Stabilisers in Polymers by
   Modern Liquid Chromatography 211
   D. MUNTEANU

Index 315

List of Contributors

S. AL-MALAIKA
Department of Chemical Engineering and Applied Chemistry, Aston University, Aston Triangle, Birmingham B4 7ET, UK

D. J. CARLSSON
Division of Chemistry, National Research Council of Canada, Ottawa, Canada K1A OR9

S. CHMELA
Polymer Institute, Centre for Chemical Research, Slovak Academy of Sciences, Bratislava, Czechoslovakia

E. T. DENISOV
Institute of Chemical Physics, Academy of Sciences, Moscow, USSR

FRANÇOIS GUGUMUS
Ciba–Geigy Ltd., CH-4002 Basle, Switzerland

DAN MUNTEANU
Chemical Research Institute, Research Centre for Plastics, Timişoara Laboratory, Str. Gării 25, R-1900 Timişoara, Romania

JAMES E. PICKETT
General Electric Company, Corporate Research and Development, Schenectady, New York 12301, USA

KLAUS SCHWETLICK
Chemistry Section, Dresden University of Technology, 8027 Dresden, Mommsenstrasse 13, GDR

Chapter 1

A Theoretical Approach to the Optimisation of Antioxidant Action

E. T. DENISOV

Institute of Chemical Physics, Academy of Sciences,
Moscow, USSR

ABSTRACT

The mechanism of inhibited oxidation of hydrocarbons and polymers depends on the reactivity of the inhibitor InH, on the oxidisability of the substance RH, on the reactivity of radicals $RO_2^.$ and $In^.$, on the stability of the hydroperoxide ROOH and on the oxidation conditions. The mechanism is determined by the key steps of the radical chain oxidation reaction and this is in turn determined by the bond energies of R—H and In—H. The rate constants of the chain propagation step ($RO_2^.$ + RH) and chain termination step ($RO_2^.$ + InH) depend on D_{R-H} and D_{In-H}. These two rate constants can be taken as universal parameters using correlation equations and expressing the rate constants of other steps of inhibited oxidation as functions of $k(RO_2^. + RH)$ and $k(RO_2^.$ + InH). Therefore the mechanism of oxidation of all hydrocarbon materials by a number of inhibitors, for which the same correlations are valid, may be considered theoretically within the framework of kinetic topology where the mechanism involved is the result of the position of the system in the space with coordinates: reactivity of RH, reactivity of InH, and reaction conditions (temperature, concentrations of reactants, etc.). This approach permits the theoretical solution of such problems as *the identification of the mechanism of inhibited oxidation of RH in systems with given structures and conditions, the estimation of oxidation rate and induction period using only rate constants $k(RO_2^{\cdot} + RH)$ and $k(RO_2^{\cdot} + InH)$, and the choice of an inhibitor with optimal retarding action. It also permits the analysis of the problem of the upper temperature limit of effective retardation for a given class of inhibitors.*

1. INTRODUCTION

A rational approach to the selection of inhibitors and the prediction of their lifetimes and effectiveness in compositions requires the knowledge of the mechanisms and kinetic parameters of oxidation of the substrate as well as the effect of the added inhibitor on the mechanism and kinetics. It is clear that the ability to decelerate oxidation depends on the nature of the oxidised compound, temperature and other reaction conditions. It may be viable to have simple algorithms for estimating the appropriateness of an inhibitor for a given composition. Compounds, InH_2 capable of reacting with the peroxyl radical to yield the inhibitor radical ($In^{\cdot}$) are widely used for the inhibition of oxidation of hydrocarbons. The inhibiting effect is determined by the reactivity of both the inhibitor itself and the radicals that it produces.

The selection of effective inhibitors is at present carried out empirically. Very often the preliminary selection of sufficiently effective inhibitors is made by the oxidation of model compounds in the presence of inhibitors. The theoretical assessment of inhibitor effectiveness may be made by computational modelling of the kinetics of oxidation, if the rate constants of all the required reactions are known. However such an approach is true only for a given inhibitor in the given system. It is therefore a problem of general and practical interest to be able adequately to transform results of inhibiting activity in one system to another system and other inhibitors in order to obtain a more general picture of the inhibiting ability of a chosen class of inhibitors. To solve this problem a special theoretical approach has been developed that permits the estimation of inhibitor activity in one system to be transformed to other conditions and provides a general analysis of the inhibiting activity of the whole class of inhibitors. This approach was named kinetic topology of the inhibited oxidation of hydrocarbons.

2 MECHANISM OF INHIBITED OXIDATION OF HYDROCARBONS

The oxidation kinetics and mechanisms of hydrocarbons (RH) as well as polyolefins have been studied in much detail.[1-4]

The early stage of oxidation of hydrocarbons, resulting in the formation of hydroperoxide, consists of the following elementary steps:

$$\text{I(initiator)} \xrightarrow{k_i} 2\text{r}^\cdot \tag{i}$$

$$\text{R}^\cdot + \text{O}_2 \xrightarrow{k_1} \text{RO}_2^\cdot \tag{1}$$

$$\text{RO}_2^\cdot + \text{RH} \xrightarrow{k_2} \text{ROOH} + \text{R}^\cdot \tag{2}$$

$$\text{r}^\cdot + \text{RH} \xrightarrow{k_2'} \text{rH} + \text{R}^\cdot \tag{2'}$$

$$\text{ROOH} \xrightarrow{k_3} \text{RO}^\cdot + \text{HO}^\cdot \tag{3}$$

or

$$\text{ROOH} + \text{RH} \xrightarrow{k_3'} \text{RO}^\cdot + \text{H}_2\text{O} + \text{R}^\cdot \tag{3'}$$

or

$$\text{ROOH} + \text{CH}_2=\text{CHX} \xrightarrow{k_3''} \text{RO}^\cdot + \text{HOCH}_2\text{CHX} \tag{3''}$$

or

$$2\text{ROOH} \xrightarrow{2k_3'''} \text{RO}_2^\cdot + \text{H}_2\text{O} + \text{RO}^\cdot \tag{3'''}$$

$$\text{R}^\cdot + \text{R}^\cdot \xrightarrow{2k_4} \text{RR (or RH + olefin)} \tag{4}$$

$$\text{R}^\cdot + \text{RO}_2^\cdot \xrightarrow{k_5} \text{ROOR} \tag{5}$$

$$\text{RO}_2^\cdot + \text{RO}_2^\cdot \xrightarrow{2k_6} \text{ROH} + \text{O}_2 + \text{R}' = 0 \quad \text{or} \quad \text{ROOR} + \text{O}_2 \tag{6}$$

Oxidation is a chain process if chain propagation by reactions (1) and (2) is faster than chain termination reactions (4)–(6). Radical $\text{R}^\cdot$ reacts vigorously with oxygen ($k_1 = 10^7$–$10^9 \, \text{l mol}^{-1} \text{s}^{-1}$); therefore, for oxygen concentrations above $10^{-4} \, \text{mol l}^{-1}$, $[\text{RO}_2^\cdot] \gg [\text{R}^\cdot]$ and the chains are terminated by reaction (6). Under these conditions the oxidation rate $v = v_i + k_2(2k_6)^{-1/2}[\text{RH}] \times v_i^{1/2}$ and the chain mechanism is realised at $v_i < \frac{1}{2} \cdot k_2^2 \times k_6^{-1}[\text{RH}]^2$.

The following reactions take place in the system upon introduction of InH which reacts with $RO_2^:$:[3-9]

$$r + InH \xrightarrow{k_7'} rH + In^\cdot \tag{7'}$$

$$RO_2^\cdot + InH \xrightarrow{k_7} ROOH + In^\cdot \tag{7}$$

$$In^\cdot + ROOH \xrightarrow{k_{-7}} InH + RO_2^\cdot \tag{-7}$$

$$RO_2^\cdot + In^\cdot \xrightarrow{k_8} \text{Products (InOOR)} \tag{8}$$

$$In^\cdot + In^\cdot \xrightarrow{2k_9} \text{Products} \tag{9}$$

$$In^\cdot + RH \xrightarrow{k_{10}} InH + R^\cdot \tag{10}$$

$$InH + ROOH \xrightarrow{k_{11}} \text{Products} \tag{11}$$

$$InH + O_2 \xrightarrow{k_{12}} In^\cdot + HO_2^\cdot \tag{12}$$

$$InOOR \xrightarrow{k_{13}} InO^\cdot + RO^\cdot \tag{13}$$

$$In^\cdot \xrightarrow{k_{14}} Q + r^\cdot \tag{14}$$

$$In_1^\cdot + In_2H \underset{k_{-15}}{\overset{k_{15}}{\rightleftharpoons}} In_1H + In_2^\cdot \tag{15}$$

Reactions (i)–(15), including the alternative reactions ('), give the principal kinetic scheme of inhibited hydrocarbon oxidation.

In one and the same system these reactions take place with different rates; therefore they differ in significance. It is important then that the key reactions which determine the 'kinetic profile' of the chosen system should be picked out.

The general kinetic scheme written above may be divided by line (9) from the basic mechanism.[10,11] The basic mechanism includes its own set of elementary key reactions. One can find also a combination of basic mechanism in real systems when key reactions of two basic mechanisms are realised with comparable rates. The key reactions of the basic mechanisms as well as equations for reaction rates of

Table 1
The mechanism and rate of inhibited oxidation of hydrocarbons

Mechanism	Key reactions	Oxidation rate, v (mol l^{-1} s^{-1})
I	r˙ + InH (7'), r˙ + RH (2')	$v_i(1 + k_7'[InH]/k_2^0[RH])^{-1}$
II	RO$_2^•$ + InH (7), In˙ + RO$_2^•$ (8)	v_i
III	RO$_2^•$ + RH (2), RO$_2^•$ + InH (7)	$k_2[RH]v_i/fk_7[InH]$
IV	In˙ + RO$_2^•$ (8), In˙ + RH (10)	$k_2[RH]^{3/2}(k_{10}v_i/fk_7k_8[InH])^{1/2}$
V	RO$_2^•$ + InH (7) InOOR → InO˙ + RO˙ (13)	$k_2[RH]v_i/fk_7[InH]$
VI	InH + O$_2$ (12)	$k_2k_{12}[RH][O_2]/fk_7[InH]$
VII	In˙ + ROOH (−7), In˙ + RO$_2^•$ (8)	$k_2[RH](k_{-7}[ROOH]v_i/fk_7k_8[InH])^{1/2}$
VIII	RO$_2^•$ + InH (7), In˙ → Q + r˙ (14)	$k_2[RH](k_{14}v_i/fk_7k_8[InH])^{1/2}$
IX	InH + ROOH (11)	$k_2k_{11}[RH][ROOH]/fk_7[InH]$

oxidation are given in Table 1. The kinetic analysis is carried out for phenols. According to their structure all phenols (InH) may be divided into the following three groups.

(a) *Group A* phenols reacting with RO$_2^•$, ROOH and O$_2$, the In˙ is active enough to react with RH and ROOH under certain conditions; the products of their reactions with RO$_2^•$ are not peroxides InOOR; In˙ do not decompose to active radicals. This group includes all InH except 2,6-di-*tert* alkyl-substituted and *ortho*- or *para*-alkoxy-substituted phenols. For such InH, mechanisms I, II, III, IV, VI, VII and IX may be valid (cf. Table 1).

(b) *Group B* phenols generating inactive In˙ which under oxidation conditions are practically unable to react with RH and ROOH; InH themselves are capable of slow reaction with ROOH and RO$_2^•$. By reacting with RO$_2^•$, aroxyls of these InH form peroxides which decompose to free radicals. Therefore such InH can react via mechanisms I, II, III and V only.

(c) *Group C* phenols with alkoxy substituents producing In˙ which dissociate to an alkyl radical which propagates the chain.

3 THE CORRELATION EQUATIONS

The number of key reactions in the general scheme is ten. Each inhibitor has its own set of rate constants for the elementary steps. The variety of mechanisms as well as the great number of steps with InH and In· impede the analysis of inhibited oxidation in general. Such analysis may be facilitated by diminishing the number of parameters characteristic of each basic mechanism. This may be done using linear correlation equations, namely correlation between the energy of activation and the enthalpy of the reaction (equations of the Polyani–Semenov type). When inhibitor InH reacts with $RO_2^·$ or ROOH, or O_2 the In—H bond is cleaved. The greater the strength of the In—H bond, the lower is the rate constant of the reaction of the phenol and the abstraction of hydrogen atoms from phenolic groups is faster when the In—H bond is weaker. On the other hand, the higher the activity of the radical In·, the stronger is the In—H bond. That is why the rate constants of key reaction (5) with InH and In· depends on the In—H bond energy, and their activation energies should be related to $D_{In—H}$. In the same way the activation energies of reactions of RH may be expressed by two parameters $D_{In—H}$ and $D_{R—H}$.[8,12] Pre-exponential factors A may be regarded as constant within the same class of reagents. The correlation equations for the activation energies of key reactions are given in Table 2. These equations are simple and valid for kinetic analysis. However, $D_{R—H}$ and $D_{In—H}$ as basic parameters have one defect, namely that they are known only for a few

Table 2
Linear correlation between activation energy E and bond energies $D_{R—H}$ and $D_{In—H}$[12]

Reaction	E (kJ mol^{-1})	$\log A$ (l mol^{-1} s^{-1})
RO· + RH (i)	$0{\cdot}30 D_{R—H} - 92{\cdot}5$	9·2
sec-$RO_2^·$ + RH (2)	$0{\cdot}55 D_{R—H} - 144$	9·0
tert-$RO_2^·$ + RH (2)	$0{\cdot}55 D_{R—H} - 144$	8·2
Group A phenols		
RO· + InH (7)	$0{\cdot}16 D_{In—H} - 42$	11
$RO_2^·$ + InH (7)	$0{\cdot}32 D_{In—H} - 94$	7·2
In· + ROOH (−7)	$274 - 0{\cdot}68 D_{In—H}$	7·2
In· + RH (10)	$0{\cdot}69(D_{R—H} - D_{In—H}) + 64{\cdot}5$	9·2
InH + ROOH (11)	$D_{In—H} - 254$	10
InH + O_2 (12)	$D_{In—H} - 218$	9·9
Group B phenols		
$RO_2^·$ + InH (7)	$0{\cdot}37 D_{In—H} - 107$	7·2
In· + ROOH (−7)	$266 - 0{\cdot}63 D_{In—H}$	7·2

Table 3
Correlation equations for the key reactions of phenol-inhibited oxidation of hydrocarbons

Reaction	$\log k \, [k \, (l \, mol^{-1} \, s^{-1})]$
$O_2 + RH$ (01)	$12 \cdot 0 - 7\,900/T + (370/T) \log k_2^*$
sec-$RO_2^{\cdot} + RH$ (2)	$9 \cdot 0 - 3\,000/T + (330/T) \log k_2^*$
tert-$RO_2^{\cdot} + RH$ (2)	$8 \cdot 2 - 3\,000/T + (330/T) \log k_2^*$
$ROOH \rightarrow RO^{\cdot} + HO^{\cdot}$ (3)	$13 - 7\,400/T$
Group A phenols	
$RO_2^{\cdot} + InH$ (7)	$7 \cdot 2 - 2\,400/T + (330/T) \log k_7^*$
$In^{\cdot} + ROOH$ (−7)	$7 \cdot 2 + 1\,260/T - (710/T) \log k_7^*$
$RO_2^{\cdot} + In^{\cdot}$ (8)	$8 \cdot 5$
$In^{\cdot} + In^{\cdot}$ (9)	$8 \cdot 5$
$In^{\cdot} + $ sec-RH (10)	$9 \cdot 2 - 1\,800/T - (520/T) \log k_7^* + (530/T) \log k_2^*$
$InH + ROOH$ (11)	$10 \cdot 0 - 8\,100/T + (570/T) \log k_7^*$
$InH + O_2$ (12)	$9 \cdot 9 - 8\,700/T + (570/T) \log k_7^*$
Group B phenols	
$RO_2^{\cdot} + InH$ (7)	$7 \cdot 2 - 2\,410/T + (330/T) \log k_7^*$
$In^{\cdot} + ROOH$ (−7)	$7 \cdot 2 + 530/T - (710/T) \log k_7^*$
$InOOR \rightarrow InO^{\cdot} + RO^{\cdot}$ (13)	$14 - 7\,225/T$
Group C phenols	
$In^{\cdot} \rightarrow Q + r^{\cdot}$ (14)	$12 \cdot 7 - 5\,810/T$

compounds. Therefore to characterise the activity of InH and RH one would prefer parameters which are known for many hydrocarbons and inhibitors. Rate constants for $RO_2^{\cdot}$ reactions with RH (k_2^*) and InH (k_7^*) at 333 K were chosen as such parameters.[13] The correlation equations for different k as a function of absolute temperature (T) and k_2^* and k_7^* are given in Table 3.

4 KINETIC TOPOLOGY OF INHIBITED OXIDATION OF HYDROCARBONS

From the kinetic data on inhibited oxidation of RH we draw the conclusion that even for a system consisting of a particular InH added to a particular RH the oxidation may take one of several possible mechanistic routes determined by a number of key reactions (cf. Table 1). If we have investigated InH in a given system so that we know all the elementary reaction rate constants involving RH, InH, $RO_2^{\cdot}$, $In^{\cdot}$,

ROOH and O_2, we will be able to calculate the RH oxidation kinetics under any conditions. If we do not know the rate constants for all the elementary reactions we may solve the problem by staging a series of experiments. However, this oversimplified approach does not give us the answers to a number of general questions; for example, how is the inhibited oxidation mechanism related to the InH and RH structures and reactivity of $RO_2^•$? What particular InH will be most effective under given conditions in a given system? Answers to these and like questions can be obtained using the following principles.[10-13]

1. The elementary reactions accompanying RH oxidation in the presence of InH may be numerous, but only a few of them determine the actual reaction mechanism and may be called key reactions. For example, the reactions involved in cumene oxidation in the presence of p-cresol at 320–380 K are (2), (7), (−7), (9), (10). However, the rates of reactions $In^• + ROOH$ (−7) and $In^• + RH$ (10) are much lower than the rate of $RO_2^• + In^•$ (8). Therefore, when $v_2 > v_i$ where v_2 pertains to reaction $RO_2^• + RH$ (2), reactions (2) and (−7) will be the crucial ones, and the oxidation rate

$$v = \frac{k_2[\text{RH}]v_i}{2k_7}[\text{ln H}]$$

will be controlled by two reactions, (2) and (7), which are the key ones. The ratio of the rate constants of these reactions, k_7/k_2, will be characteristic of InH effectiveness. For chain termination by the reaction of RH with $RO_2^•$ the following mechanisms, each consisting of a set of elementary steps, can be singled out as principal (see Table 1).

2. If we use a different InH or a different RH, or change the oxidation conditions, the mechanism may change too. Therefore, to enable a systematic approach to be made, it may be helpful to consider the range of operating conditions of the alternative inhibited oxidation mechanisms. For one InH and one RH it will be a set of conditions under which the mechanism in question persists. But the concept of the range of operating conditions may be broadened if we extend it to a range of oxidising compounds and a range of inhibitors.

3. The range of operating conditions of each mechanism (its 'domain of existence') depends on the relative rates of a large number of elementary steps. For instance, the following seven inequalities must be realised in order to channel the reaction according to

mechanism III where the chains are terminated via reaction (7):

$$k_2[\text{RH}] > k_7[\text{InH}] \quad k_{-7}[\text{ROOH}] < k_8[\text{RO}_2^{\cdot}]$$
$$k_{10}[\text{RH}] < k_8[\text{RO}_2^{\cdot}] \quad v_i > (k_{11}[\text{ROOH}] + k_{12}[\text{O}_2])[\text{InH}]$$
$$k_{13}[\text{InOOR}] < v_i \quad k_{14} < k_8[\text{RO}_2^{\cdot}]$$

In real situations all the elementary reactions (1)–(14) proceed but many of them go at a relatively low rate. This multiparameter relationship may be simplified if we use correlation equations. The range of InH must be broken down into groups, each falling under a specific set of correlation equations. According to their structure all phenols InH may be divided into the following three groups.

Group A consists of phenols in which the In$^{\cdot}$ are active enough to react with RH and ROOH under certain conditions. The products of their reactions with RO$_2^{\cdot}$ are not peroxides and do not decompose to active radicals. For such InH, mechanisms I, II, III, IV, VI, VII and IX may be valid.

Group B consists of phenols generating inactive In$^{\cdot}$ that are essentially unable to react with RH and ROOH, although the InH themselves are capable of reacting slowly with ROOH and O$_2$. By reaction between RO$_2^{\cdot}$ and aroxyls, peroxides are formed that decompose to free radicals. Such InH can react by only mechanisms I, II, III and V.

Group C consists of phenols with alkoxy substituents producing In$^{\cdot}$ which dissociate to alkyl radicals that propagate the chain. Such InH can react by mechanisms I–IX.

All reactions in which InH and In$^{\cdot}$ take part depend on the In—H bond dissociation energy. Therefore within each group of InH the activation energies of each elementary step may be expressed in terms of one parameter, viz. $D_{\text{In—H}}$. On the other hand, the activation energies for reactions in which RH takes part may be expressed in terms of the R—H bond dissociation energy. It should not be overlooked however, that tertiary RO$_2^{\cdot}$ react somewhat more slowly than either secondary or primary RO$_2^{\cdot}$.

4. It is therefore possible to reduce the variety of structural factors, of which the relative impact on oxidation is determined by the rate constants of a total of nine reactions, to only two parameters, namely, $k_2^* = k_2$ and $k_7^* = k_7$, both at 333 K (Table 3).

5. The domain of each mechanism may now be expressed as a certain volume in three-dimensional space: k_2^*–k_7^*–conditions (T, v_i,

Table 4
Equations for domain boundary strips at $v_i = 10^{-7}$ mol l^{-1} s^{-1}, [RH] = 10 mol l^{-1}, [InH] = 10^{-3} mol l^{-1}, [ROOH] = 10^{-3} mol l^{-1}

Mechanism		Boundary equation
I–II (2′, 7′)	$k'_2[\text{RH}] = k'_7[\text{InH}]$	$T = 7 \cdot 12 D_{\text{R—H}} - 3 \cdot 80 D_{\text{In—H}} - 1194$
Group A phenols		
II–III (2, 7)	$k_2[\text{RH}] = fk_7[\text{InH}]$	$T = 5 \cdot 26 D_{\text{R—H}} - 3 \cdot 05 D_{\text{In—H}} - 477$
III–IV (8, 10)	$k_8 v_i = k_7 k_{10}[\text{RH}][\text{InH}]$	$T = 2 \cdot 69 D_{\text{R—H}} - 1 \cdot 44 D_{\text{In—H}} - 115$
III, IV–VI (7, 12)	$fk_{12}[\text{InH}][\text{O}_2] = v_i$	$T = 4 \cdot 40 D_{\text{In—H}} - 959$
III–VII (−7, 8)	$k_8 v_i = k_7 k_{-7}[\text{ROOH}][\text{InH}]$	$T = 963 - 1 \cdot 92 D_{\text{In—H}}$
Group B phenols		
III–VIII (8, 14)	$k_8 v_i = k_7 k_{14}[\text{InH}]$	
III–IX (7, 11)	$fk_{11}[\text{InH}][\text{ROOH}] = v_i$	

InH, RH, ROOH, O$_2$). The domain will be defined by the ranges of k_2^*, k_7^* and experimental conditions (e.g. temperature). One domain will be separated from another by a boundary 'strip', i.e. such ranges of k_2^* and k_7^* and conditions under which signs of both mechanisms will be evident. The boundary conditions between any two domains may be formulated as the appropriate inequalities and expressed in terms of $D_{\text{R—H}}$ and $D_{\text{In—H}}$[12] (Table 4). This presentation allows us to make certain conclusions. Firstly, we may now pinpoint the chief factors which control the mechanism. They are $D_{\text{R—H}}$, $D_{\text{ArO—H}}$ and T. Concentrations of RH and ArOH are affecting but not controlling the mechanism. Such factors as v_i, [O$_2$] and [ROOH] will affect the mechanism only under specific conditions. Secondly, from the analysis of the analytical formulae it follows that there are structural ranges responsible for each particular mechanism. It appears that it is not always possible by merely varying the conditions to implement a particular mechanism for a given pair, RH–InH. For example, mechanism III (Table 1) may be realised only for such pairs for which $D_{\text{R—H}} < 0 \cdot 63 D_{\text{ArO—H}} + 142$ kJ mol^{-1}.[3]

As mentioned above, the correlation equations based on k_2^* and k_7^* are more convenient for practical use.[13] The boundary strips between domains of basic mechanisms are given in Table 5. Figures 1 and 2 show domains of inhibited oxidation as a function of InH activity and RH oxidisability at different T. The boundary strip has a certain finite width, so that the transition from one mechanism to another is smooth.

Table 5
Equations for domain boundary strips and their presentation in the $T = X/Y$ form

Mechanism	X	Y
Group A phenols		
II–III	$600 + 330 (\log k_7^* - \log k_2^*)$	$1.8 + \log([RH]/f[InH])$
III–IV	$4200 + 90 \log k_7^* - 530 \log k_2^*$	$7.9 + \log([RH][InH]/v_i)$
III–VI	$8700 - 570 \log k_7^*$	$9.9 + \log(f[InH]O_2/v_i)$
III–VII	$1140 + 380 \log k_7^*$	$5.9 + \log([ROOH][InH]/v_i)$
III–VIII	$8210 - 330 \log k_7^*$	$11.4 + \log([InH]/v_i)$
VII–IX	$8100 - 570 \log k_7^*$	$10 + \log(f[InH][ROOH]/v_i)$
Group B phenols		
II–III	$600 + 330 (\log k_7^* - \log k_2^*)$	$1.8 + \log([RH]/f[InH])$
III–V	7225	$14 + \log([InH]/v_i)$

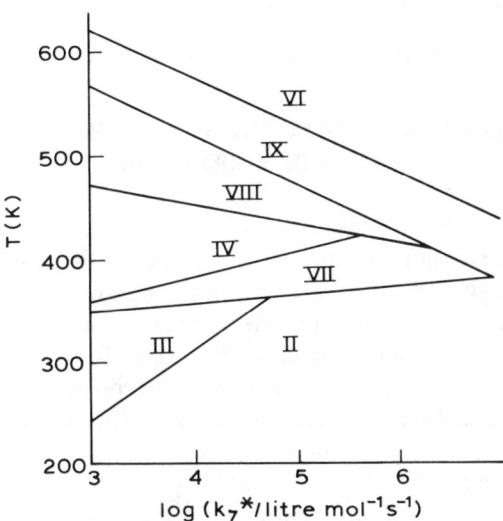

Fig. 1. Graphic representation of the domains of realisation of different mechanisms of phenol-inhibited hydrocarbon oxidations in T vs $\log k_7^*$ coordinates with $k_2^* = 1 \, l \, mol^{-1} \, s^{-1}$, $[RH] = 10 \, mol \, l^{-1}$, $f = 2$, $[InH] = [ROOH] = [O_2] = 10^{-3} \, mol \, l^{-1}$ and $v_i = 10^{-7} \, mol \, l^{-1} \, s^{-1}$. The mechanism numbering is given in Table 1.

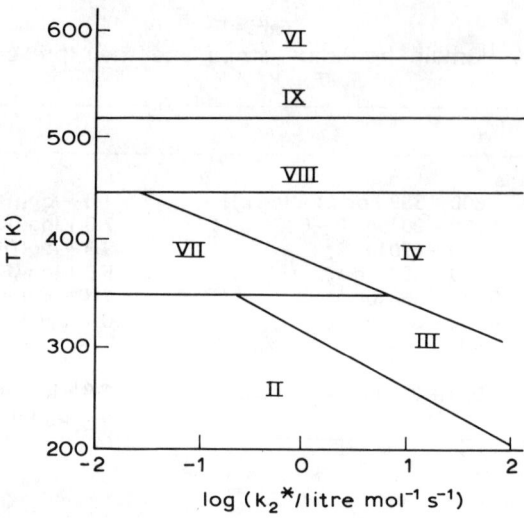

Fig. 2. Graphic representation of the domains of realisation of different mechanisms of phenol-inhibited oxidation of hydrocarbons in T vs $\log k_2^*$ coordinates with $k_7^* = 10^{-4}$ l mol^{-1} s^{-1}, [RH] = 10 mol l^{-1}, $f = 2$, [InH] = [ROOH] = [O$_2$] = 10^{-3} mol l^{-1}, and $v_i = 10^{-7}$ mol l^{-1} s^{-1}. The mechanism numbering is given in Table 1.

5 PARAMETRIC EQUATIONS FOR THE RATE OF INHIBITED OXIDATION

As we have seen, the kinetics of inhibited oxidation are determined by key reactions, and they change when we change RH, or InH, or the conditions of oxidation. The rate constants of all elementary steps may be expressed through two parameters, namely k_2^* and k_7^*. This gives us the possibility of devising universal formulae for estimation of the oxidation rate v as a function of only two rate constants k_2^* and k_7^* and temperature[13] (see Table 6). The discrepancies between experimental and estimated values of v are illustrated by the following examples[13] of tetralin oxidation inhibited by C_6H_5OH.[10]

		v (l mol^{-1} s^{-1})	
Mechanism	T	Experimental	Calculated
IV	343	2.7×10^{-6}	7.9×10^{-6}
VII	330	6.2×10^{-7}	1.6×10^{-6}

Table 6
Equations for oxidation rate with phenols of type A

Mechanism	$\log v \ [v(\text{mol } l^{-1} \ s^{-1})]$
III	$1\cdot8 - 600/T + (330/T)(\log k_2^* - \log k_7^*) + \log([\text{RH}]v_i/f[\text{InH}])$
IV	$5\cdot8 - 2\,700/T + (600/T)\log k_2^* - (420/T)\log k_7^* + 1\cdot5\log[\text{RH}]$ $+ \frac{1}{2}\log(v_i/f[\text{InH}])$
VII	$4\cdot8 - 1\,170/T + (330/T)\log k_2^* - (520/T)\log k_7^* + \log[\text{RH}]$ $+ \frac{1}{2}\log([\text{ROOH}]v_i/f[\text{InH}])$
VIII	$7\cdot5 - 4\,700/T + (330/T)\log k_2^* - (170/T)\log k_7^* + \frac{1}{2}\log([\text{RH}]^2 v_i/f[\text{InH}])$

6 THE OPTIMUM INHIBITOR

The inhibiting effect is characterised by the duration and extent of reaction inhibition (effectiveness). The latter, for chain termination, may be expressed in terms of the chain length $\gamma = v/v_i$. That InH which under specific conditions provides for the minimum chain length is the most effective. From the data of Table 6 it follows that the inhibiting effectiveness depends on the oxidation mechanism, types of reagents (RH and InH) and conditions (T, v_i, reagent concentrations). In Table 6 all the parameters relating to reactions which involve InH and In· are in terms of k_7^*. From consideration of the equations for v under different mechanisms it follows that the smaller v is, the higher is the k_7^*. Hence the InH characterised by the maximum k_7^* will be the most effective one in any case.

Another important characteristic of InH is the duration of its inhibiting period τ which, provided InH is consumed only in chain terminations, is determined by its initial concentration $[\text{InH}]_0$, coefficient f, and initiation rate v_i. Under these conditions the rate of consumption of InH, $v_{\text{InH}} = v_i/f$. Side reactions, e.g. reactions of InH with O_2 and ROOH, will reduce the effective inhibition period. The effectiveness of InH as an inhibiting agent is higher, the smaller the contribution of side reactions to the total consumption of InH. Therefore the optimum InH will be the one which assures the minimum chain length γ with a relatively low loss of InH in side reactions. The important factor is the relative fraction of InH lost in side reactions, $\omega = fv'_{\text{InH}}/v_i$, where v'_{InH} is the rate of InH consumption in side reactions. Suppose, for effective InH, $\omega \leq 0\cdot25$. Since k_{11} and k_{12} increase sympathetically with k_7^* (cf. Table 3), the optimum InH is therefore expected to satisfy contradictory requirements: it must be most active in chain termination steps but reacts as slowly as possible with O_2 and ROOH. If the main side route of InH consumption is its reaction with O_2, the condition $\omega \leq 0.25$ will mean that $4k_{12}[\text{InH}][O_2] \leq v_i$ which, after substitution of the correlation equation from Table 3 and necessary manipulations, takes the form[13]

$$\log k_7^* \leq 15.3 - [18.4 - 1.8 \log(v_i/f[\text{InH}][O_2])] \times 10^{-3} T$$

Obviously the InH which has the maximum k_7^* and satisfies the above inequality will be the best one. It follows from eqn (4), that the optimum k_7^* value will be lower, the higher the temperature and InH and O_2 concentrations. At $v_i = 10^{-7}$ mol l^{-1} s^{-1}, $[\text{InH}] = [O_2] = 10^{-3}$ mol l^{-1} and $f = 2$, the optimum $\log k_7^* = 8$ ($T = 350$ K); 7 (400 K); 6 (450 K) and 5 (500 K). For ArOH, $k_7^* \leq 10^7$ mol l^{-1} s^{-1}.

If it is the reaction of InH with ROOH which is the main route of InH consumption, the inequality, upon substitution of the correlation equation and transformation, takes the form[13]

$$\log k_7^* \leq 14.2 - [18.6 - 1.8 \log(v_i/f[\text{InH}][\text{ROOH}])] \times 10^{-3} T$$

It follows from the equation that at higher temperatures the range of suitable InH is narrower. Under $v_i = 10^{-7}$ mol l^{-1} s^{-1}, $f = 2$, $[\text{InH}] = [\text{ROOH}] = 10^{-3}$ mol l^{-1}, the optimum values of $\log k_7^*$ are 6.9 (350 K); 5.8 (400 K); 4.8 (450 K) and 3.8 (500 K).

In the general case InH reacts both with O_2 and ROOH. Therefore to use an InH as effectively as possible, the total rate of these reactions must be relatively low, i.e.

$$f(k_{11}[\text{ROOH}] + k_{12}[O_2])[\text{InH}] \leq 0.25 v_i$$

The competition between reactions (11) and (12) depends on T, $[O_2]$ and $[\text{ROOH}]$

$$\log(v_{11}/v_{12}) = \frac{600}{T} + \log \frac{[\text{ROOH}]}{[O_2]}$$

At temperatures typical for oxidation (350–450 K) and $[\text{ROOH}] \geq 0.1[O_2]$, the prevalent reaction is that between ArOH and ROOH.[11]

7 EFFECTIVE INHIBITION TEMPERATURES

It is important to know the temperatures at which a compound of a given class may be used as an effective inhibitor. In the case of InH the problem may be solved on the quantitative level if we rely on the approach and equations presented in this paper. The inhibitors which terminate chains remain effective as long as they assure short enough reaction chains γ_{max}, where γ_{max} is the maximum possible chain length in a system. On the other hand, for InH to be effective, k_7^* should not, as we have shown above, be greater than some optimum value $k_{7\,opt}^*$. Using the correlations of Table 3 and the formulae for $k_{7\,opt}$ (see above) one may obtain T_{max}, the temperature above which inhibition by a given mechanism stops being effective. For example, in the case of mechanism III, $\gamma = k_2[RH]/fk_7[InH]$ and, putting $\gamma = \gamma_{max}$, taking the logarithm and substituting the equations of Table 3, we obtain

$$\log \gamma_{max} = 1 \cdot 8 \frac{600}{T} + \frac{300}{T} \log(k_2^*/k_7^*) + \log([RH]/[InH])$$

Substituting the formulae for $k_{7\,opt}^*$ in this equation [for side reaction (11)] we come up with the formula given in Table 7. This table[13] contains the formulae for estimating T_{max} under different oxidation mechanisms for any k_2^* and the k_7^* of the optimum InH, assuming the reaction with ROOH as the main side route. Evaluation of T_{max} for n-alkane oxidation under the conditions [InH] = [ROOH] =

Table 7
(a) Formulae for the upper temperature limit T_{max} of inhibiting action of phenols of group A presented in the form $T_{max} = X/Y$ at v_i = constant

Mechanism	X	Y
III	$600 + 330 \log(k_7^*/k_2^*)$	$1 \cdot 8 - \log \gamma_{max} + \log([RH]/f[InH])$
IV	$2\,700 + 420 \log k_7^* - 600 \log k_2^*$	$5 \cdot 7 - \log \gamma_{max} + 1 \cdot 5 \log[RH]$ $- 0 \cdot 5 \log(f[InH]v_i)$
VII	$1\,170 + 520 \log k_7^* - 330 \log k_2^*$	$4 \cdot 8 - \log \gamma_{max} + \log[RH]$ $+ 0 \cdot 5 \log([ROOH]/f[InH]v_i)$
VIII	$4\,700 - 330 \log k_2$	$7 \cdot 5 - \log \gamma_{max} + \log[RH]$ $- 0 \cdot 5 \log(f[InH]v_i)$

(b) Formulae for T_{max} with k_7^* expressed by equation (10), $\gamma_{max} = 10$

Mechanism	X	Y
III	$5\,290 - 330 \log k_2^*$	$6 \cdot 9 + \log[RH] - 0 \cdot 6 \log(v_i/[ROOH])$ $- 0 \cdot 4 \log(f[InH]v_i)$
IV	$8\,660 - 600 \log k_2^*$	$14 \cdot 5 + 1 \cdot 5 \log[RH] - 1 \cdot 3 \log v_i$ $+ 0 \cdot 3 \log(f[InH]) + 0 \cdot 8 \log[ROOH]$
IV	$9\,180 - 530 \log k_2^*$	$13 \cdot 7 + \log[RH] - 1 \cdot 4 \log v_i$ $+ 0 \cdot 9 \log(f[InH][ROOH])$
VII	$8\,550 - 330 \log k_2^*$	$13 \cdot 5 + \log[RH] + 1 \cdot 4 \log[ROOH]$ $+ 0 \cdot 4 \log(f[InH]) - 1 \cdot 4 \log v_i$

10^{-3} mol l^{-1}, $v_i = 10^{-7}$ mol l^{-1} s^{-1}, gives $T_{max} = 440$ K (mechanism III); 445 K [IV, reactions (2), (7), (8), (10)]; 520 K [IV, reactions (9), (10)]; 400 K (VII) and 425 K (VIII). Referring to Fig. 1 we see that for n-alkanes in the presence of ROOH mechanism VII is preferable to mechanisms II and IV because of the very strong R–H bond. Therefore the most confident of all estimates will be T_{max} for mechanism VII (Table 7). Apparently all such estimates are approximate; the correlation equations themselves are approximate. For instance in the case of oxidation by mechanisms III and for an error in the difference of measured activation energies ($E_2 - E_1$) of 6 kJ mol^{-1}, the error in T_{max} will be ± 30 K ($T_{max} = 440 \pm 30$ K).

8 KINETICS OF INHIBITED AUTOXIDATION OF HYDROCARBONS

An important characteristic of the autoxidation of a hydrocarbon is that its product (hydroperoxide) is an initiator[1-3] which causes a progressively increasing initiation rate in the course of the reaction. The rate of acceleration depends in turn on the rate of chain oxidation, i.e. there is a kind of a positive feedback between the autoinitiation and autoxidation reactions. A similar feedback exists in the oxidation of other organic compounds, too.

Firstly, the more effectively does the inhibitor terminate the chains, the slower is its consumption and the longer the effective inhibition period τ, whereas in an initiated chain reaction, with $v_i =$ constant, the τ is independent of the inhibitor effectiveness. Secondly, the autoxidation of RH can be inhibited not only with compounds which act to terminate the chains but also with compounds which decompose ROOH.[1,2,6] Thirdly, critical phenomena are often observed in inhibited autoxidation experiments[1] which must be attributed to the above-mentioned feedback effect.

Since ROOH is decomposed during autoxidation, the oxidation may follow either of two regimes: non-steady state, or quasi-steady state with respect to ROOH.

Under non-steady conditions ROOH is stable and almost no decomposition is perceptible during the induction period, i.e. its decomposition rate constant $k_\Sigma \ll \tau^{-1}$. Obviously such a regime is associated with specific conditions of inhibited oxidation, as well as with the structure and reactivity of RH, ROOH and InH.

Since oxidation of RH and consumption of InH are interrelated processes, the O_2 absorption rate may be quantitatively expressed in terms of the rate of consumption of InH in the system using the following equations (v_{InH} is the rate of consumption of InH):

$$v_i = v_{i_0} + k_3[\text{ROOH}] \simeq k_3[\text{ROOH}]; \quad v_{InH} = v_i/f \quad (9)$$

$$v = k_2[\text{RH}][\text{RO}_2^\cdot] + k_7[\text{InH}][\text{RO}_2^\cdot] \quad (10)$$

For every possible mechanism of inhibited oxidation one may correlate [$RO_2^\cdot$] with [InH] and [ROOH], express it mathematically and, after solving a set of two differential equations describing absorption of oxygen and consumption of InH, express the absorbed quantity of oxygen in terms of the InH consumed:

$$\Delta[O_2] = [\text{ROOH}] = \int_0^t v \, dt; \quad \tau = f \int_{[\text{InH}]_0}^0 \frac{d[\text{InH}]}{v_i} \quad (11)$$

For each particular mechanism the correlation will have its special form. It should be noted that all oxygen is not always converted to hydroperoxide. For example, because of isomerisation occurring during solid-phase oxidation of polyolefins, only a portion of the absorbed O_2 is converted to ROOH.[14] In this case the formulae for [ROOH] and [O_2] should be corrected accordingly.

$$\begin{array}{c}\diagdown\quad\diagup\text{OOH}\\ \diagup\text{C}\diagdown\quad\cdot\text{C}\diagdown\\ \quad\text{CH}_2\diagup\end{array} \longrightarrow \begin{array}{c}\quad\text{O}^\cdot\\ \diagdown\quad|\quad\diagup\\ \diagup\text{C}\diagdown\quad\diagdown\\ \quad\text{CH}_2\diagup\end{array}\begin{array}{c}\text{OH}\\ |\\ \text{C}-\\ |\end{array} \quad (12)$$

Since the rate of inhibitor consumption, $v_{InH} = v_i/f$, and v_i tend to increase during oxidation, the InH consumption kinetics are substantially non-linear. During the early stages of oxidation $v_{InH} = v_{i_0}/f$, but as more ROOH accumulates, v_{InH} increases and becomes maximal by the end of the induction period.

At a sufficiently high temperature, or in the presence of an ROOH decomposer, ROOH will rapidly dissociate and therefore the oxidation regime will quickly become quasi-steady as regards the ROOH concentration, with the decomposition rate being equal to the rate of its formation. The change from a non-steady to the quasi-steady condition is related to the induction period τ, which depends on the InH type and concentration. The transition from one inhibitor to another often manifests itself in transitions from one type of autoxidation process to another and various critical phenomena.

Table 8
Formulae for γ, $[ROOH]_s$ and $[InH]_{cr}$ at inhibited quasi-steady autoxidation of RH ($\beta = k_3/k_\Sigma$)

Mechanism	γ	$[InH]_{cr}$ $(mol\ l^{-1})$	$[ROOH]_s$ $(mol\ l^{-1})$
III	$\dfrac{k_2[RH]}{fk_7[InH]}$	$\dfrac{\beta k_2[RH]}{fk_7}$	$\dfrac{v_{i_0}\beta\gamma}{k_3(1-\beta\gamma)}$
IV	β^{-1}	—	$\dfrac{\beta^2 k_2^2 k_{10}[RH]^3}{fk_3 k_7 k_8[InH]}$
VII	$\dfrac{k_2 k_7^{1/2}[RH]}{(fk_3 k_7 k_8[InH])^{1/2}}$	$\dfrac{\beta^2 k_2^2 k_{-7}[RH]}{fk_3 k_7 k_8}$	$\dfrac{[InH]_{cr}v_{i_0}}{fk_3([InH]-[InH]_{cr})}$
VIII	β^{-1}	—	$\dfrac{\beta^2 k_2^2 k_{14}[RH]^2}{fk_3 k_7 k_8[InH]}$

What we mean by critical effects in inhibited autoxidation of RH is that under a certain critical InH concentration, $[InH]_{cr}$, there takes place a sharp change in the τ versus $[InH]$ relationship, i.e. $d\tau/d[InH]$ for $[InH] > [InH]_{cr}$ is much higher than $d\tau/d[InH]$ for $[InH] < [InH]_{cr}$. Critical effects may arise when inhibited oxidation proceeds via mechanisms III and VII (see Table 8) and $v_{i_0} < k_3[ROOH]$, so that decomposition of ROOH is sufficiently rapid and the condition $k_\Sigma \tau > 1$ for $[InH] > [InH]_{cr}$ is satisfied ($\beta = k_3/k_\Sigma$). As we have said above, the critical phenomena are due to the feedback effect in inhibited

Table 9
Formulae for induction period of inhibited autoxidation of RH

Mechanism	τ/τ_0	x
III	$1 - x^{-1}(1 + \ln x)$	$x = fk_7[InH]_0/\beta k_2[RH]$
	$\log x = -1.8 + 600/T + (330/T)\log(k_7^*/k_2^*) + \log(f[InH]_0/\beta[RH])$	
IV	$1 - x^{-1}\ln(1 + x)$	$x = fk_7 k_8[InH]_0 v_{i_0}/\beta^2 k_2^2 k_{10}[RH]^3$
	$\log x = -11.5 + 5\,400/T + (850/T)\log k_7^* - (1\,200/T)\log k_2^*$ $+ \log(f[InH]_0 v_{i_0}/\beta^2[RH]^3)$	
VII	$1 - x^{-1}[1 + \ln(2 - 2\sqrt{x} + x) + \arctan(\sqrt{x} - 1)]$	$x = fk_3 k_7 k_8[InH]_0/\beta^2 k_2^2 k_7[RH]^2$
	$\log x = -9.7 + 2\,340/T + (1\,040/T)\log k_7^* - (670/T)\log k_2^*$ $+ \log(fk_3[InH]_0/\beta^2[RH]^2)$	
VIII	$1 - x^{-1}\ln(1 - x)$	$x = fk_7 k_8[InH]_0 v_{i_0}/\beta^2 k_2^2 k_{14}[RH]^2$
	$\log x = -15 + 9\,400/T + (330/T)\log k_7^* - (670/T)\log k_2^*$ $+ \log(f[InH]_0 v_{i_0}/\beta^2[RH]^2)$	

oxidation and occur when both the formation and decomposition rates are similarly dependent on ROOH; for example they may be proportional to [ROOH]—which is just the case in routes III and VII. If the oxidation rate $v \simeq [ROOH]^n$ with $n < 1$, and the decomposition rate is proportional to [ROOH], the critical effects will never take place. Table 8 contains formulae for chain lengths γ, $[InH]_{cr}$ and the quasi-steady hydroperoxide concentration $[ROOH]_s$ for different inhibited oxidation mechanisms.[15] It is seen that the critical effects take place when the chain reaction proceeds via routes III and VII, whereas no critical affects can be possible with other mechanisms. The formulae for induction period τ are given in Table 9.[15] These formulae permit τ to be estimated when only k_2 and k_7 are known.

The problem of optimum inhibitor definition under autoxidation conditions is more difficult than during oxidation with $v_i = $ constant due to variation of v_i during the course of the reaction. The optimum inhibitor was defined[15] as the inhibitor that gives the longest induction period for a given mechanism of inhibition. Such an inhibitor will be consumed in the oxidising substrate at a minimum rate when $[InH] = 0.5[InH]_0$. Formulae for optimum inhibitor concentration depend on the mechanism of oxidation and are given in Table 10.

When the rate of hydroperoxide formation in oxidising RH is much

Table 10
Equations for the selection of the optimum inhibitor from Group A phenols

Mechanism	$\log k_7 [k_7 (1\ mol^{-1}\ s^{-1})$ at 333 K]
$v_i = $ constant, InH reacts with ROOH	
III–VIII	$14 \cdot 2 - 18 \cdot 6 \times 10^{-3} T + 1 \cdot 8 \times 10^{-3} T \log(v_i/f[InH][ROOH])$
$v_i = k_3 [ROOH]$, quasi-stationary regime	
III	$14 \cdot 2 - 17 \cdot 5 \times 10^{-3} T + 1 \cdot 8 \times 10^{-3} T \log(k_3/(1 + f)[InH]_0)$
IV	Maximum k_7
VII	Maximum k_7
VIII	$14 \cdot 2 - 17 \cdot 5 \times 10^{-3} T + 1 \cdot 8 \times 10^{-3} T \log(k_3/(1 + f)[InH]_0)$
$v_i = k_3 [ROOH]$, non-steady conditions	
III	$14 \cdot 2 - 17 \cdot 5 \times 10^{-3} T + 1 \cdot 8 \times 10^{-3} T \log(k_3/(1 + f)[InH]_0)$
IV	$14 \cdot 2 - 17 \cdot 5 \times 10^{-3} T + 1 \cdot 8 \times 10^{-3} T \log(k_3/(1 + f)[InH]_0)$
VII	$15 \cdot 2 - 17 \cdot 5 \times 10^{-3} T + 1 \cdot 8 \times 10^{-3} T \log(k_3/(1 + f)[InH]_0)$
VIII	$13 \cdot 6 - 17 \cdot 5 \times 10^{-3} T + 1 \cdot 8 \times 10^{-3} T \log(k_3/(1 + f)[InH]_0)$

Table 11
Formulae for hydroperoxide formation in the non-steady autoxidation regime

Mechanism	[ROOH]	a
II	$[InH]_0(1 - x - a \ln x)$	$k_2[RH]/k_7[InH]_0$
III	$-[InH]_0 a \ln x$	$k_2[RH]/k_7[InH]_0$
IV	$a[InH]_0^{1/3}(1 - \sqrt[3]{x})^{2/3}$	$(9fk_2^2 k_{10}/k_3 k_7 k_8)^{1/3}[RH]$
V	$a[InH]^{1/2}(1 - \sqrt{x})$	$2k_2[RH](fk_{-7}/k_3 k_7 k_8)^{1/2}$
VI	$[InH]_0 \left\{ a \ln \dfrac{1}{x} + b(1 - x) \right\}$	$k_2[RH]/k_7[InH]_0$
		$(b = 2k_2 e_{11} k_{11}[RH]/fk_3 k_7)$
VII	$-[InH]_0 a \ln x$	$k_2[RH]/k_7[InH]_0$
VIII	$-[InH]_0 a \ln x$	$k_2[RH](1 + 2e_{13})/k_7[InH]_0$
IX	$a[InH]_0^{1/3}(1 - \sqrt[3]{x})^{2/3}$	$(9fk_2^2[RH]^2 k_{14}/k_3 k_7 k_8)^{1/3}$

faster than that of decomposition, the oxidation proceeds in non-stationary regime. In this case concentrations of ROOH formed and InH consumed are interconnected as InH is consumed by the rate $v_{InH} = v_i/f$, $v_i \simeq k_3[ROOH]$ and $v_{ROOH} = v = k_2[RH][RO_2^{\cdot}]$. For example, in the case of mechanism IV, one obtains

$$-\frac{d[ROOH]}{d[InH]} = fk_2[RH]^{3/2}(k_{10}/fk_3 k_7 k_8[InH][ROOH])^{1/2} \quad (13)$$

The result of integration is given in Table 11, which gives formulae for hydroperoxide formation and parameter a as a function of k_2^* and k_7^*. These formulae permit the analytical solution of the problem of optimum inhibitor concentration under non-stationary conditions of autoxidation (see Table 10).

9 CONCLUSIONS

As we have seen, the inhibiting effect of phenols depends on the rate constants of their reactions with peroxyl radicals k_7, but it is not this reaction alone which controls the inhibitor effectiveness. Also important are reactions of inhibitor radicals In$^{\cdot}$, including In$^{\cdot}$ + In$^{\cdot}$ (9), In$^{\cdot}$ + ROOH (−7) and In$^{\cdot}$ + RH (10). Kinetic analysis of the situation has shown that, depending on the conditions and activity of InH, RH and In$^{\cdot}$, some reactions are more important than others and may be called 'key' reactions. Using the parametric relationship between the rate constants of reactions involving InH, In˙ and $RO_2^{\cdot}$, the variety of reactions in such systems may generally be reduced to the following two key reactions: $RO_2^{\cdot} + InH$ (7) and $RO_2^{\prime} + RH$ (2). This enables one to tackle the problem of estimating the effectiveness of inhibition of different RH with a whole class of inhibitors under different conditions as well as to choose the optimum InH, to estimate the temperature ranges of effective inhibition, and to opt for the inhibition mechanism depending on the conditions and InH structure.

The approach stated above and the formulae given may be used not only in the study of pure hydrocarbon oxidation but also for their mixtures. For oxidation of hydrocarbon fuels, it was shown that the kinetics of oxidation of a mixture of hydrocarbons is the same as for individual hydrocarbons. The same situation applies in the case of inhibited oxidation of hydrocarbon mixtures, because the same parameters (v_{i_0}, $k_2 k_6^{-1/2}$, k_3, β, k_7/k_2 etc.) characterise the oxidation both of mixtures and of individual hydrocarbons. The difference is that these are kinetic composite parameters in the case of mixtures.

The approach to oxidation of hydrocarbon polymers is complicated by the following circumstances. Firstly, rate constants of all bimolecular reactions depend on molecular mobility in the given polymer matrix.[16] This dependence is explained by the fact that different orientations of reagents have different energies in the polymer matrix that influence its transition state. As molecular mobility increases with the temperature increase, the conditions approach those of liquid hydrocarbons. Secondly, the diffusion of reagents in polymers is slower than in liquids, and as it is the rate of diffusion that determines the rate constants of such reactions as (6), (8), (9), this is why the borderlines between domains of inhibited oxidation mechanisms depend on the polymer matrix and its molecular mobility. However, one can expect that displacement of borderlines will not be large in comparison with hydrocarbons, due to dependence of borderlines on ratios of rate constants that change much less than the absolute values. The calculated T_{max} of effective inhibitor retardation in hydrocarbon oxidation is apparently close to that for hydrocarbon polymers.

REFERENCES

1. Emanuel, N. M., Denisov, E. T. & Maizus, Z. K., *Liquid-Phase Oxidation of Hydrocarbons*. Plenum Press, New York, 1967.

2. Scott, G., *Atmospheric Oxidation and Antioxidants*. Elsevier, Amsterdam, 1965.
3. Denisov, E. T., *Uspekhi Khimii*, **42** (1973) 361.
4. Mill, T. & Hendry, D. G., in *Comprehensive Chemical Kinetics*, vol. 16. Elsevier, Amsterdam, 1980, p. 70.
5. Howard, J. A., *Adv. Free Radical Chem.*, **4** (1972) 49.
6. Scott, G., in *Developments in Polymer Stabilisation—4*, ed. G. Scott, Applied Science Publishers, London, 1981, p. 1.
7. Mahoney, L. R., *Angew. Chem.* **81** (1968) 555.
8. Beliakov, V. A., Shanina, E. L., Roginskii, V. A. & Miller, V. B., *Izvestija AN SSSR, Ser. khim.*, (1975) 2685.
9. Rubtsov, V. I., Roginskii, B. A., Miller, M. B. & Zaikov, G. E., *Kinetika i Kataliz*, **21** (1980) 612.
10. Denisov, E. T., *Oxid. Commun.*, **6** (1984) 309.
11. Denisov, E. T., *Neftekhimiya*, **22** (1982) 448.
12. Denisov, E. T., *Khim. Fiz.*, **2** (1983) 229.
13. Denisov, E. T., *Khim. Fiz.*, **3** (1984) 1114.
14. Denisov, E. T., in *Developments in Polymer Stabilisation—5*, ed. G. Scott, Applied Science Publishers, London, 1982, p. 23.
15. Denisov, E. T., *Khim. Fiz.*, **4** (1985) 67.
16. Denisov, E. T., *Makromol. Chem., Suppl.* **8** (1984) 63.

Chapter 2

Mechanisms of Antioxidant Action of Phosphite and Phosphonite Esters

KLAUS SCHWETLICK
Dresden University of Technology, Dresden, GDR

ABSTRACT

Phosphite and phosphonite esters are used on a large scale for the stabilisation of polymers against degradation during processing and long-term applications. They function as antioxidants by various mechanisms depending on their structure, the nature of the substrate to be stabilised and the reaction conditions.

All phosphites and phosphonites are hydroperoxide-decomposing secondary antioxidants. Their reactivity in hydroperoxide reduction decreases with increasing electron-acceptor ability and bulk of the groups bound to phosphorus in the order phosphonites > alkyl phosphites > aryl phosphites > hindered aryl phosphites.

Five-membered cyclic phosphites are capable of decomposing hydroperoxides catalytically due to the formation of acidic hydrogen phosphates by hydrolysis and peroxidolysis in the course of reaction. The o-hydroxyphenyl phosphates formed in this way from o-phenylene phosphites are excellent chain breaking antioxidants.

Aryl phosphites, particularly these derived from sterically hindered phenols, can act as chain-breaking primary antioxidants by reduction of peroxyl radicals to alkoxyl radicals. The latter react further with the phosphites by substitution, releasing aroxyl radicals which terminate the radical chain oxidation. The chain-breaking antioxidant activity of aryl phosphites is lower than that of hindered phenols, because the rate constants of their reaction with peroxyl radicals and their stoichiometric inhibition factors are lower than those of phenols. The stoichiometric inhibition factors decrease from one to zero with decreasing concentration of the phosphite and increasing oxidisability of the substrate. Therefore, phosphites themselves are active chain-breakers only at rather high concentrations, predominantly in substrates of low oxidisability at lower temperatures.

In oxidising media at higher temperatures, however, hydrolysis of phosphites and phosphonites takes place in addition to oxidation. The phenols so formed synergised by the parent phosphorus compounds and their hydrolysis products, are responsible for the high antioxidative activity of aryl phosphites and phosphonites, especially the hindered compounds, in oxidations at higher temperatures.

1 INTRODUCTION

Esters of phosphorous and phosphonic acids are used on a large scale as non-discolouring antioxidants for the stabilisation of organic materials against degradation during fabrication, processing and long-term applications. In the stabilisation of polymers, such as rubber, polyolefins, ABS and PVC, they represent a class of compounds whose importance is increasing and is now comparable with that of the sterically hindered phenols. In Scheme 1 are listed the main phosphites, **I–VII**, and a phosphonite **VIII** industrially used today as well as some cyclic derivatives, **IX–XI**, which have been studied repeatedly in the literature.

Generally, phosphorus antioxidants are used in combination with hindered phenols and other stabilisers, but the sterically hindered aryl phosphites and phosphonites which have become available in recent times are, under some conditions, active by themselves and replace phenols especially in the processing stabilisation of polyolefins. The share of phosphorus compounds in the total consumption of antioxidants has increased in the last few years, whereas that of phenols has stagnated.

In spite of their great practical importance, the detailed mechanisms of antioxidant action of organic phosphorus compounds and, in

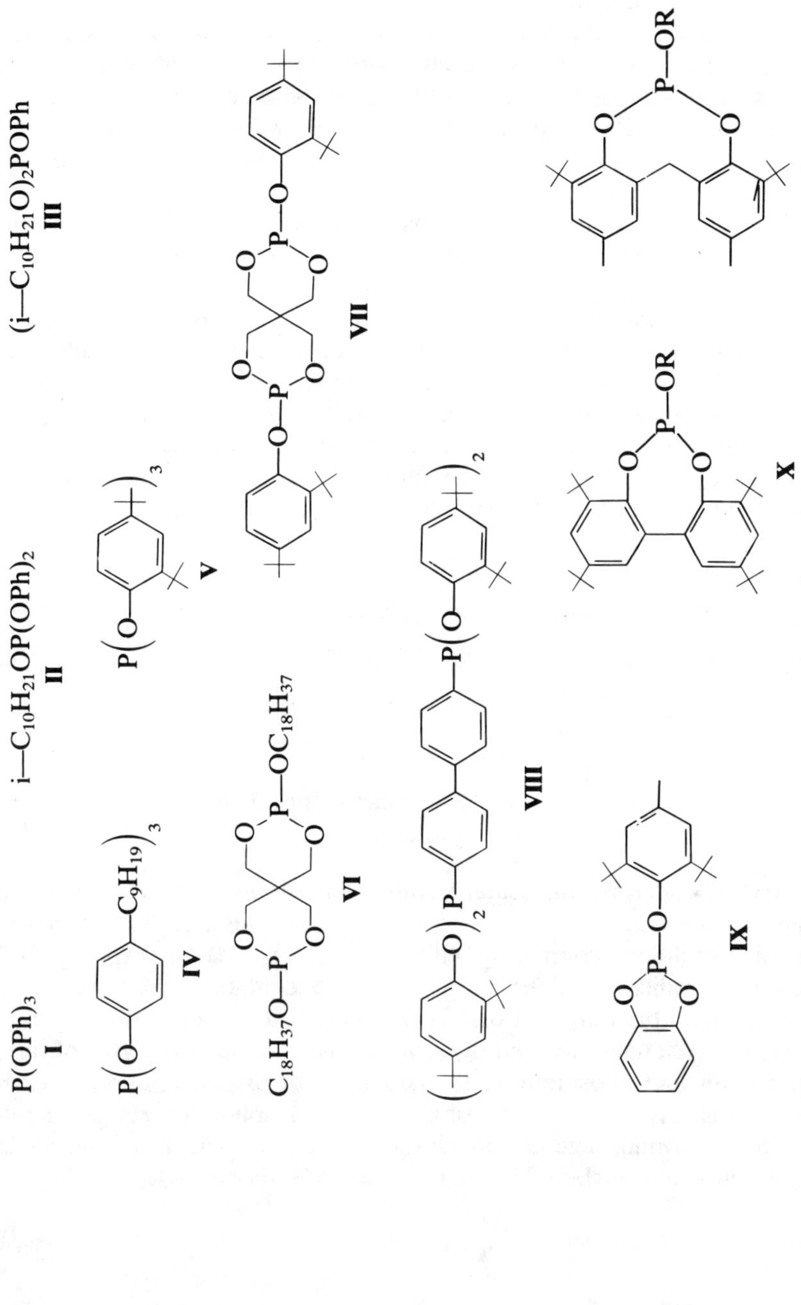

P(OPh)₃　　i—C₁₀H₂₁OP(OPh)₂　　(i—C₁₀H₂₁O)₂POPh
I　　　　　　**II**　　　　　　　　**III**

Scheme 1

particular, the relationships between chemical structure and antioxidant activity, have been studied far less intensively than those of phenols. The influence of diverse reaction conditions and the nature and role of transformation products of these inhibitors in the stabilisation process have begun to be elucidated only in recent times. These investigations started in the USSR, where Kirpichnikov's and Levin's groups, followed by Mukmeneva and Pobedimskii, pioneered work in this field; this was later extended by Rysavý & Sláma, by Humphris & Scott, by us and by others.

The efficiency and mechanisms of action of organophosphorus stabilisers have been reviewed by the Soviet authors[1,2] and by ourselves.[3,4]

Phosphites and phosphonites can function as antioxidants by several mechanisms depending on their structure, the nature of the substrate to be stabilised, and the reaction conditions.

All phosphites and phosphonites are preventive antioxidants decomposing hydroperoxides in a non-radical way and so suppressing the chain-branching step (3) in the familiar radical chain mechanism of autoxidation depicted in a simplified version in Scheme 2.

$$(RH) \rightarrow R^\cdot \quad (i)$$

$$R^\cdot + O_2 \rightarrow ROO^\cdot \quad (1)$$

$$ROO^\cdot + RH \rightarrow ROOH + R^\cdot \quad (2)$$

$$ROOH + RH \rightarrow RO^\cdot + H_2O + R^\cdot \quad (3)$$

$$RO^\cdot + RH \rightarrow ROH + R^\cdot \quad (4)$$

$$2ROO^\cdot \rightarrow \text{inactive products} \quad (5)$$

Scheme 2

Aryl phosphites, particularly those based on sterically hindered phenols, are able to act primarily as chain-breaking antioxidants under certain conditions competing with steps (2) and (4) by trapping $RO_2^\cdot$ and $RO^\cdot$ radicals and so terminating the oxidation chain reaction. Strong chain-breaking action, furthermore, can result from reaction products (phenols) formed by hydrolysis and peroxidolysis of aryl phosphites and phosphonites in the course of oxidation under special conditions. Additionally, phosphites and phosphonites act as metal-complex forming agents, blocking metal ions which cause chain initiation and branching by reaction with ROOH and RH.

Because of these various action mechanisms the antioxidant activity of phosphorus compounds is not a fixed unchangeable property but depends strongly on the oxidation conditions and the nature of the oxidisable material.

2 HYDROPEROXIDE-DECOMPOSING ANTIOXIDANT ACTION OF PHOSPHITES AND PHOSPHONITES

2.1 The Stoichiometric Reaction of Phosphites and Phosphonites with Hydroperoxides

It is well known that organic phosphites readily reduce hydroperoxides to give alcohols and the corresponding phosphate esters.[5,6] The reaction proceeds with a 1:1 stoichiometry by a non-radical mechanism, probably a S_N2 mechanism at the O—O bond with P as the nucleophile,

$$ROOH + P(OR')_3 \rightarrow RO^- + HO\overset{+}{-}P(OR')_3 \rightarrow ROH + O{=}P(OR')_3$$

(6)

obeying a second-order rate law, first-order with respect to both reactants. Rate constants have been determined for the reactions of various phosphites with 1,1-diphenylethyl hydroperoxide,[7] tert-butyl hydroperoxide,[8-10] cyclohexenyl hydroperoxide[11] and cumyl hydroperoxide[11-19] in solution.

Some selected values and activation parameters taken from our own work, under identical conditions and with systematic variation of phosphite structure, are presented in Table 1. The rate constants cover a range of three orders of magnitude. The reactivity of the phosphites towards hydroperoxides is governed mainly by polar and steric effects of the groups bound to phosphorus, decreasing with increasing electron-acceptor ability and bulk of substituent groups in the order

Alkyl phosphites > aryl phosphites > hindered aryl phosphites

Thus, the simple saturated alkyl and cycloalkyl phosphites are most reactive, whereas the fully hindered aryl and cyclic arylene phosphites react most slowly. Altogether, the rate constants are rather high, so that, especially at somewhat higher temperatures, the decomposition of hydroperoxides by phosphites is a very fast reaction.

This also holds for the decomposition of polymer hydroperoxides. The reaction of hindered aromatic phosphites with solid polyethylene

Table 1
Stoichiometric reaction of phosphites with cumyl hydroperoxide in chlorobenzene ($[P]_0 = [ROOH]_0 = 0.2$ M)[17,18]

Phosphite	$10^3 \times k$ (30°C) ($M^{-1} s^{-1}$)	$\Delta H^\ddagger$ ($kJ\,mol^{-1}$)	$\Delta S^\ddagger$ (25°C) ($J\,K^{-1}\,mol^{-1}$)
P(OEt)$_3$ [a]	350		
P(OiPr)$_3$ [a]	760		
P(OtBu)$_3$	220		
P(OPh)$_3$	31	87	+12
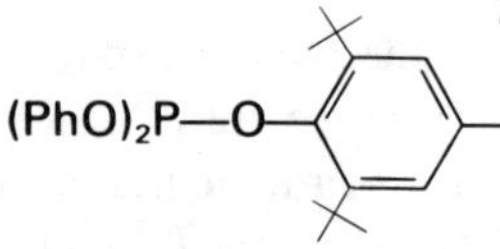 (PhO)$_2$P—O—(2,4,6-tri-tBu-phenyl) [b]	6.5		
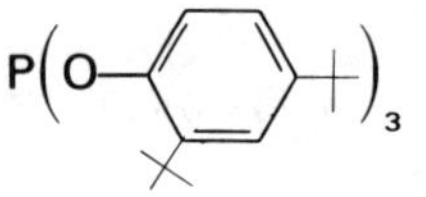 P(O—(4-tBu-phenyl))$_3$	4.9		
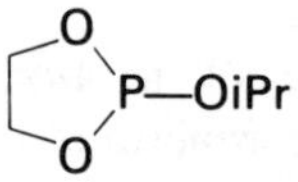 (ethylenedioxy)P—OiPr	330		
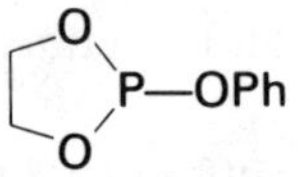 (ethylenedioxy)P—OPh	65		
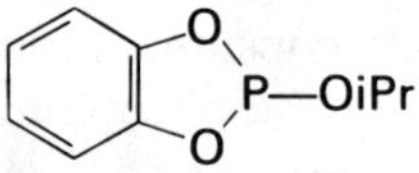 (catecholato)P—OiPr	190		
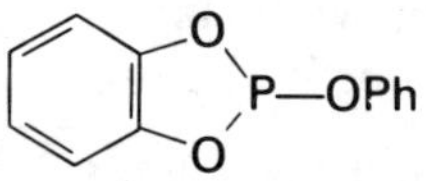 (catecholato)P—OPh	99		
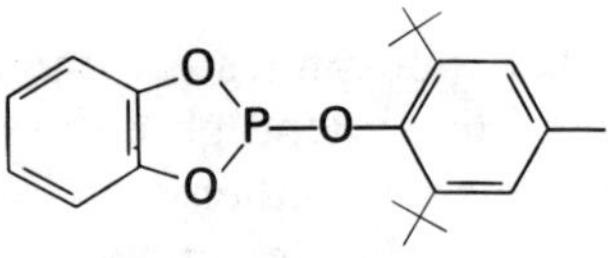 (catecholato)P—O—(2,4,6-tri-tBu-phenyl)	3.8	69	−62
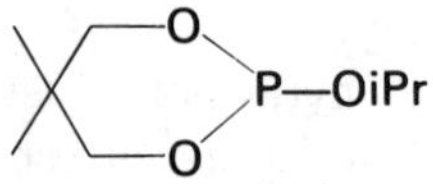 (neopentylenedioxy)P—OiPr	120		

Table 1—*contd.*

Phosphite	$10^3 \times k$ (30°C) ($M^{-1} s^{-1}$)	$\Delta H^{\ddagger}$ (kJ mol^{-1})	$\Delta S^{\ddagger}$ (25°C) (J K^{-1} mol^{-1})
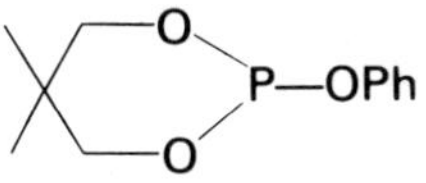 (cyclic phosphite) P—OPh	50		
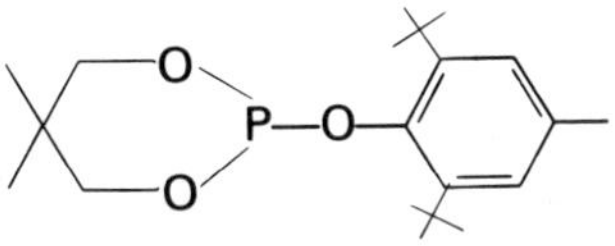 (cyclic phosphite) P—O-aryl	37	52	−105
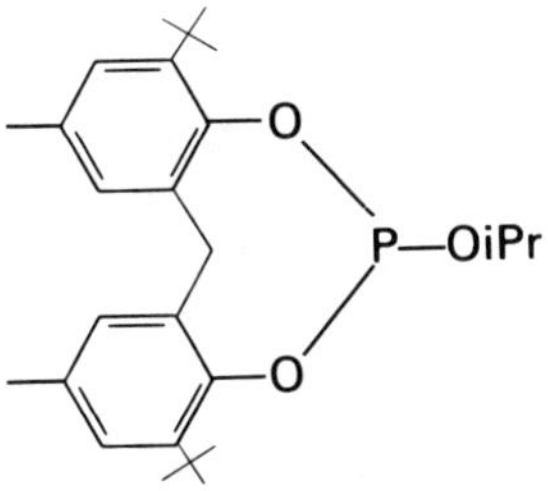 (dibenzyl cyclic) P—OiPr	8·3		
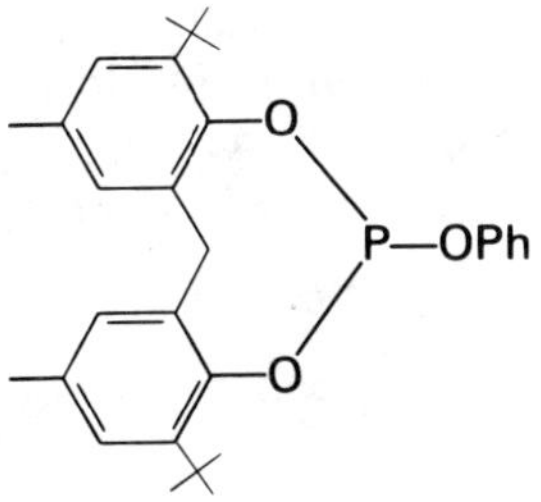 (dibenzyl cyclic) P—OPh	4·6	75	−42
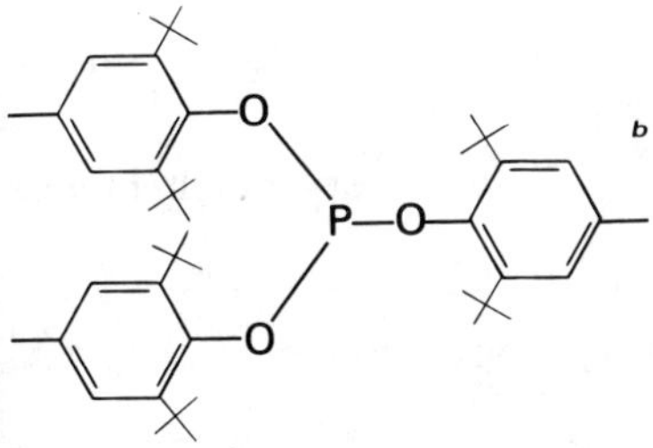 (dibenzyl cyclic) P—O-aryl[b]	0·83	102	+35
Ph—P(OPh)$_2$[c]	600		

(continued)

Table 1—contd.

Phosphite	$10^3 \times k$ (30°C) $(M^{-1} s^{-1})$	$\Delta H^\ddagger$ $(kJ\, mol^{-1})$	$\Delta S^\ddagger$ (25°C) $(J\, K^{-1}\, mol^{-1})$
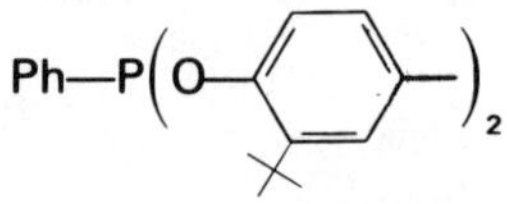 Ph—P(O—C₆H₃(X)—)₂	1500		
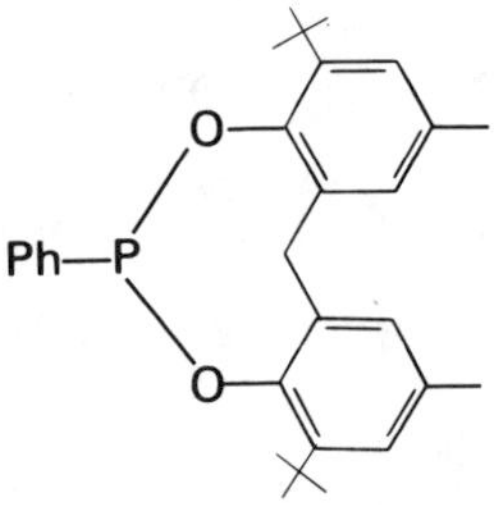 Ph—P (cyclic with two O—C₆H₃(X)—CH₂— linkages)	13		

^{a} $[P]_0 = [ROOH]_0 = 0\cdot1$ M.
^{b} $[ROOH]_0 = 0\cdot4$ M.
^{c} $[P]_0 = [ROOH]_0 = 0\cdot05$ M.

hydroperoxide at 50–80°C has been shown to proceed initially with rate constants of the same order of magnitude as with the low-molecular-weight hydroperoxides in the liquid phase.[20] At higher conversions, however, a drastic decrease of rate was observed, most pronounced at low phosphite and high hydroperoxide concentrations.[20,21]

Phosphonites react with hydroperoxides in an analogous way to phosphites giving the corresponding phosphonates,

$$ROOH + Ar\!-\!P(OR')_2 \rightarrow ROH + \underset{O}{\overset{Ar}{\diagup\!\!\!\diagdown}} P(OR')_2$$

but their hydroperoxide-decomposing reactivity is very much higher than that of phosphites (Table 1).[18,22]

2.2 The Catalytic Decomposition of Hydroperoxides by Phosphites and Phosphonites

In addition to and following the stoichiometric reaction, five-membered cyclic phosphites (1,3,2-dioxaphosphole derivatives) are

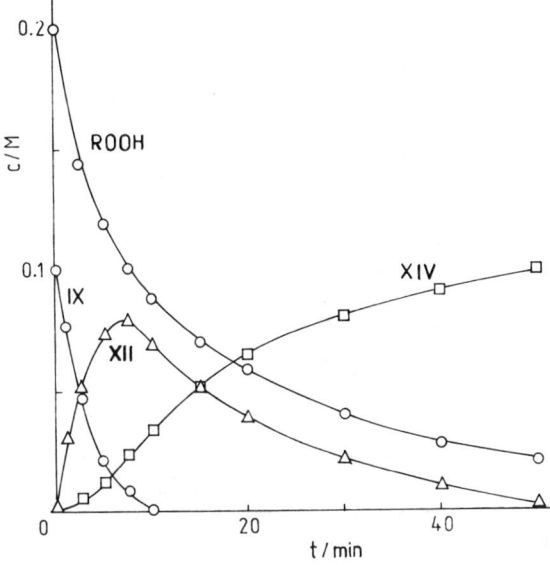

Fig. 1. Products of reaction of 2,6-di-*tert*-butyl-4-methylphenyl *o*-phenylene phosphite with cumyl hydroperoxide in chlorobenzene at 50°C.[15]

capable of reacting with alkyl hydroperoxides *catalytically*, so that more than a stoichiometric amount of hydroperoxide is decomposed by these phosphites.[13] This catalytic mode of reaction has especially been studied with cumyl hydroperoxide, where it leads to decomposition products, phenol and acetone, which are easily detectable indicators for the catalytic pathway.[10,13,17,19,23,24]

The catalytic decomposition is preceded by a fast reaction stage in which the catalytically active species is formed from the phosphite and hydroperoxide. We have observed by HPLC and ^{31}P-NMR spectroscopy two phosphates during the course of reaction of 2,6-di-*tert*-butyl-4-methylphenyl *o*-phenylene phosphite (**IX**) with cumyl hydroperoxide (Fig. 1).[15] The first (**XII**), going through a maximum of concentration, is the one formed by stoichiometric oxidation of **IX** with hydroperoxide. It is transformed into the second phosphate (**XIV**) in two ways, rapidly by reaction with water (from the dehydration of cumyl alcohol and from moisture) and more slowly with the hydroperoxide giving a peroxyphosphate (**XIII**) which is unstable and decomposes to **XIV** and phenol and acetone (Scheme 3). The proportion of these two

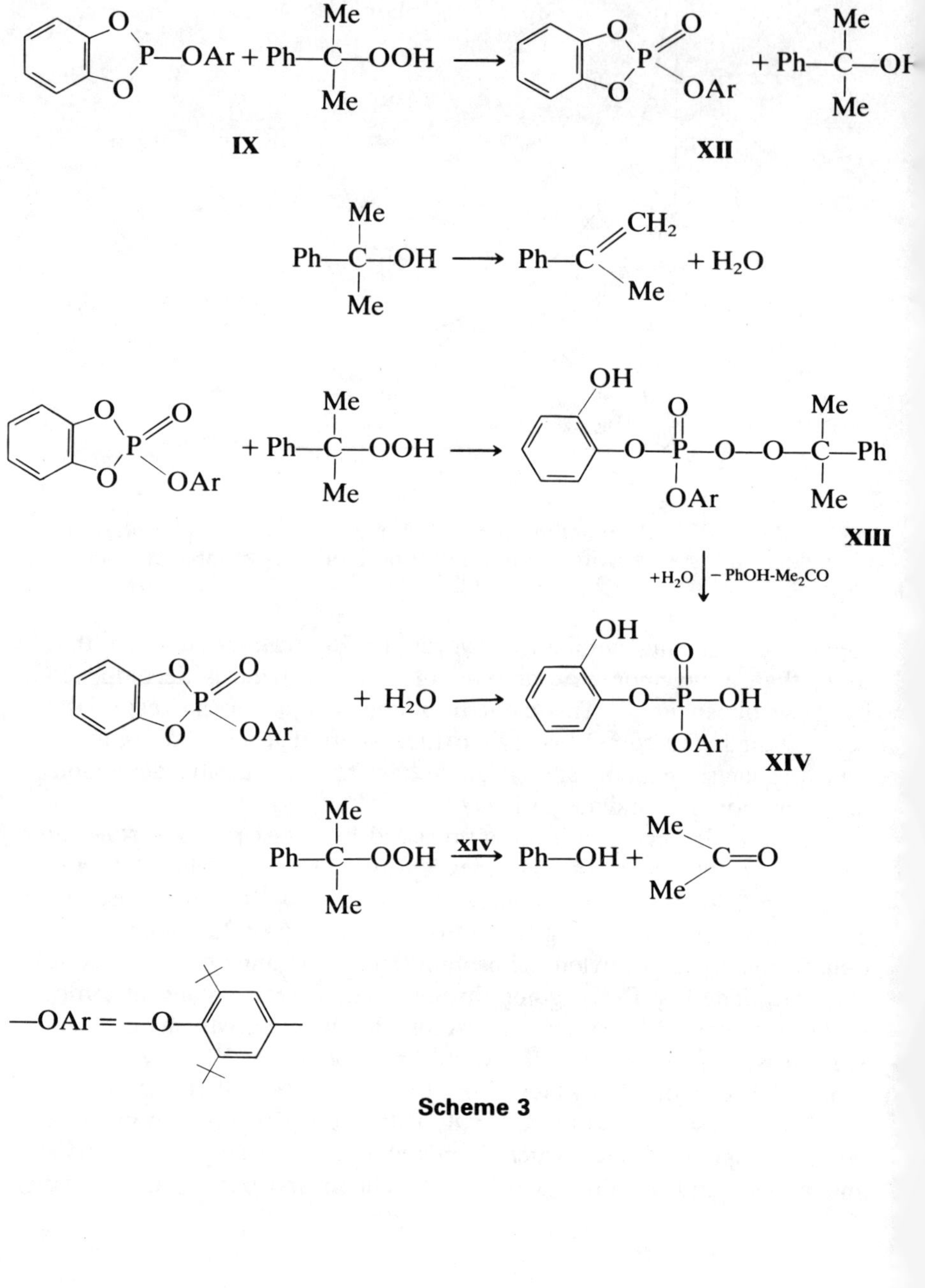

Scheme 3

routes depends on the water/hydroperoxide ratio. The open-chained hydrogen phosphate **XIV** is the acidic catalyst in the presence of which cumyl hydroperoxide is decomposed to phenol and acetone.[15]

XIV is also formed in the reaction of the *o*-phenylene phosphite **IX** with other hydroperoxides, e.g. *tert*-butyl and tetralyl hydroperoxide.[15] It is, as a substituted phenol, a very efficient chain-breaking antioxidant and responsible for the inhibiting activity of **IX** in the later stages of the oxidation of hydrocarbons (cf. Section 3).

Cyclic *o*-phenylene phosphonites react with hydroperoxides similarly to phosphites, giving initially the corresponding *o*-phenylene phosphonates which decompose hydroperoxides catalytically.[22]

Unlike the *o*-phenylene phosphites of type **IX** which catalytically decompose cumyl hydroperoxide even at −20°C, ethylene phosphites, e.g. **XV**, are not able to react with the peroxide in a molar ratio [ROOH]/[P] > 1 at temperatures below 50°C and in the absence of water.[17] A catalytic decomposition, however, is observed at 75°C. The analogous reaction products to those with **IX** (Scheme 3) are formed, but the open-chained hydroxyethyl phosphates **XVII** are unstable under the reaction conditions and condense to give polyphosphate esters.[17]

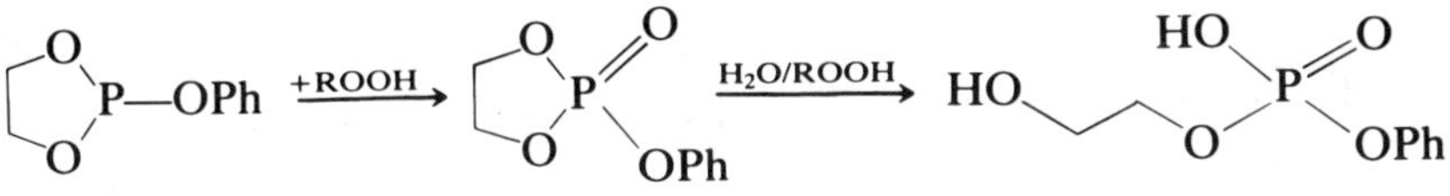

$$\text{XV} \xrightarrow{+\text{ROOH}} \text{XVI} \xrightarrow{H_2O/ROOH} \text{XVII}$$

⟶ polyphosphate esters

The catalytic decomposition of hydroperoxides by cyclic ethylene and phenylene phosphites proceeds more slowly than the stoichiometric reaction at the same phosphite concentrations. This is evident by comparing Table 1 with Table 2, which contains some second-order rate constants for the catalytic stage of the decomposition of cumyl hydroperoxide. With respect to phosphite structure, the following order holds for the reactivity in the catalytic hydroperoxide decomposition:

Non-sterically hindered aromatic phosphites > aliphatic phosphites > hindered aromatic phosphites

which is significantly different from that in the stoichiometric reaction.

Table 2
Catalytic decomposition of cumyl hydroperoxide by cyclic phosphites in chlorobenzene[17,18]

Phosphite	T (°C)	$[P]_0$ (M)	$[ROOH]_0$ (M)	$10^3 \times k$ (M s^{-1})
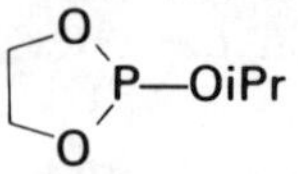 ethylenedioxy-P—OiPr	75	0·05	0·25	0·42
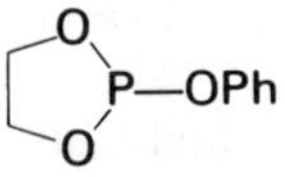 ethylenedioxy-P—OPh	75	0·05	0·2	0·84
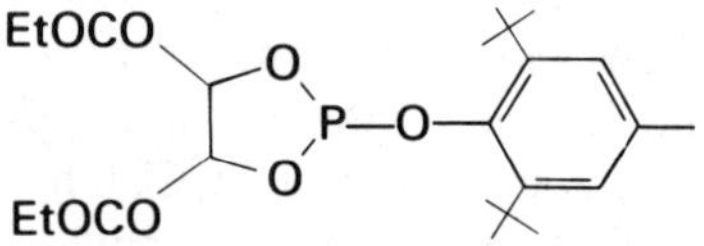 (EtOCO)-substituted dioxaphospholane P—O—Ar	75	0·05	0·25	0·22
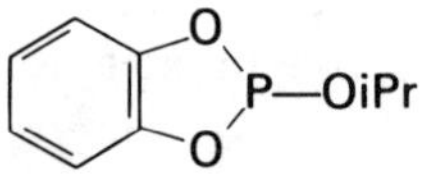 benzodioxaphosphole P—OiPr	30	0·01	0·2	1·5
	75	0·01	0·25	39
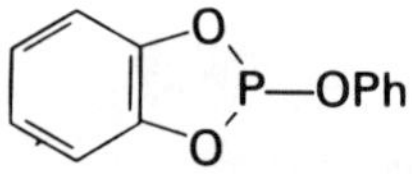 benzodioxaphosphole P—OPh	75	0·01	0·15	120
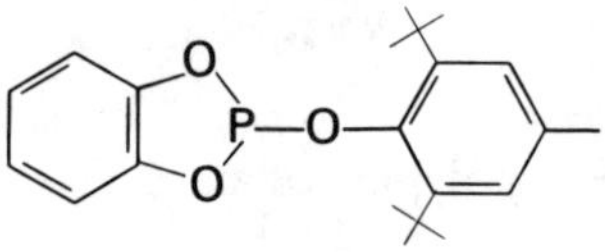 benzodioxaphosphole P—O—Ar	75	0·05	0·25	10

In contrast to some claims,[13,24] six- and higher-membered cyclic phosphites are not capable of decomposing hydroperoxides in a catalytic manner.[17,19] This parallels the well-known circumstances in the hydrolysis of cyclic phosphate esters, where only the five-membered derivatives exhibit an extraordinary reactivity.[25-27]

Certain phosphite derivatives, such as arylamidophosphites,[28] thiophosphites,[18,19] phosphorus acid esters of thiobisphenols,[29] and some diaryl hydrogen phosphites,[30] are capable of decomposing hydroperoxides catalytically.

2.3 Radical Side Reactions

The heterolytic mechanism of hydroperoxide decomposition by phosphites is accompanied by a homolytic reaction in a minor proportion. The reacting phosphite–hydroperoxide system exhibits chemiluminescence[31] and ^{31}P-CIDNP[32-34] and initiates radical oxidation and polymerisation reactions.[23,24] Intermediate radicals can be detected by ESR spectroscopy[35] and can be trapped by NO and stable nitroxyl radicals.[12,36] In the case of cumyl hydroperoxide, the formation of acetophenone as by-product can be considered a good indicator for the radical side reaction.[17,24] From the trapping experiments and the acetophenone yield, the extent of the homolytic pathway has been estimated to be below 1% for alkyl phosphites and up to 5% for aryl phosphites at 20°C.[12,36] A mechanism for this radical side reaction starting with an electron transfer from the hydroperoxide to the phosphite, has been postulated (Scheme 4).[8]

$$ROOH + P(OR')_3 \rightarrow R\dot{\bar{O}}\bar{O}H + \overset{+}{\cdot}P(OR')_3 \rightarrow RO^{\cdot} + HO\overset{+}{P}(OR')_3$$
$$HO\overset{+}{P}(OR')_3 \rightarrow HPO(OR')_2 + R'O^{\cdot}$$
$$RO^{\cdot} \rightarrow PhCOMe + Me^{\cdot} \quad (R = PhCMe_2)$$

Scheme 4

This mechanism explains the formation of aryloxyl radicals (if $R' = Ar$)[35] and acetophenone as by-product in the decomposition of cumyl hydroperoxide by aryl phosphites.[17,24] The formation of phosphonates, observed by ^{31}P-CIDNP, obviously also results from the intermediate phosphoranyl radicals with participation of the solvent.[34]

3 CHAIN-BREAKING ANTIOXIDANT ACTION OF PHOSPHITES AND PHOSPHONITES

It is now generally accepted that certain phosphites, particularly the sterically hindered aryl phosphites, are also capable of acting as chain-terminating antioxidants[1-4] in addition to their hydroperoxide-decomposing action, although in the past some workers denied the direct reaction of those phosphites with peroxyl radicals and related their observed activity to their ability to release phenols in the course of autoxidation.[15,37]

In order to act as a chain-breaking antioxidant, a stabiliser must fulfil two basic requirements: it must be able to compete effectively with the substrate RH for the chain-propagating $RO_2^{\cdot}$ radicals and it must form an efficient chain-terminating agent in this reaction with $RO_2^{\cdot}$ radicals. Hindered aryl phosphites always meet the second requirement. They react with alkylperoxyl radicals to give hindered aryloxyl radicals which are capable of terminating the autoxidation chain reaction:

$$ROO^{\cdot} + P(OAr)_3 \rightarrow ROO\dot{P}(OAr)_3 \rightarrow RO^{\cdot} + O{=}P(OAr)_3 \qquad (7)$$
$$RO^{\cdot} + P(OAr)_3 \rightarrow RO\dot{P}(OAr)_3 \rightarrow ROP(OAr)_2 + {}^{\cdot}OAr \qquad (8)$$
$$ROO^{\cdot} + ArO^{\cdot} \rightarrow \text{inactive products} \qquad (9)$$

On the other hand, alkyl phosphites react with $RO_2^{\cdot}$ radicals to form the corresponding alkyl radicals $R^{\cdot}$ which do not terminate but propagate the oxidation:

$$ROO^{\cdot} + P(OR')_3 \rightarrow ROO\dot{P}(OR')_3 \rightarrow RO^{\cdot} + O{=}P(OR')_3 \qquad (7)$$
$$RO^{\cdot} + P(OR')_3 \rightarrow RO\dot{P}(OR')_3 \rightarrow R^{\cdot} + O{=}P(OR')_3 \qquad (10)$$
$$R^{\cdot} + O_2 \rightarrow ROO^{\cdot}$$

In both cases the phosphite is oxidised by $RO_2^{\cdot}$ radicals in a first reaction step giving the phosphate and $RO^{\cdot}$ radicals. Crucial for the antioxidative efficiency of phosphites, therefore, is the second step, their reaction with the *alkoxyl* radicals formed in the first step. Only those phosphites which react with alkoxyl radicals by *substitution* to give isomeric phosphite-releasing chain-terminating aryloxyl radicals can act as primary antioxidants. In this case the intermediate alkoxyphosphoranyl radicals undergo α-*scission*, whereas β-*scission* leads to *oxidation* of the phosphite and chain-propagating alkyl radicals:

$$RO^{\cdot} + {>}P{-}OAr \longrightarrow RO{-}\overset{|}{\underset{|}{P}}{-}OAr \underset{\beta}{\overset{\alpha}{\rightleftarrows}} \begin{array}{l} RO{-}P{<} + ArO^{\cdot} \quad \text{Substitution} \\ {>}P{\overset{O}{\underset{OAr}{\nearrow}}} + R^{\cdot} \quad \text{Oxidation} \end{array}$$

The diverse reactions of phosphoranyl radicals which depend on their structure are understood rather well and have been reviewed in a more comprehensive context.[38,39]

3.1 The Inhibition of Autoxidation by Phosphites and Phosphonites

The chain-breaking antioxidant efficiency of phosphites can best be studied in the inhibition of hydrocarbon oxidation by phosphites *at low temperatures* (<100°C), where the homolytic decomposition of hydroperoxides formed does not yet occur and, therefore, inhibition is also not caused by the hydroperoxide-decomposing action of the phosphites. From the kinetics of such inhibited oxidations the two quantities characterising the effectiveness of a chain-breaking antioxidant can be evaluated; namely, the rate constant, k_7, of its reaction with peroxyl radicals and the stoichiometric inhibition factor f, the number of peroxyl radicals trapped by one molecule of the antioxidant. The constant k_7 determines the rate of the inhibited oxidation and the effective antioxidant concentration; f determines the length of the inhibition period.

The inhibition of the oxidation of hydrocarbons by organic phosphites at low temperatures has been studied by several authors, with partially contradictory results and interpretations.[15,23,40,41]

In Fig. 2 is depicted the fate of the phosphorus species studied by ^{31}P-NMR spectroscopy in the course of cumene oxidation inhibited by the sterically hindered 2,6-di-*tert*-butyl-4-methylphenyl neopentylene phosphite at 65°C.[41] The phosphite is consumed at a rate slightly higher than the rate of initiation ($r_i = 2 \cdot 4 \times 10^{-6}$ M s^{-1}), giving the

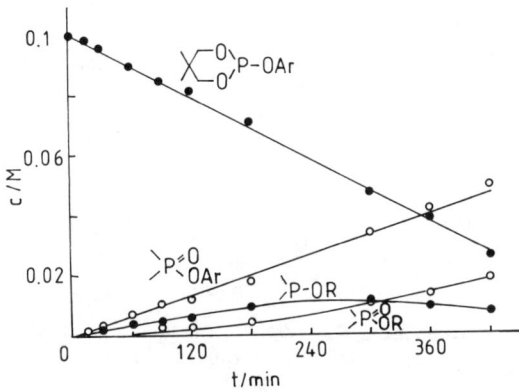

Fig. 2. Transformation of 2,6-di-*tert*-butyl-4-methylphenyl neopentylene phosphite in the AIBN-initiated oxidation of cumene in chlorobenzene at 65°C ([P]$_0$ = 0·1 M, [RH] = 1·71 M, [AIBN]$_0$ = 0·1 M).[41]

corresponding phosphate and the substituted alkyl phosphite which is further oxidised to the substituted alkyl phosphate. The ratio of oxidation to substitution products of the parent phosphite,

$$\left[\begin{array}{c}\diagup\\ \diagdown\end{array}\!\!P\!\!\diagup_{OAr}^{O}\right]\Big/\left(\left[\begin{array}{c}\diagup\\ \diagdown\end{array}\!\!POR\right]+\left[\begin{array}{c}\diagup\\ \diagdown\end{array}\!\!P\!\!\diagup_{OR}^{O}\right]\right)$$

is larger than 1, as might be expected according to the mechanism, reactions (7)–(9), indicating, as does the higher consumption rate of the phosphite, that additional processes must be involved in the oxidation of the parent phosphite. The reaction of the phosphite with hydroperoxides formed in the course of oxidation [reaction (6)] is the main process involved.

In the oxidation of tetralin inhibited by 2,6-*tert*-butyl-4-methylphenyl *o*-phenylene phosphite (**IX**), the situation is similar (Fig. 3),[41] but the hydroxylphenyl phosphate **XIV** appears additionally, due to ring opening of the intermediate phosphate **XII** by peroxidolysis and hydrolysis according to Scheme 3.

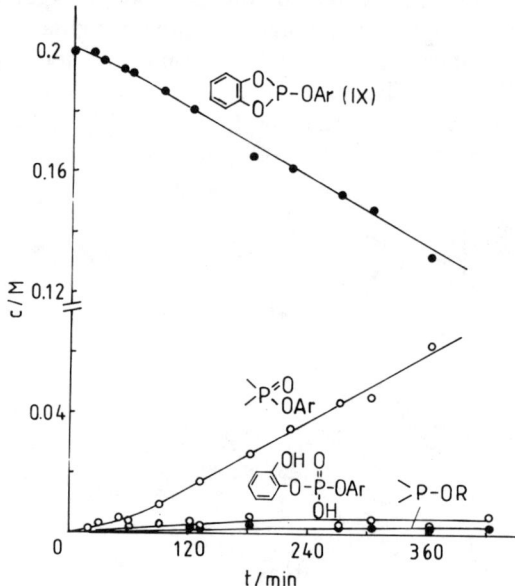

Fig. 3. Transformation of 2,6-di-*tert*-butyl-4-methylphenyl *o*-phenylene phosphite in the AIBN-initiated oxidation of tetralin in chlorobenzene at 65°C ([P]$_0$ = 0·2 M, [RH] = 1·75 M, [AIBN]$_0$ = 0·1 M).[41]

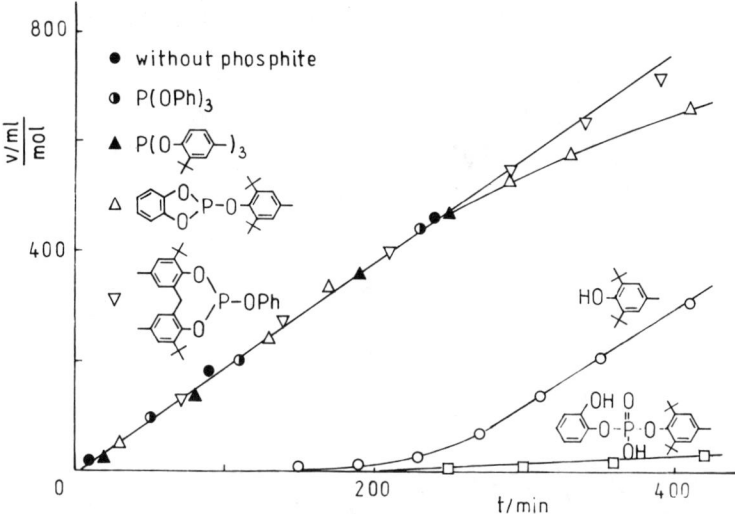

Fig. 4. Oxidation of tetralin in the presence of phosphites in chlorobenzene at 60°C ($[P]_0 = 5 \times 10^{-4}$ M, [AIBN] = 5×10^{-3} M, [RH] = 3·73 M).[68]

The oxygen uptake of tetralin at 60°C in the presence of 5×10^{-4} M antioxidant is shown in Fig. 4.[41] At this low concentration, none of the phosphites used retarded the oxidation, whereas the hindered phenols, 2,6-di-*tert*-butyl-4-methylphenol (BHT) and the hydroxyphenyl phosphate **XIV**, caused a pronounced induction period. The behaviour of the *o*-phenylene phosphite **IX** is exceptional in that it acted antioxidatively in the later stages of the reaction. As mentioned above, this is because this cyclic phosphite is transformed into the hydroxyphenyl phosphate **XIV** in the course of oxidation.[10,17]

At higher concentrations, phosphites with open-chain sterically hindered aryl groups inhibit the autoxidation of tetralin and cumene (Fig. 5). The non-hindered aryl phosphites are inactive initially, but become active in the later stages of oxidation. Sterically hindered eight-membered cyclic arylene phosphites are also inefficient. Altogether, the antioxidative efficiency of all phosphites is less than that of the hindered phenol, BHT.

It is remarkable that phenylphosphonites including the hindered aryl ones, do not initially inhibit the oxidation of cumene and tetralin at 65°C.[18,42] In the further course of oxidation, however, inhibition becomes evident as in the case of phenyl phosphites referred to above.

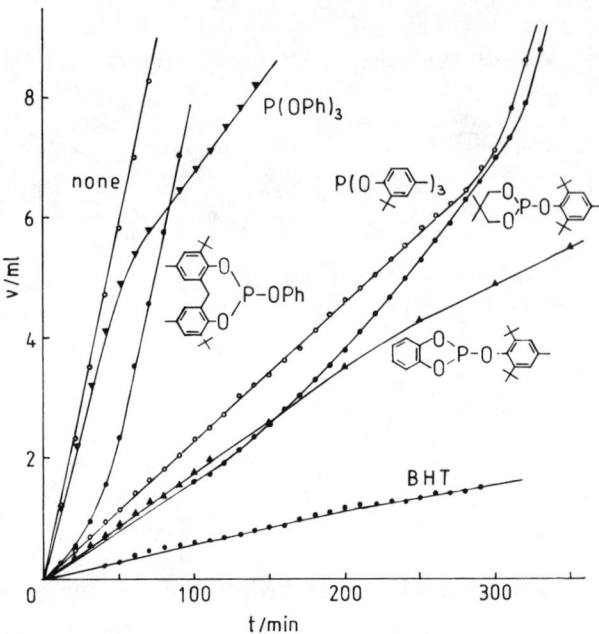

Fig. 5. Oxidation of cumene in the presence of phosphites in chlorobenzene at 65°C ($[P]_0 = 0.05$ M, $[AIBN] = 0.1$ M, $[RH] = 1.71$ M).[41]

Table 3 reveals that, at equal phosphite concentrations, the induction periods are longer and the rates of inhibited oxidation are lower in cumene than in tetralin, i.e. the antioxidative efficiency of phosphites depends on the nature of the hydrocarbon substrate; it is higher, the less oxidisable the substrate.[41] Stoichiometric inhibition factors, f, and rate constants of the reaction of phosphites with peroxyl radicals, k_7, evaluated from the inhibited oxidations of these two hydrocarbons are also listed in Table 3. For the sterically hindered aryl phosphites k_7 is in the order of 2×10^2 to 1×10^3 l mol^{-1} s^{-1} at 65°C, which is 10^2 to 10^3 times smaller than some other values reported in the literature,[40] but in good agreement with the values determined from the inhibition of alkyl phosphite oxidations by these phosphites[48] and with a value extrapolated directly from ESR-measurements of the rate of *tert*-butyl peroxyl radicals with phosphites at lower temperatures.[43] These values of k_7 are about 50 times smaller than the corresponding values for hindered phenols, reflecting the smaller retarding effects of phosphites on the rates of inhibited oxidations in comparison with phenols.

The induction periods and stoichiometric inhibition factors for phosphites are also smaller ($f < 1$) than those of phenols (BHT: $f = 2$). The dependencies of the induction periods and stoichiometric factors on the concentration of the phosphites are striking. The length of induction period is not directly proportional to the phosphite concentration but decreases at a greater rate than the antioxidant concentration. Its dependence on $[P]_0$ is approximately linear but below some apparent critical antioxidant concentration no induction period can be observed (Fig. 6).[41] This is also reflected in the dependence of the stoichiometric factor, f, on the antioxidant concentration: f is not a constant, as in the case of phenols, but decreases with decreasing phosphite concentration from 1 at very high concentrations to zero at very low concentrations (Fig. 7). These observations may be rationalised kinetically by taking into account the destruction of the phosphite according to eqn. (6) in the course of oxidation,[41] which leads to the following expressions: for the induction period,

$$t_{ind} = \frac{1}{r_i}\left\{[P]_0 - \frac{k_2[RH]}{2k_7}\ln\left(1 + \frac{2k_7[P]_0}{k_2[RH]}\right)\right\} \quad (11)$$

for the stoichiometric inhibition factor,

$$f = 1 - \frac{k_2[RH]}{2k_7[P]_0}\ln\left(1 + \frac{2k_7[P]_0}{k_2[RH]}\right) \quad (12)$$

and for the critical antioxidant concentration,

$$[P]_{cr} \approx \frac{k_2[RH]}{2k_7} \quad (13)$$

The destruction of the phosphite by hydroperoxide is more pronounced the more hydroperoxide is formed in the reaction, e.g. the lower the phosphite concentration, the more easily is the hydrocarbon oxidised. That means that the chain-breaking antioxidant efficiency of phosphites depends not only on their intrinsic antioxidative properties (k_7) but also on the nature of the substrate to be stabilised. The higher the oxidisability of the substrate (k_2), the shorter the induction period, the lower the stoichiometric factors of inhibition and the higher the effective antioxidant concentration of the phosphite needed for inhibition. Therefore, in the easily oxidisable aralkyl hydrocarbons, the required minimum phosphite concentrations are rather high. The same should be true for the analogous polymers (polystyrene, polydienes,

Table 3
AIBN-initiated oxidation of cumene and tetralin inhibited by phosphites and BHT, respectively, in chlorobenzene at 65°C ([P]$_0$] = 0·05 M, [AIBN] = 0·1 M, [cumene] = 1·71 M, [tetralin] = 1·75 M)[41]

Antioxidant	Cumene				Tetralin			
	t_{ind} (min)	f	$10^6 \times r_0$ (M s^{-1})	k_7 (M^{-1} s^{-1})	t_{ind} (min)	f	$10^6 \times r_0$ (M s^{-1})	k_7 (M^{-1} s^{-1})
None	0	0	10·5		0	0	19·8	
P(OPh)$_3$	0	0	9·3		0	0	12·8	
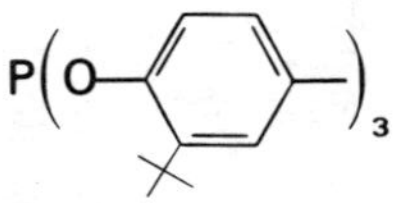 [structure: P(Ar)$_3$]	300	0·72	3·33	300	95	0·26	5·46	640
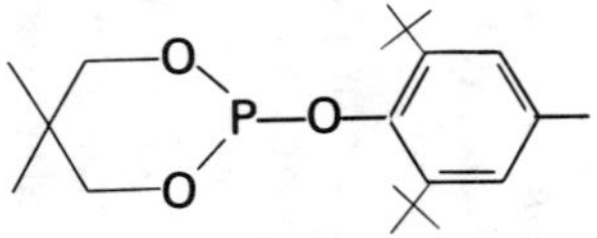 [structure: phosphite]	310	0·75	3·04	500	200	0·55	4·26	1420
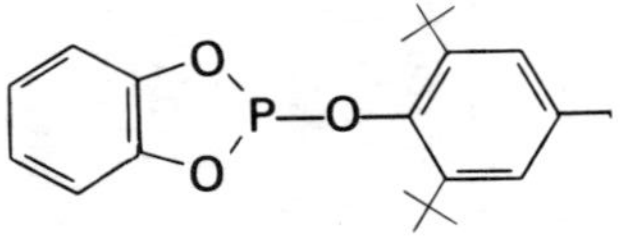 [structure: catechol phosphite]			3·16	400			5·03	810

7·6	25	3·52	230	2·43	1·5 × 10⁴
		0·13	45	2·0	
				2·0	
				2·56	3·0 × 10⁴

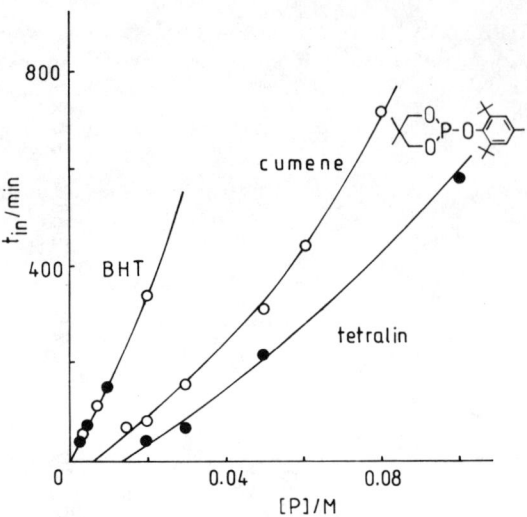

Fig. 6. Induction periods in the AIBN-initiated oxidation of cumene and tetralin inhibited by 2,6-di-*tert*-butyl-4-methylphenyl neopentylene phosphite and BHT, respectively (conditions as in Fig. 5).[41]

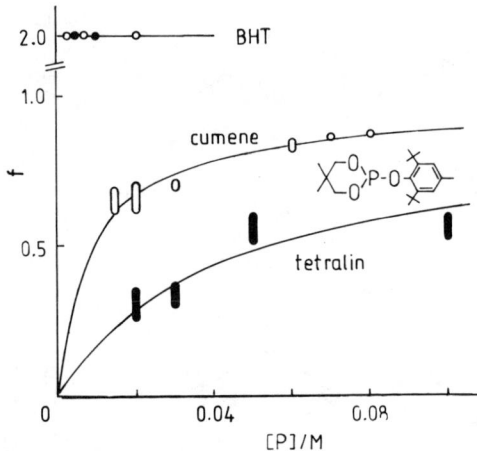

Fig. 7. Stoichiometric inhibition factors of 2,6-di-*tert*-butyl-4-methylphenyl neopentylene phosphite and BHT, respectively, in the AIBN-initiated oxidation of cumene and tetralin. Curves best fitted with eqn (12) giving $k_7 = 370$ $M^{-1}s^{-1}$ for cumene and 1×10^3 $M^{-1}s^{-1}$ for tetralin (conditions as in Fig. 5)[41]

polyacrylonitrile, polyethers and their copolymers), whereas in saturated aliphatic hydrocarbons and polyolefins these minimum concentrations should be lower.

Unfortunately, there are few kinetic studies of phosphite-inhibited oxidations of polymers at low temperatures to prove these suggestions; only the oxidation of solid isotactic polypropylene at 85°C has been reported. Here, the hindered tris(2-*tert*-butyl-4-methylphenyl) phosphite as well as the non-hindered triphenyl phosphite are effective as inhibitors[40] above some 'critical' antioxidant concentration, but these concentrations are of the same order (and not lower) as in the case of the higher oxidisable liquid aralkanes.[44] This may reflect the lower rate constants, k_7, of the reaction of the phosphites with polymer peroxyl radicals in the solid state due to restricted diffusion compared with the liquid phase. Therefore, the antioxidant activity of phosphites in the solid polymer is lower than in liquid hydrocarbons.[40]

3.2 Reactions of Phosphites and Phosphonites with Alkylperoxyl Radicals

The structural features determining the chain-breaking antioxidant ability of organophosphorus compounds have been elucidated in detail by studying their reactions with alkylperoxyl radicals and alkoxyl radicals. These investigations confirm and expand the results obtained in the inhibition experiments mentioned so far.

Peroxyl radicals, generated by thermal decomposition of azo-compounds in the presence of oxygen,

$$R-N=N-R \rightarrow 2R^\cdot + N_2$$
$$R^\cdot + O_2 \rightarrow ROO^\cdot$$

were reacted with phosphites and phosphonites.[45–50]

Trialkyl phosphites react with O_2 in the presence of azo-bis-isobutyronitrile (AIBN) in a rapid radical chain reaction to give the corresponding phosphates[45,47–49]

$$P(OR')_3 + \tfrac{1}{2}O_2 \rightarrow O=P(OR')_3$$

according to eqns (7) and (10). Dialkyl arylphosphonites behave analogously giving the corresponding phosphonates:[50]

$$PhP(OR')_2 + \tfrac{1}{2}O_2 \rightarrow Ph\overset{\overset{\displaystyle O}{\|}}{P}(OR')_2$$

Therefore alkyl phosphites and phosphonites cannot act as chain-breaking antioxidants.

From the rates of oxidation, the rate constants k_7 for the reactions of trialkyl phosphites with 2-cyanopropyl-2-peroxyl radicals have been calculated (Table 4).[49] They are in the order of $10^3\,l\,mol^{-1}\,s^{-1}$ at 65°C,

Table 4
Products and kinetic parameters of reactions of phosphites and phosphonites with oxygen/AIBN in chlorobenzene at 65°C ($[P]_0 = 0.2$ M, [AIBN] = 0.1 M, $r_i = 2.4 \times 10^{-6}\,M\,s^{-1}$)[42,49]

	$>$P—OR' (mole fraction)	$>$P(=O)OR'	$>$P—OCMe$_2$CN (mole fraction)	r/r_i	k_7 ($M^{-1}\,s^{-1}$)
P(OEt)$_3$	1		0	1 500[a]	1 840
P(OiPr)$_3$	1		0	1 080[a]	1 330
P(OtBu)$_3$	1		0	580[a]	730
P(OPh)$_3$	0.96		0.04	12	
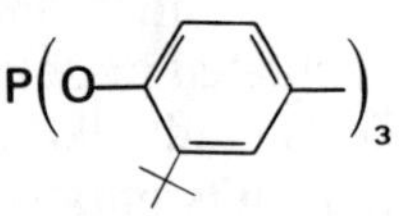 P(O—C$_6$H$_4$—)$_3$	0.7		0.3	1.8	
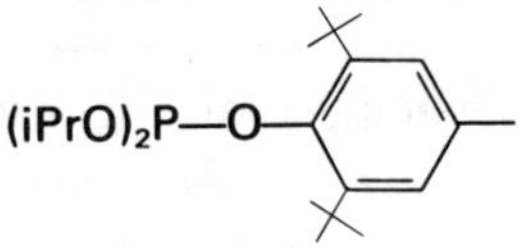 (iPrO)$_2$P—O—Ar	0.7		0.3	1.7	
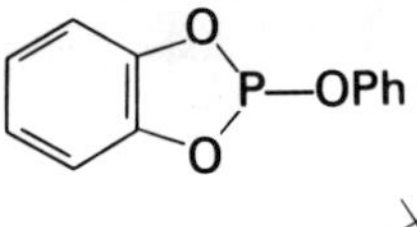 catechol-P—OPh	0.8		0.2[b]	4	
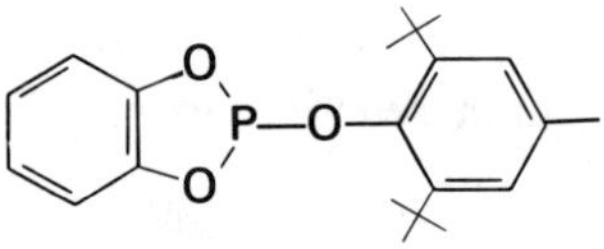 catechol-P—O—Ar	0.5		0.5	1.2	
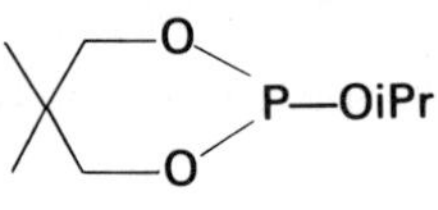 neopentyl-P—OiPr	1		0	790	970
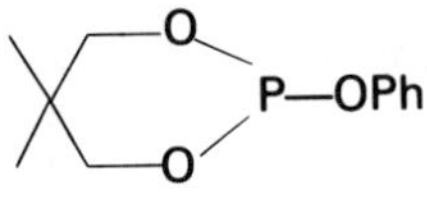 neopentyl-P—OPh	1		0	55	660

Table 4—contd.

$>\!\!P\!\!-\!\!OR'$	$>\!\!P(\!=\!\!O)\!-\!\!OR'$ (mole fraction)	$>\!\!P\!\!-\!\!OCMe_2CN$ (mole fraction)	r/r_i	$\dfrac{k_7}{(M^{-1}\,s^{-1})}$
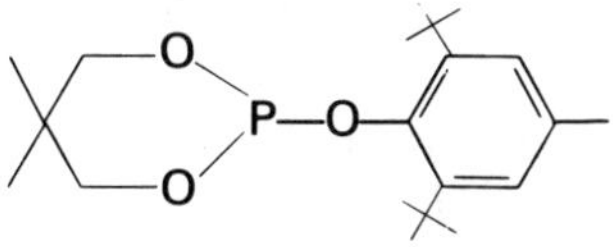 [cyclic phosphite with aryl-O group]	0·6	0·4	1·4	
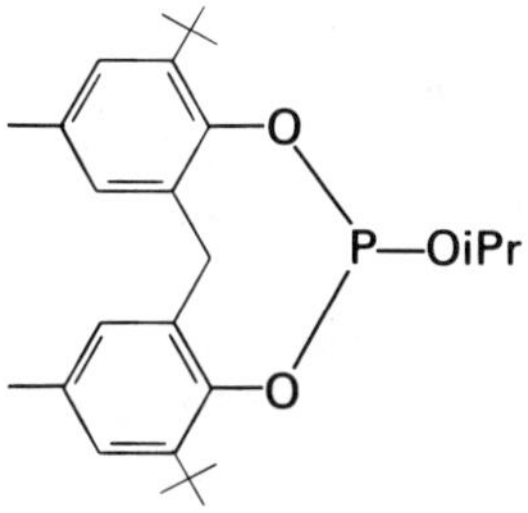 [bis-aryloxy P—OiPr]	1	0	8	100
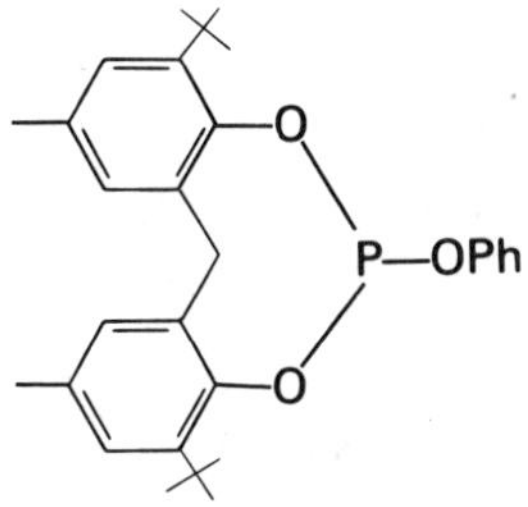 [bis-aryloxy P—OPh]	1	0	6	80
Ph—P(OiPr)$_2$	1	0	22	120
Ph—P(OPh)$_2$	1	0	28	150
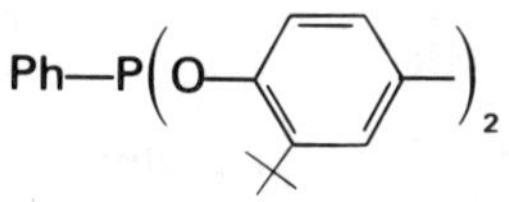 Ph—P(O—Ar)$_2$	0·95	0·05	4·8	
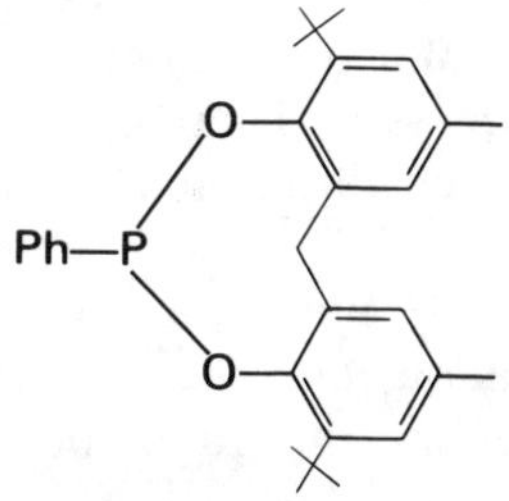 Ph—P[bis-aryloxy cyclic]	1	0	5·1	27

[a] $[AIBN] = 1{\cdot}0 \times 10^{-3}$ M, $r_i = 2{\cdot}4 \times 10^{-8}$ M s^{-1}.
[b] Ring-opening products.

similar to that expected by extrapolating the rate constants determined directly by kinetic ESR spectroscopy[43] at lower temperatures. The rate constant k_7 decreases with increasing bulk of R'.

Non-sterically hindered aryl and arylene phosphites and arylphosphonites exhibit the same behaviour in the AIBN-initiated oxidation as the alkyl esters. They are oxidised to give the corresponding phosphates[46,48,49] and phosphonates,[42] respectively (see Table 4).

In the reaction of triphenyl phosphite a small amount (4 mol %) of the substitution product, ROP(OPh)$_2$, is found besides the oxidation product, OP(OPh)$_3$.[49] In this case the intermediate phosphoranyl radical, ROṖ(OPh)$_3$, undergoes α-scission in addition to β-splitting. Because phenoxyl radicals are formed in the α-scission, triphenyl phosphite should act as chain-terminating antioxidant to a certain degree.

Phosphites with exocyclic sterically hindered aryl groups, e.g. **V**, **VII**, **IX**, react with oxygen/AIBN only very slowly. Chain propagation is almost completely suppressed ($r/r_i \approx 1$).[1,49] In addition to the oxidation product of the phosphite, the substitution product and its phosphate are also formed (Table 4):[49]

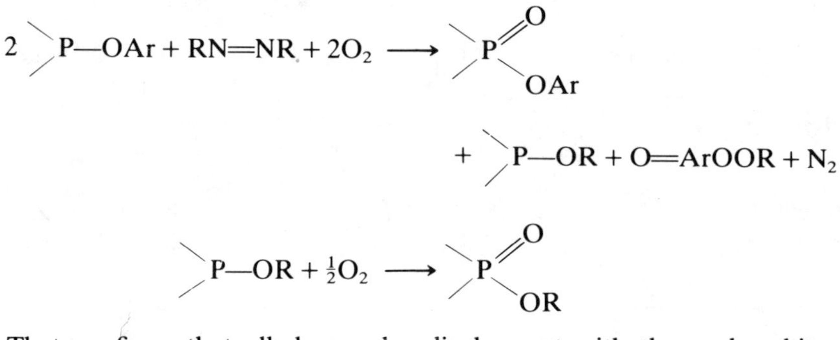

$$2 \ \diagdown\!\!P\!\!-\!\!OAr + RN\!\!=\!\!NR + 2O_2 \longrightarrow \ \diagdown\!\!P\diagup\!\!\!\diagdown\!\!\substack{O \\ OAr}$$

$$+ \ \diagdown\!\!P\!\!-\!\!OR + O\!\!=\!\!ArOOR + N_2$$

$$\diagdown\!\!P\!\!-\!\!OR + \tfrac{1}{2}O_2 \longrightarrow \ \diagdown\!\!P\diagup\!\!\!\diagdown\!\!\substack{O \\ OR}$$

That confirms that alkylperoxyl radicals react with these phosphites according to eqns (7) and (8) by α-scission of the intermediate alkoxyphosphoranyl radicals releasing sterically hindered aryloxyl radicals which effectively trap peroxyl radicals [eqn (9)]. Therefore, sterically hindered aryl phosphites can act as chain-terminating antioxidants.

In the reaction of mixtures of alkyl and sterically hindered aryl phosphites with oxygen/AIBN, the chain oxidation of the alkyl phosphite is inhibited by the aryl phosphite. From the rates of consumption of the phosphites, the rate constants, k_7, of the reactions of hindered aryl phosphites with 2-cyanopropyl-2-peroxyl radicals are accessible.[49] Again, they are in the order of $(1-3) \times 10^2 \, l \, mol^{-1} \, s^{-1}$ at 65°C and somewhat lower than those of alkyl phosphites.

Unlike the open-chained derivatives, sterically hindered cyclic arylene phosphites, e.g. dibenzo[d,g]-1,3,2-dioxaphosphocins derived from methylenbisphenol '2246', with non-hindered exocyclic aryl groups are completely oxidised in the reaction with oxygen/AIBN (Table 4). Substitution products by ring opening are not formed. Remarkably, hindered aryl phenylphosphonites also react with AIBN/O_2 mainly by oxidation; cyanoisopropyl phenylphosphonites formed by displacement of a hindered aryloxyl radical are observed only in minor proportions (Table 4).[42]

Because of this predominant oxidation, sterically hindered aryl phosphonites and cyclic arylene phosphites are less effective chain-breaking antioxidants in low-temperature oxidations than the acyclic hindered aryl phosphites (see Section 3.1).

3.3 Reactions of Phosphites and Phosphonites with Alkoxyl Radicals

The reactions of alkoxyl radicals with organic phosphites and phosphonites have been extensively studied[5,46,51-58] and reviewed,[38,39] emphasising the structure and reaction modes of the intermediate alkoxyphosphoranyl radicals. With respect to the mechanism of antioxidant action of phosphorus compounds, product studies with *tert*-butyloxyl radicals generated from di-*tert*-butyl peroxide and di-*tert*-butyl peroxalate (DTBPO) are most relevant.[46,54,58]

With *tert*-butyloxyl radicals from DTBPO,

$$tBuO\text{—}OCOCOO\text{—}OtBu \rightarrow 2tBuO\cdot + 2CO_2$$

aliphatic phosphites react to form the corresponding phosphates according to eqn (10) (Table 5), in accordance with previous results and confirming that these phosphites are not suitable as chain-breaking antioxidants.

Aryl phosphites including the non-sterically hindered ones, react with *tert*-butoxyl radicals at low temperatures (50°C) exclusively by substitution according to eqn (8) to give aroxyl radicals and *tert*-butyl phosphites which are easily oxidised to *tert*-butyl phosphates. Phosphites with several aroxyl groups react until all aroxyls are displaced,

Table 5
Products of reaction of phosphites with di-*tert*-butyl peroxalate in chlorobenzene at 50°C ($[P]_0 = 0.2$ M, $[DTBPO]_0 = 0.25$ M)[58]

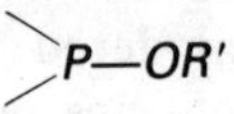

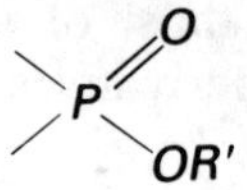

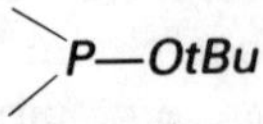

>P—OR'	>P(=O)OR'	>P—OtBu	Phosphonates
		(mole fraction)	
P(OEt)$_3$	1·0	0	—
P(OiPr)$_3$	0·84	0	—
P(OPh)$_3$	0	1·0	—
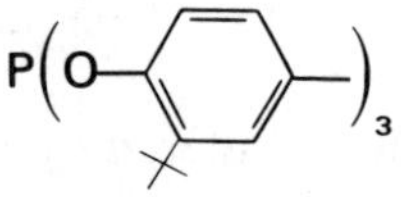 P(O—C$_6$H$_3$(tBu)$_2$)$_3$	0	0·3	0·6
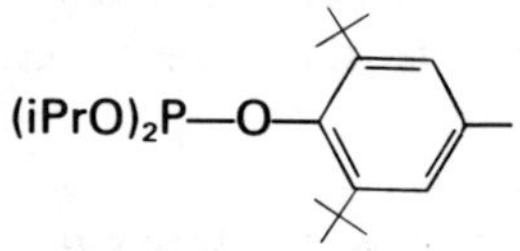 (iPrO)$_2$P—O—C$_6$H$_3$(tBu)$_2$	0	0·5	0·5
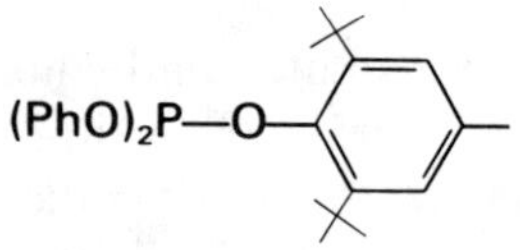 (PhO)$_2$P—O—C$_6$H$_3$(tBu)$_2$	0	0·5	0·45
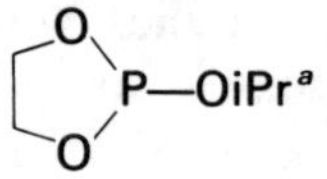 (1,3,2-dioxaphospholane)P—OiPr[a]	0·44	0	—
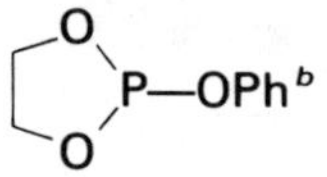 (1,3,2-dioxaphospholane)P—OPh[b]	0·2	0·8	—
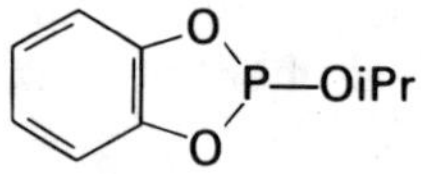 (benzo-1,3,2-dioxaphospholane)P—OiPr	0·8	0·2	—
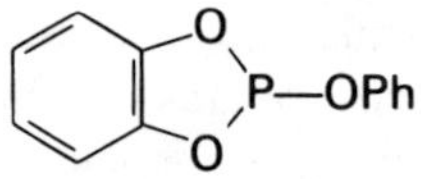 (benzo-1,3,2-dioxaphospholane)P—OPh	0	0·9	—

Table 5—*contd.*

>P—OR'	>P(=O)(OR')	>P—OtBu (mole fraction)	Phosphonates
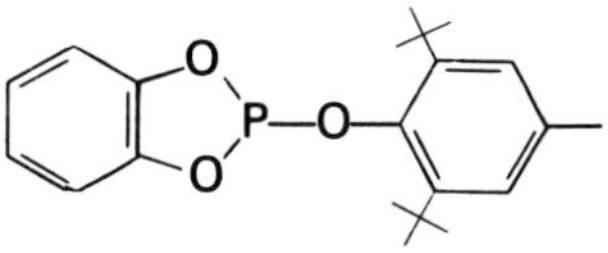 catechol cyclic phosphite with O-aryl (di-t-butyl)	0	0·65	0·35
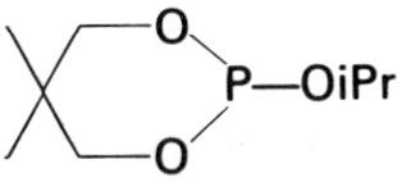 neopentyl cyclic P—OiPr	0·88	0	—
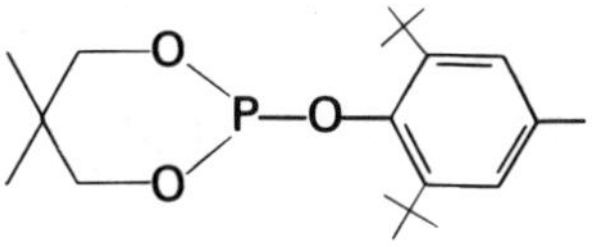 neopentyl cyclic P—O-aryl (di-t-butyl)	0	0·8	0·2
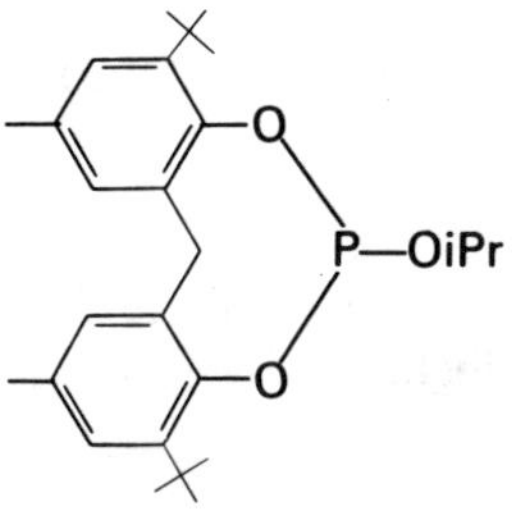 bis-phenol methylene cyclic P—OiPr	+	+	+
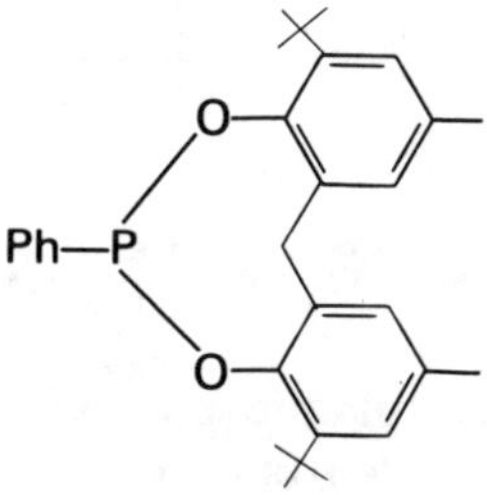 bis-phenol methylene cyclic Ph—P	1·0	0	0

[a] $[DTBPO]_0 = 0·2$ M, 55°C.
[b] $[DTBPO]_0 = 0·1$ M, 55°C.

after which the *tert*-butyl phosphite is oxidised:

$$P(OPh)_3 \xrightarrow[-PhO^\cdot]{+tBuO^\cdot} tBuOP(OPh)_2 \xrightarrow[-PhO^\cdot]{+tBuO^\cdot} (tBuO)_2POPh$$

$$\xrightarrow[-PhO^\cdot]{+tBuO^\cdot} (tBuO)_3P \xrightarrow[-tBu^\cdot]{+tBuO^\cdot} (tBuO)_3P=O$$

At higher temperatures (>50°C) the *tert*-butyl esters formed are unstable and decompose to hydrogen phosphites by elimination of isobutylene:[46,57]

$$tBuOP(OPh)_2 \rightarrow H_2C=CMe_2 + O=PH(OPh)_2 \qquad (14)$$

Furthermore, it seems that β-scission of the intermediate phosphoranyl radicals also takes place, so that aryl phosphates are observed as well as hydrogen phosphites.[57]

In the reaction of 4-methylaryl phosphites with *tert*-butoxyl radicals, phosphonates are always observed as by-products, often in high yields,[58] e.g.

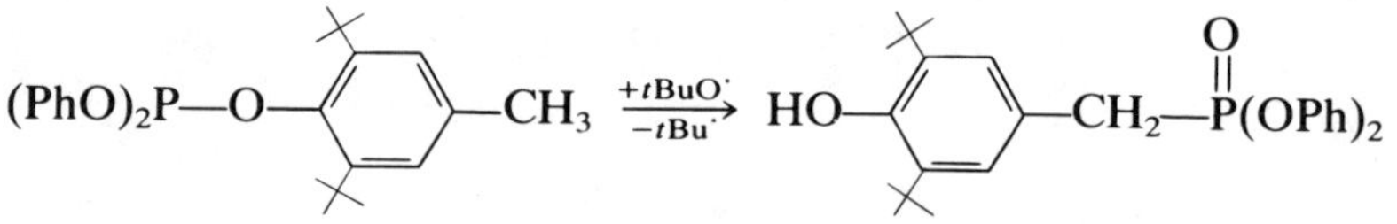

(PhO)$_2$P—O—C$_6$H$_2$(tBu)$_2$—CH$_3$ $\xrightarrow[-tBu^\cdot]{+tBuO^\cdot}$ HO—C$_6$H$_2$(tBu)$_2$—CH$_2$—P(O)(OPh)$_2$

The same phosphonates are formed in the reaction of mixtures of aryl phosphates and 4-methylphenols, for instance BHT, with DTBPO:[58]

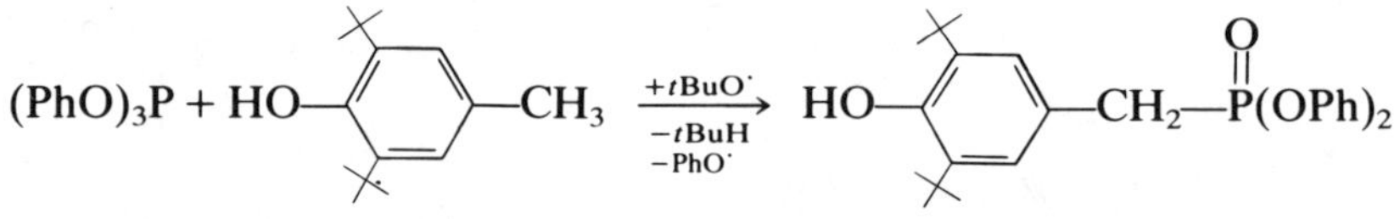

(PhO)$_3$P + HO—C$_6$H$_2$(tBu)$_2$—CH$_3$ $\xrightarrow[\substack{-tBuH \\ -PhO^\cdot}]{+tBuO^\cdot}$ HO—C$_6$H$_2$(tBu)$_2$—CH$_2$—P(O)(OPh)$_2$

Additional phosphonates result from radical reactions of these hydroxyarylphosphonates at the phenolic moiety.

In contrast to acyclic aryl phosphites, cyclic aromatic phosphites including sterically hindered ones, react with *tert*-butoxyls predominantly by oxidation and not substitution, as is typical for aliphatic phosphites. The intermediate arylene phosphoranyl radicals mainly undergo β- and not α-scission by ring opening.[58]

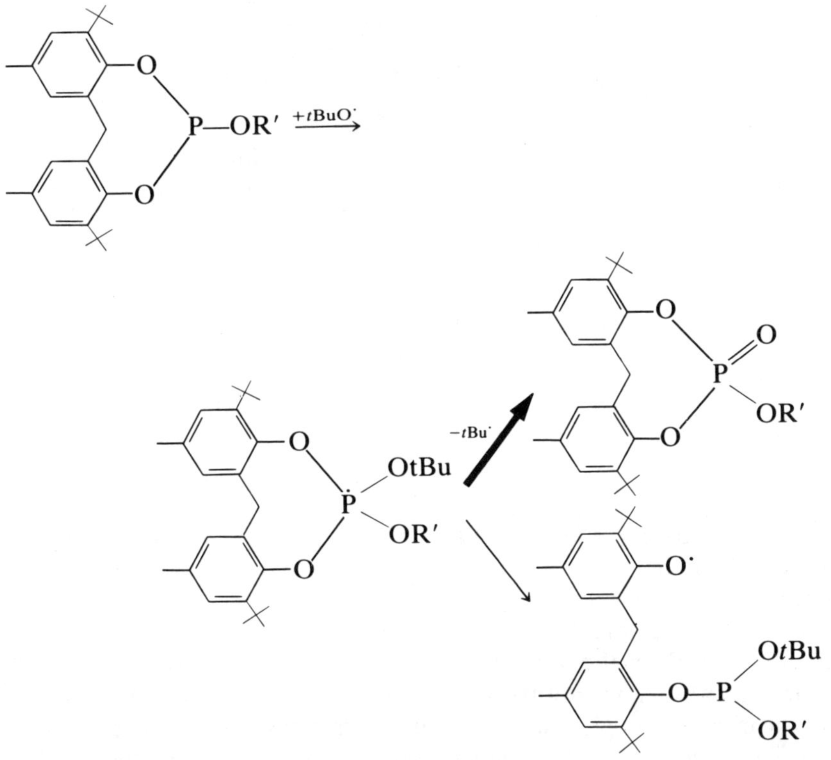

Likewise, hindered aryl phenylphosphonites react with *tert*-butoxyl radicals mainly by oxidation (β-scission of the intermediate phenylphosphoranyl radicals) giving the corresponding phenylphosphonates[18,54] (Table 5). This special behaviour of hindered aryl phosphonites and of hindered cyclic arylene phosphites towards alkoxy radicals accounts for their poor inhibiting ability in the low-temperature oxidation of hydrocarbons mentioned in Section 3.1.

Returning to the reactions of the acyclic aryl phosphites, it is remarkable that phenyl and other non-hindered aryl phosphites react with *tert*-butyloxyl radicals in a different manner from cumyloxyl, tetralyloxyl and cyanoisopropyloxyl radicals mentioned before. With the latter group, oxidation to the corresponding phenyl phosphates occurs by β-scission of the intermediate phosphoranyl radicals [eqn (10)], whereas with the former group, *tert*-butyl phosphites are built up by α-scission and substitution of a phenoxyl radical [eqn (8)]:

$$\text{Me---}\underset{\underset{\text{Me}}{|}}{\overset{\overset{\text{Me}}{|}}{\text{C}}}\text{---O}^\cdot + \text{>P---OPh} \longrightarrow \text{Me---}\underset{\underset{\text{Me}}{|}}{\overset{\overset{\text{Me}}{|}}{\text{C}}}\text{---O---}\overset{|}{\underset{|}{\text{P}^\cdot}}\text{---OPh}$$

$$\longrightarrow \text{Me---}\underset{\underset{\text{Me}}{|}}{\overset{\overset{\text{Me}}{|}}{\text{C}}}\text{---O---P}\diagup\diagdown + {}^\cdot\text{OPh}$$

$$\text{Ph---}\underset{\underset{\text{Me}}{|}}{\overset{\overset{\text{Me}}{|}}{\text{C}}}\text{---O}^\cdot + \text{>P---OPh} \longrightarrow \text{Ph---}\underset{\underset{\text{Me}}{|}}{\overset{\overset{\text{Me}}{|}}{\text{C}}}\text{---O---}\overset{|}{\underset{|}{\text{P}^\cdot}}\text{---OPh}$$

$$\longrightarrow \text{Ph---C}^\cdot\diagup^{\text{Me}}_{\text{Me}} + \text{O=}\underset{|}{\overset{\text{Me}}{|}}\text{P---OPh}$$

Obviously, this is because the $PhCMe_2-O$ and $NCCMe_2-O$ bonds are weaker than the Me_3C-O bond and C–O scission successfully competes with P–O scission in the phosphoranyl radical.

Thus the chain-breaking efficiency of non-hindered aryl phosphites should depend on the substrate to be stabilised. They should exhibit some chain-breaking antioxidant activity only in saturated aliphatic hydrocarbons and polyolefins, whereas in compounds and polymers bearing radical-stabilising substituents, such as polystyrenes, polydienes, polyacrylonitrile, polyethers, etc., their chain-breaking efficiency should be low.

4 ANTIOXIDANT ACTION OF PHOSPHITES AT HIGHER TEMPERATURES

Polymers can be exposed to temperatures up to 200–300°C during manufacture, processing and special end-uses.

In autoxidations at higher temperatures, homolytic decomposition of the hydroperoxides formed takes place [reaction (3)] giving rise to reactive radicals which cause auto-initiation and chain branching of the oxidation reaction. Under such conditions, the hydroperoxide-decomposing ability of phosphites and phosphonites becomes effective in suppressing chain branching and so inhibiting oxidation. For aliphatic phosphites this is the only way of acting antioxidatively. This is responsible for the relatively low efficiency of these phosphites when used alone, on the one hand, and for their synergistic action in combinations with phenols, on the other.

Aromatic phosphites may act by chain breaking, in addition to hydroperoxide decomposition, at higher temperatures, by reaction with peroxyl and alkoxyl radicals, thus releasing aryloxyl radicals. The two modes of antioxidant action have different susceptibilities to changes of temperature: the ratio of direct radical trapping to hydroperoxide decomposition should decrease with increasing temperature, because the rate of the reaction of alkylperoxyl with the substrate [reaction (2)] increases very much faster with rising temperature than the rate of reaction of peroxyl radicals with the phosphite [reaction (7)] a process of low activation energy.[43] But according to eqn (12), low k_7/k_2 ratios give rise to low f factors and, therefore, to low chain-breaking antioxidant activity. Additionally, the ratio of α- to β-splitting in the intermediate alkoxyphosphoranyl radical RO—P˙—OAr should decrease with increasing temperature,[54] i.e. substitution of the phosphite, which leads to termination, is less likely at higher temperatures than oxidation. Thus, the chain-breaking efficiency of phosphites, which is already lower than that of hindered phenols at low temperatures, should decrease further with increasing temperature.

In practice it has been observed, however, that in the oxidation of polyethylene[64–66] and in the oxidation of polypropylene at 160–200°C, alkylated aryl and arylene phosphites, the hindered phenyl phosphites in particular, exhibit a rather high antioxidant activity resembling or even surpassing that of the parent hindered phenols.[9,15,23,59–63,67,68] This also holds for the hindered eight-membered cyclic phosphites (dibenzo-dioxaphosphocins) and for the hindered aryl phosphonites[18] which were only poor radical-trapping agents in the initial stages of low-temperature oxidations (cf. Section 3.1). Some other chain-terminating action connected with the hindered phenolic moiety of the molecule, therefore, must be responsible for powerful antioxidant activity at higher temperatures.

It has repeatedly been suggested, but not unequivocally proved, that phenols formed by hydrolysis of the parent phosphites are the species responsible for the antioxidant activity of phosphites in high-temperature oxidations.[15,37,69,70] Other authors have considered the hydrolysis of phosphates formed from the parent phosphites to be the source of phenols.[19]

Water, always present in autoxidations, is also present in those inhibited by phosphorus compounds, resulting from dehydration of alcohols formed in reaction (6) at somewhat higher temperatures, and from thermal decomposition of hydroperoxides [reaction (3)]. It is well known, too, that alkyl and non-hindered aryl phosphites and phosphonites easily hydrolyse even at ambient temperatures. Hydrolysis of hindered aryl phosphites is more restrained, and it does not occur under the conditions of autoxidation at low temperatures but at higher temperatures it may become relevant.

The formation of phenols and phosphorous acid has been reported to take place in phosphite-inhibited thermal oxidations.[69–71] ^{31}P-NMR and HPLC studies of the fate of aryl phosphites and phosphonites in the autoxidation of cumene and tetralin at 150°C clearly revealed oxidation to phosphates and hydrolysis to hydrogen phosphites and phosphonites respectively, did take place from the beginning of the

reaction; the hydrogen phosphites and phosphonites were further hydrolysed to phosphorous and phosphonic acids, respectively, whereas the phosphates and phosphonates were surprisingly stable under the reaction conditions (Fig. 8).[42] Therefore the phenols released in the hydrolysis are the essential inhibiting species in these oxidations, the activity of which may be synergistically enhanced by the hydrogen phosphites and phosphonites and by the phosphorous acid simultaneously formed. Diaryl hydrogen phosphites are themselves antioxidatively active and, as does H_3PO_3, give synergistic mixtures with hindered phenols in higher-temperature oxidations.[65,70,71]

In the autoxidation of less oxidisable hydrocarbons, for example polyolefins, hydrolysis and some direct radical trapping by aryl phosphites may take place simultaneously. The substituted phosphites formed according to reaction (8) should eliminate olefins at higher temperatures as mentioned in Section 3.3, eqn (14), so giving rise to the same hydrogen phosphites as in hydrolysis.

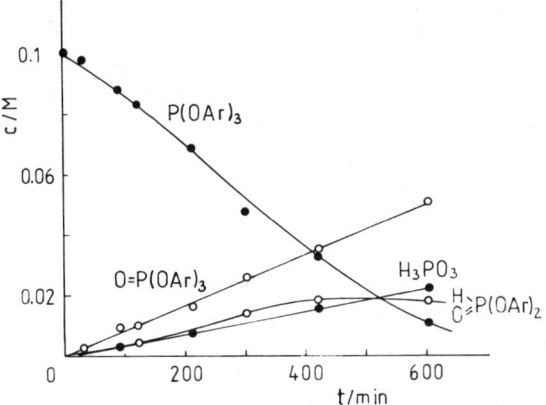

Fig. 8. Transformation of tris-(2-tert-butyl-4-methylphenyl) phosphite in the thermal oxidation of tetralin at 150°C.[42]

The five-membered o-phenylene phosphites of type **IX** seem to represent a special case also in high-temperature inhibition. Here, hydrolysis and peroxidolysis of the intermediate *phosphates* **XII** by ring opening to hydroxyphenyl phosphates **XIV** (Scheme 3) is so fast that it takes place even at ambient temperatures. The same processes occurring at higher temperatures would account for the extraordinary antioxidant activity of these cyclic phosphites[23,59,61,62] and also that of the corresponding phosphates[23] in polypropylene oxidation. Acyclic aryl phosphates do not exhibit this behaviour, they are not hydrolysed in the oxidising media and are inactive as antioxidants.[42]

5 CONCLUSIONS

The basic mechanisms by which organic phosphites and phosphonites operate as antioxidants now appear to be quite clear. They are hydroperoxide decomposition (stoichiometric and catalytic) and radical trapping by themselves and by phenols formed in their hydrolysis. The efficiency of these different modes of operation and their contribution to the total antioxidant action are also well understood, depending on the structure of the phosphorus compound, on the oxidisability of the substrate and on the reaction conditions. These mechanisms have predominantly been elucidated in the oxidations of low-molecular-weight hydrocarbons and very little is yet known about these relations in polymers of different structure and under different reaction conditions. In this field, further experimental studies are needed to verify forecasts extrapolated from the behaviour of low-molecular-weight compounds in order to provide a better understanding of the action of this important class of antioxidants in polymer manufacture, processing and end-use.

REFERENCES

1. Pobedimskii, D. G., Mukmeneva, N. A. & Kirpichnikov, P. A., *Developments in Polymer Stabilisation—2*, ed. G. Scott. Applied Science Publishers, London, 1980, p. 125.
2. Kirpichnikov, P. A., Mukmeneva, N. A. & Pobedimskii, D. G., *Usp. Khim.*, **52** (1983) 1831.
3. Schwetlick, K., *Pure Appl. Chem.*, **55** (1983) 1629.
4. Schwetlick, K., König, T., Rüger, C., Pionteck, J. & Habicher, W. D., *Polym. Degrad. Stab.*, **15** (1986) 97.
5. Walling, C. & Rabinowitz, M., *J. Am. Chem. Soc.*, **81** (1959) 1243.
6. Denney, D. B., Goodyear, W. F. & Goldstein, *J. Am. Chem. Soc.*, **82** (1960) 1393.
7. Kirpichnikov, P. A., Mukmeneva, N. A., Pudovik, A. N. & Kolyubakina, N. S., *Dokl. Akad. Nauk SSSR*, **164** (1965) 1050.
8. Pobedimskii, D. G. & Buchachenko, A. L., *Izv. Akad. Nauk SSSR, Ser. Khim.* (1968) 1181.
9. Ryšavý, D. & Sláma, Z., *Chem. Prum.*, **18** (1968) 20.
10. Schwetlick, K., Rüger, C. & Noack, R., *J. Prakt. Chem.*, **324** (1982) 697.
11. Chebotareva, E. G., Pobedimskii, D. G., Kolyubakina, N. S., Mukmeneva, N. A., Kirpichnikov, P. A. & Akhmadullina, A. G., *Kinet. Katal.*, **14** (1973) 891.
12. Kirpichnikov, P. A., Pobedimskii, D. G. & Mukmeneva, N. A., *Khim. Primen. Forfororg. Soedin., Tr. Konf.*, 1972 (1974) 215.
13. Humphris, K. J. & Scott, G., *J. Chem. Soc., Perkin Trans. 2* (1973) 826.
14. Zaichenko, L. P., Babel', V. G., Smirnov, P. A. & Proskuryakov, V. A., *Izv. Vyssh. Uchebn. Zaved., Khim. Khim. Tekhnol.*, **19** (1976) 1387.
15. Ryšavý, D. & Sláma, Z., *Angew. Makromol. Chem.*, **9** (1969) 129.
16. Rüger, C., Dissertation B, Technische Universität Dresden, 1982.
17. Rüger, C., König, T. & Schwetlick, K., *J. Prakt. Chem.*, **326** (1984) 622.
18. König, T., Dissertation, Technische Universität Dresden, 1986.
19. Holcik, J., Koenig, J. L. & Shelton, J. R., *Polym. Degrad. Stab.*, **5** (1983) 373.
20. Pobedimskii, D. G., Kurbatov, V. A., Kirpichnikov, P. A., Nasybullin, Sh. A. & Denisov, E. T., *Vysokomol. Soedin., Ser. A*, **18** (1976) 2650.
21. Kurbatov, V. A., Balandina, N. A. & Pobedimskii, D. G., *Vysokomol. Soedin., Ser. B*, **24** (1982) 421.
22. Tepper, R., Diploma Thesis, Technische Universität Dresden, 1981.

23. Humphris, K. J. & Scott, G., *Pure Appl. Chem.*, **36** (1973) 163.
24. Humphris, K. J. & Scott, G., *J. Chem. Soc., Perkin Trans.* 2 (1973) 831, 617.
25. Haake, P. C. & Westheimer, F. H., *J. Am. Chem. Soc.*, **83** (1961) 1102.
26. Kaiser, E. T. & Kudo, K., *J. Am. Chem. Soc.*, **89** (1967) 6725.
27. Gorenstein, D. G., Luxon, B. A., Findlay, J. B. & Momii, R., *J. Am. Chem. Soc.*, **99** (1977) 4170.
28. Rüger, C., Arnold, D. & Schwetlick, K., *J. Prakt. Chem.*, **324** (1982) 706.
29. König, T., Schwetlick, K., Kudelka, I. & Pospisil, J., *Polym. Degrad. Stab.*, **15** (1986) 97.
30. Cherkasova, C. A., Chebotareva, E. G., Gol'dfarb, E. I., Pobedimskii, D. G., Mukmeneva, N. A. & Kirpichnikov, P. A., *Neftekhimiya*, **21** (1981) 728.
31. Pobedimskii, D. G. & Belyakov, V. A., *Kinet. Katal.*, **10** (1969) 64.
32. Pobedimskii, D. G., Kirpichnikov, P. A., Samitov, Yu. Y. & Goldfarb, E. I., *Org. Magn. Resonance*, **5** (1973) 503.
33. Pershin, A. D., Pobedimskii, D. G., Kurbatov, V. A. & Buchachenko, A. L., *Izv. Akad. Nauk SSSR, Ser. Khim.* (1975) 581.
34. König, T., Grossmann, G., Schwetlick, K. & Rüger, C., *J. Prakt. Chem.*, **328** (1986) 763.
35. Tkač, A., Rüger, C. & Schwetlick, K., *Collect. Czechoslov. Chem. Commun.*, **45** (1980) 1182.
36. Pobedimskii, D. G., Levin, P. I. & Chelnokova, Z. B., *Izv. Akad. Nauk SSSR, Ser. Khim.* (1969) 2066.
37. Khloplyankina, M. S., Karpuchin, O. N., Buchachenko, A. L. & Levin, P. I., *Neftekhimiya*, **5** (1965) 49.
38. Roberts, B. P., *Adv. Free Radical Chem.*, **6** (1980) 225.
39. Bentrude, W. G., *Acc. Chem. Res.*, **15** (1982) 117.
40. Pobedimskii, D. G., & Kirpichnikov, P. A., *J. Polym. Sci., Polym. Chem. Ed.*, **18** (1980) 815.
41. Schwetlick, D., König, T., Pionteck, J., Sasse, D. & Habicher, W. D., *Polym. Degrad. Stab.*, **22** (1988) 357.
42. Schwetlick, K., Pionteck, J., Winkler, A., Hähner, U., Kroschwitz, H. & Habicher, W. D., *Polym. Degrad. Stab.* (in press).
43. Furimsky, E. & Howard, J. A., *J. Am. Chem. Soc.*, **95** (1973) 369.
44. Pobedimskii, D. G. & Kirpichnikov, P. A., *J. Polym. Sci., Polym. Chem. Ed.*, **18** (1980) 1587.
45. Floyd, M. B. & Boozer, C. A., *J. Am. Chem. Soc.*, **85** (1963) 984.
46. Bentrude, W. G., *Tetrahedron Lett.* (1965) 3543.
47. Ogata, Y. & Yamashita, M., *J. Chem. Soc., Perkin Trans.* 2 (1972) 730.
48. Kurbatov, V. A., Gren, G. P., Pavlova, L. A., Kirpichnikov, P. A. & Pobedimskii, D. G., *Kinet. Katal.*, **17** (1976) 329.
49. Schwetlick, K., Pionteck, J., König, T. & Habicher, W. D., *Eur. Polym. J.*, **23** (1987) 383.
50. Ogata, Y., Yamashita, M. & Toshinao, T., *Bull. Chem. Soc. Japan*, **45** (1972) 2223.
51. Walling, C. & Schmidt Pearson, M., *J. Am. Chem. Soc.*, **86** (1964) 2262.

52. Kochi, J. K. & Krusic, P. J., *J. Am. Chem. Soc.*, **91** (1969) 3944.
53. Krusic, P. J., Mahler, W. & Kochi, J. K., *J. Am. Chem. Soc.*, **94** (1972) 6033.
54. Bentrude, W. G., Hansen, E. R., Khan, W. A., Min, T. B. & Rogers, P. E., *J. Am. Chem. Soc.*, **95** (1973) 2286.
55. Watts, G. B., Griller, D. & Ingold, K. U., *J. Am. Chem. Soc.*, **94** (1972) 8784.
56. Davies, A. G., Griller, D. & Roberts, B. P., *J. Chem. Soc., Perkin Trans. 2* (1972) 933, 2224.
57. Levin, Ya. A., Ilyasov, A. V., Goldfarb, E. I. & Vorkunova, E. I., *Org. Magn. Resonance*, **5** (1973) 497.
58. Schwetlick, K., König, T., Rüger, C. & Pionteck, J., *Z. Chem.*, **26** (1986) 360.
59. Levin, P. J., Kirpichnikov, P. A., Lukovnikov, A. F. & Khloplyankina, M. A., *Vysokomol. Soedin.*, **5** (1963) 1152.
60. Mikhailov, N. V., Tokareva, L. G. & Popov, A. G., *Vysokomol. Soedin.*, **5** (1963) 188.
61. Levin, P. I., *Zh. Fiz. Khim.*, **38** (1964) 672.
62. Levin, P. I. & Bulgakova, T. A., *Vysokomol. Soedin.*, **6** (1964) 700.
63. Chelnokova, Z. B., Zimin, Yu. B. & Levin, P. I., *Vysokomol. Soedin., Ser. B*, **10** (1968) 126.
64. Kirpichnikov, P. A., Kolyubakina, N. S., Mukmeneva, N. A., Mukmenov, E. T. & Vorkunova, E. I., *Vysokomol. Soedin., Ser. B*, **12** (1970) 189.
65. Akhmadullina, A. G., Mukmeneva, N. A., Kirpichnikov, P. A., Kolyubakina, N. S. & Pobedimskii, D. G., *Vysokomol. Soedin., Ser. A.* **16** (1974) 370.
66. Pobedimskii, D. G., Kurbatov, V. A., Kirpichnikov, P. A., Nasybullin, Sh. A. & Denisov, E. T., *Vysokomol. Soedin., Ser. A*, **18** (1976) 2650.
67. Lebedeva, L. P. & Levin, P. I., *Vysokomol. Soedin., Ser. B*, **24** (1982) 379.
68. Rüger, C., König, T. & Schwetlick, K., *Acta Polym.*, **37** (1986) 435.
69. Bass, S. I. & Medvedev, S. S., *Zh. Fiz. Khim.*, **36** (1962) 2537.
70. Novoselova, L. V., Zubtsova, L. I., Babel', V. G. & Proskuryakov, V. A., *Zh. Prikl. Khim.*, **46** (1973) 1329.
71. Zaichenko, L. P., Babel', V. G. & Proskuryakov, V. A., *Zh. Prikl., Khim.*, **47** (1974) 1168, 1354; **49** (1976) 465.

Chapter 3

Antioxidant Mechanisms of Derivatives of Dithiophosphoric Acid

S. AL-MALAIKA
Aston University, Birmingham, UK

ABSTRACT

The antioxidant action of dithiophosphates in a wide range of substrates proceeds predominantly by peroxide-decomposing and radical-trapping mechanisms. A range of analytical techniques, including ^{31}P-NMR, are employed to elucidate the nature of the transformation products of dithiophosphoric acid and its derivatives and hence the mechanism of their action.

1 INTRODUCTION

Dialkyl dithiophosphoric acids (DRDPA; I) and their derivatives have been shown to possess antioxidant properties in a wide range of substrates and under quite different conditions. The acids themselves were shown to be effective inhibitors of hydrocarbon oxidation.[1] Dithiophosphoryl disulphides (DRDS; II) and polysulphides have been used[2,3] as vulcanisation accelerators and as effective sulphur-donor curing agents for rubbers. Amides of dithiophosphoric acids were shown[2] to exhibit antiozonant, antioxidant, thermal and light stabilising properties. Many oil-soluble derivatives of the higher-molecular-weight dialkyl and diaryl dithiophosphates, such as zinc and barium complexes, have been shown[2,4,5] to improve extreme-pressure properties, to decrease corrosion and to act as antioxidants and antisludging agents in detergents and lubricating oils. The solubility of these compounds in hydrocarbons depends on the length and branching of the alkyl or arylalkyl groups. Nickel complexes of dithiophosphoric acid (MDRP; III) have been used as antiozonants,[6] and light and thermal stabilisers for polyolefins,[7-14] whereas barium diaryl dithiophosphate has been recommended[15] for stabilisation of halogen-containing polymers, e.g. polyvinyl chloride.

Many derivatives of dithiophosphoric acid have been shown to exhibit strong synergistic effects when used in combination with other antioxidants in polymer systems. For example, nickel dithiophosphates and thiophosphoryl disulphides synergise effectively with thermal antioxidants (e.g. hindered phenols) and UV absorbers to give very highly stabilised polymers.[10,16]

$$(RO)_2P\begin{matrix}S\\SH\end{matrix} \qquad (RO)_2P\begin{matrix}S\\S\end{matrix}\begin{matrix}S\\S\end{matrix}P(OR)_2 \qquad (RO)_2P\begin{matrix}S\\S\end{matrix}M\begin{matrix}S\\S\end{matrix}P(OR)_2$$

DRDPA, I DRDS, II MDRP, III

R = alkyl, arylalkyl, etc.; M = a metal ion

In common with other sulphur-containing compounds, derivatives of dithiophosphoric acid function mainly as peroxide decomposers and radical scavengers.[17] Dithiophosphates are also known to act by other stabilisation mechanisms, e.g. metal deactivation, quenching of singlet oxygen and UV absorption. The relative importance of each of these mechanisms depends on the structure of the sulphur compound, the substrate and the environment. The antioxidant action of dithiophosphates has been reviewed previously.[17,18] However, much less is known about the exact nature of intermediates formed during the antioxidant action of these compounds. Although there is an appreciable amount of literature on dithiophosphates, many of the data are not immediately useful because of widely different experimental conditions used in different laboratories. The aim of this review is to examine critically some of the early work on the chemistry of dithiophosphates and to present some recent detailed studies on the nature of transformation products formed during the antioxidant action of some of the more important derivatives and metal complexes of dithiophosphoric acid under controlled and comparable experimental conditions.

2 GENERAL REACTIONS AND CHARACTERISTICS OF DITHIOPHOSPHATES

Dialkyl dithiophosphoric acids are prepared by reaction of phosphorus pentasulphide with the appropriate alcohol, a reaction which was recognised more than a century ago.[19] The acid is rather unstable to the oxidative–hydrolytic action of aqueous reagents on exposure to the atmosphere.[20] Scheme 1 shows some of the more important reactions of dithiophosphoric acid.[21] The acid was shown[22] to be a highly acidic 'mercaptan' which is readily oxidised to the corresponding disulphide (DRDS; **II**), and which adds with ease to olefins (irrespective of the presence or absence of peroxides in the olefin) to give esters (e.g. **IV**).

Transesterification of the acid by a higher-boiling alcohol (R'OH) yields an acid, **V**, which contains the alkoxy group of the homologous alcohol R'OH.[23] Reaction with alkylperoxyl radicals give the dithiophosphoryl radical (**VI**).[1] Dithiophosphoric acids, **I**, were shown[24] to be excellent reducing agents; for example, they reduce sulphoxides to the corresponding sulphides while themselves being oxidised quantitatively to thiophosphoryl disulphides. Reaction of the acids with inorganic bases and metal salts gives the corresponding metal dithiophosphates [see reaction (a) in Scheme 1]. Dithiophosphoric acids were shown[25,26] to be more acidic than the structurally related dithiocarbamic and xanthic acids. This high acidity was suggested[25] to be responsible for the higher stability of metal dithiophosphates [e.g. the nickel complex NiDRP (**III**)] towards acids and decomposition to sulphides when compared with the analogous dithiocarbamates and xanthates.[25]

The effect of the structure of the ligands on the antioxidant properties of metal complexes (e.g. **III**, M = Zn) was investigated[27] by comparing the antioxidant activity of the zinc salts of various phosphoric acids [dithiophosphoric acid **I**), monothiophosphoric acid (**VII**), and phosphoric acid (**VIII**) in addition to those of the corresponding phosphonic acids, **IX–XI**]. The antioxidant effectiveness was found to follow the order **I** ≅ **IX** > **VII** ≅ **X** ≫ **VIII** ≅

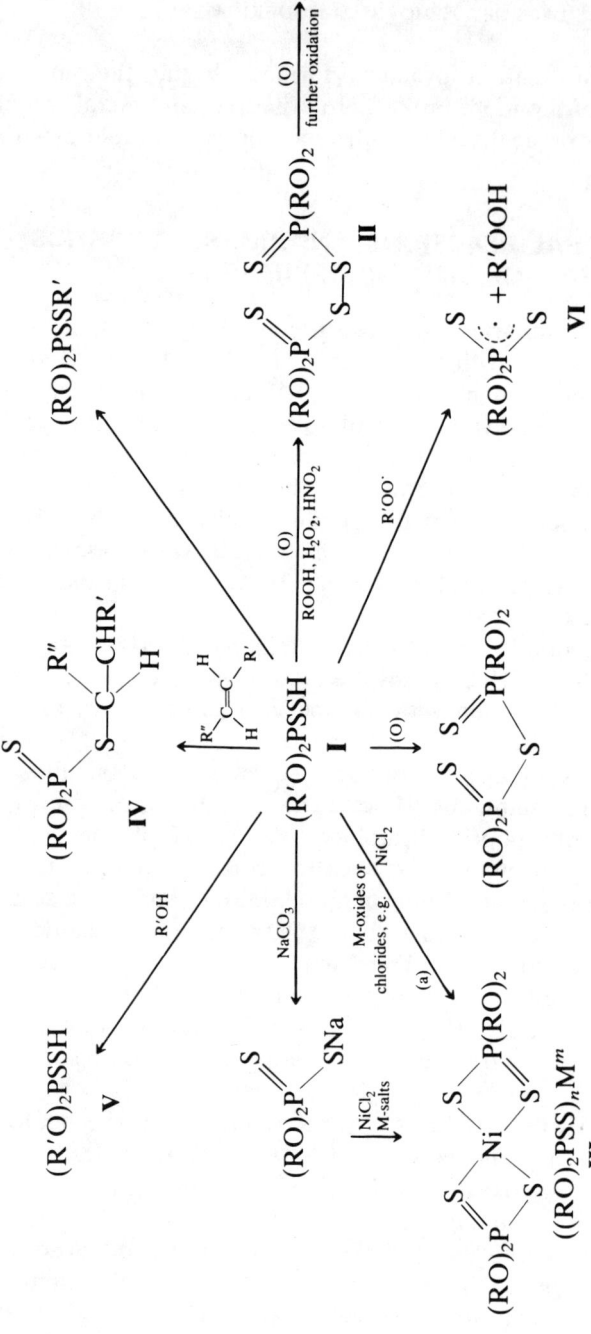

Scheme 1. Reactions of dithiophosphoric acid

XI, indicating the importance of the >P-S and $\text{>P}\overset{\overset{\text{S}}{\|}}{-}$ groups. The importance of the sulphur element in the structure of the ligand is further supported by the fact that fully oxygenated acid derivatives (e.g. **VIII** and **XI**) do not inhibit the oxidation of hydrocarbons.[27] Moreover, it was found that the antioxidant effectiveness of the phosphates **I**, **VII** and **VIII**, is, in general, higher than that of the corresponding phosphonates (**IX–XI**).

```
   R—O                R—O                R—O
     \                  \                  \
      P—S—              P—O—              P—O—
     /‖                 /‖                 /‖
   R—O S              R—O S              R—O O
      I                 VII                VIII

   R—C                R—C                R—C
     \                  \                  \
      P—S—              P—O—              P—O—
     /‖                 /‖                 /‖
   R—C S              R—C S              R—C O
     IX                  X                  XI
```

3 DITHIOPHOSPHATES AS RADICAL TRAPPING AGENTS

The chain-breaking activity of the dithiophosphates was first demonstrated in the mid 1960s. Colclough & Cunneen[28] and Ivanov & Shopov[29] demonstrated the formation of inactive products from reactions of metal dithiolates with alkylperoxyl radicals. It was shown[28] that zinc dithiophosphate and other metal dithiolates, e.g. zinc complexes of dithiocarbamic and xanthic acids, decrease or inhibit the rate of oxidation of squalene at 60°C in the presence of a free-radical initiator such as azobis-isobutyronitrile (AIBN). A mechanism involving the deactivation of alkylperoxyl radicals by the metal chelate via an electron-transfer step [reaction (1)] was proposed,[28] although no evidence was given to support this mechanism. A more detailed study of the reaction of various metal dithiolates with alkylperoxyl radicals was subsequently carried out by Burn.[30,31] Zinc, potassium, lead, iron(III), nickel, cadmium and copper(I) dithiophosphates were all found to act as radical traps. Burn[30] isolated thiosphosphoryl disulphide (DRDS; **II**) (10% yield) as a product of the reaction of a zinc

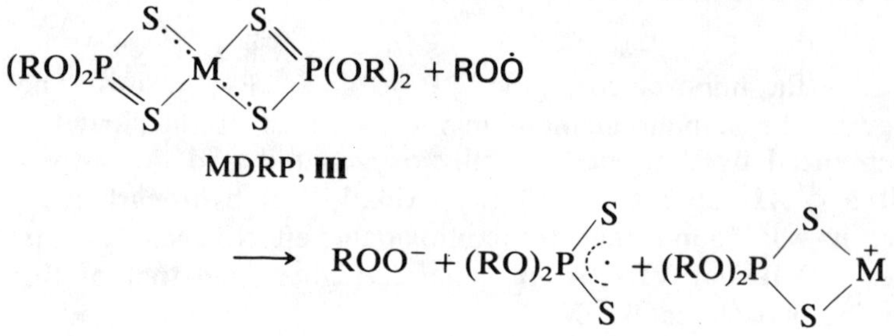

$$(RO)_2P\underset{S}{\overset{S}{\diamond}}M\underset{S}{\overset{S}{\diamond}}P(OR)_2 + RO\dot{O}$$

MDRP, **III**

$$\longrightarrow ROO^- + (RO)_2P\underset{S}{\overset{S}{\diamond}} + (RO)_2P\underset{S}{\overset{S}{\diamond}}\overset{+}{M} \quad (1)$$

dithiophosphate (MDRP, $M = Zn$) with alkylperoxyl radicals, although recently it was claimed[32] that a much higher concentration of disulphide (60%) is formed from this reaction. Burn suggested that both the sulphur atom and the metal centre are important for the deactivation of alkylperoxyl radicals. The proposed mechanism [reaction (2)] involves an initial electron transfer from the sulphur atom to the peroxyl radical while the function of the metal atom provides an easy route for heterolysis of the peroxyl-containing intermediate;[30] see reaction (2). Howard and coworkers[1,33] also proposed that the reaction

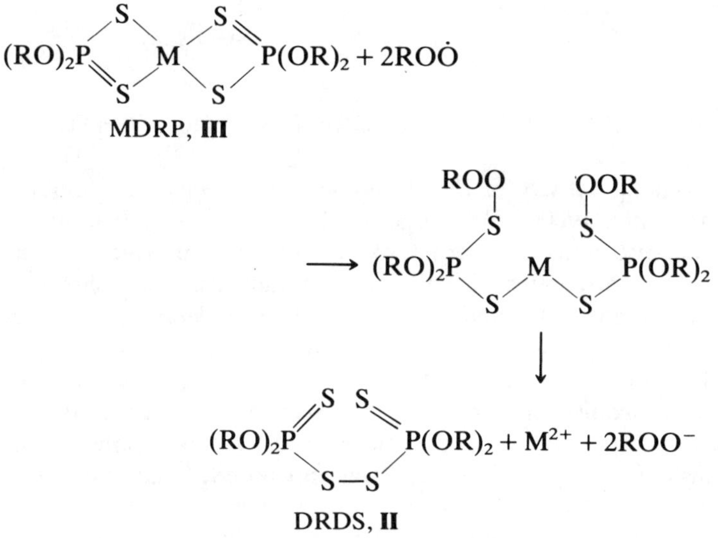

$$(RO)_2P\underset{S}{\overset{S}{\diamond}}M\underset{S}{\overset{S}{\diamond}}P(OR)_2 + 2RO\dot{O}$$

MDRP, **III**

$$\longrightarrow (RO)_2P\underset{S}{\overset{S-ROO}{\diamond}}M\underset{S}{\overset{S-OOR}{\diamond}}P(OR)_2$$

$$\downarrow$$

$$(RO)_2P\underset{S-S}{\overset{S\,\,\,S}{\diamond}}P(OR)_2 + M^{2+} + 2ROO^- \quad (2)$$

DRDS, **II**

of alkylperoxyl radicals with dithiophosphates occurs at the metal centre either by an electron-transfer mechanism or by a process which involves a Zn(III) species as a transition state or transient intermediate [reaction (3)].

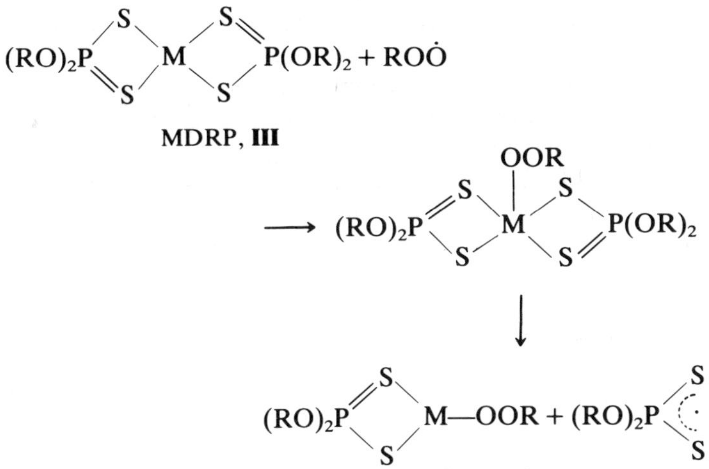

$$(RO)_2P\overset{S}{\underset{S}{\diagup\diagdown}}M\overset{S}{\underset{S}{\diagdown\diagup}}P(OR)_2 + RO\dot{O}$$

MDRP, III

$$\longrightarrow (RO)_2P\overset{S}{\underset{S}{\diagup\diagdown}}\overset{OOR}{\underset{}{M}}\overset{S}{\underset{S}{\diagdown\diagup}}P(OR)_2$$

$$\downarrow \qquad (3)$$

$$(RO)_2P\overset{S}{\underset{S}{\diagup\diagdown}}M\text{—OOR} + (RO)_2P\overset{S}{\underset{S}{\diagdown}}$$

VI

The dithiophosphoryl radical (VI) formed in the above reaction then undergoes further reaction with alkylperoxyl to give inert products. When compared with metal dithiophosphates, the metal free thiophosphoryl moieties (e.g. the disulphide, II) are inefficient radical deactivators.[32,34-37]

The coefficient of inhibition, f, defined as the number of alkylperoxyl radicals trapped per dithiophosphate moiety, is a useful indicator of the antioxidant lifetime. The value of f has been determined by different methods; the induction period method which was shown not to be very reliable, and a thermal-initiation method where a value of f was shown to be close to 1 in most cases for different dithiophosphates.[30,38] For the zinc complex, the value of f lies in the range 0·7–3 as measured by the thermal method, whereas a much larger value ($\cong$4·5) was obtained from photo-initiated autoxidation studies.[39]

4 DITHIOPHOSPHATES AS PEROXIDOLYTIC AGENTS

Kennerley & Patterson were the first to study the cumene hydroperoxide (CHP)-decomposing activity of metal dithiolates and other sulphur-containing compounds; they proposed a catalytic decomposition mechanism.[40] They further proposed that the parent sulphur-containing compounds were not the real inhibitors but rather the product(s) of their reaction with CHP. In spite of the earlier work of Hock & Lang,[41] who suggested a catalytic decomposition of hydroperoxides under the influence of electrophilic reagents E, e.g. mineral acids or Lewis acids, with acetone and phenol as products of heterolytic decomposition of CHP [reaction (4)], Kennerley & Patterson considered it unlikely that the antioxidant action of peroxide decomposers was associated with their conversion to strong acids, and the identity of the active product(s) remained unknown.

$$R_2-\underset{R_3}{\underset{|}{\overset{R_1}{\overset{|}{C}}}}-OOH + E \longrightarrow R_2-\underset{R_3}{\underset{|}{\overset{R_1}{\overset{|}{C}}}}-O^+ + EOH^-$$

$$\downarrow \qquad (4)$$

$$R_1-O-\underset{R_3}{\underset{|}{\overset{R_2}{\overset{|}{C^+}}}} \longrightarrow O=C\overset{R_2}{\underset{R_3}{\diagdown}} + R_1OH + E$$

In the early 1960s Holdsworth et al.[42] were the first to propose that sulphur oxides, e.g. SO_2, are the true catalysts for the ionic decomposition of hydroperoxides by dithiocarbamates (e.g. ZnDRC). Due to the similarity in the behaviour of ZnDRC with the structurally related dithiophosphate (ZnDRP), they proposed that SO_2 was a common peroxide-decomposing species produced from these metal complexes. Subsequent work on reactions of ZnDRP with CHP,[43,44] however, failed to detect SO_2. Scott & Husbands[45] showed later that SO_2 does in fact only undergo stoichiometric reaction with hydroperoxides and this was found to be associated with the generation of free radicals. SO_3, on the other hand, was shown[45] to be a powerful catalyst for the decomposition of CHP to phenol and acetone, the ionic decomposition products. Extensive mechanistic studies on dithiophosphates and related sulphur-containing compounds, which were carried out in the last decade by many investigators,[17,18,34,44,46–54] had confirmed the early findings that this class of compounds undergoes a complex series of oxidation reactions involving free radicals (pro-oxidants) leading to sulphur acids as the main catalyst for decomposition of hydroperoxides. This will be discussed in the following sections. Moreover, metal dithiophosphates, **III** (e.g.

ZnDRP and NiDRP) have been shown[34,43,44,54–56] to exhibit multistep hydroperoxide decomposition curves which represent both radical and ionic mechanisms; the importance of each stage was determined by the (initial hydroperoxide/metal complex) ratio used. Generally, the use of a catalytic amount of the peroxide leads to decomposition by an ionic mechanism in the final stage. Stoichiometric amounts, on the other hand, lead to decomposition during the first stage of the reaction in a homolytic process. The initial stage of the decomposition of hydroperoxide is generally responsible for the pro-oxidant effect observed during oxidation of hydrocarbons in the presence of hydroperoxides with some metal complexes.

5 THE ANTIOXIDANT ACTION OF METAL-FREE DERIVATIVES OF DITHIOPHOSPHORIC ACID

5.1 Thiophosphoryl Disulphides (DRDS; II)

Examination of the antioxidant role of thiophosphoryl disulphides is important since they have been identified as a transformation product from reactions of many metal dithiophosphate complexes (e.g. Zn, Ni) with hydroperoxides.[27,29,34,43,57] It was demonstrated that heterolytic decomposition of hydroperoxides predominate at all molar ratios of the peroxide to thiophosphoryl disulphide (DRDS) whilst the radical-scavenging mechanisms contribution is only minor.[17,35,36] Questions, however, still remain as to the details of the mechanism of action, such as the extent of contribution of each of the above processes to the overall mechanism, and whether the disulphide itself, or its oxidation products, are the main catalyst for the ionic decomposition of hydroperoxides. The experimental conditions used in different laboratories for carrying out these mechanistic studies are quite varied, giving rise to different data and interpretations and apparent discrepancies in the details of the mechanisms.

In order to elucidate the role of the disulphide as a transformation product during the oxidation of metal dithiophosphate complexes, we have undertaken a mechanistic study of the reaction of disulphides with hydroperoxides at high temperatures (110–180°C) by monitoring the kinetics of appearance and disappearance of oxidation/transformation products by a variety of complementary experimental techniques, e.g. oxygen absorption, hydroperoxide decomposition and product analysis using GLC, IR spectroscopy and ^{31}P-NMR.

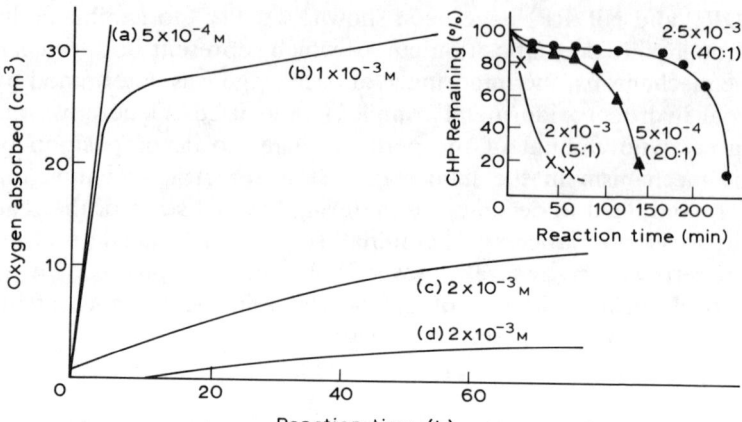

Fig. 1. Effect of thiophosphoryl disulphide, DRDS, R = isobutyl, on the oxidation of decalin at 130°C in the presence of 1×10^{-2} M CHP. Inset shows the decomposition of CHP (1×10^{-2} M) by DRDS in chlorobenzene at 110°C. The molar concentrations of the disulphide are given on the curves.

In the presence of different molar ratios of DRDS/CHP, oxidation of decalin at 130°C occurs immediately as shown (Fig. 1) by the rapid uptake of oxygen leading to a second much slower oxidation stage.[54] The extent of oxidation in the first stage decreases with increasing the disulphide concentration. This initial pro-oxidant stage indicates the inability of the disulphide to inhibit the hydrocarbon oxidation in the presence of excess hydroperoxides. However, the presence of the second autoretarding oxidation stage suggests clearly that, under these conditions, the DRDS must be oxidised to more powerful catalysts during the first step and these are responsible for the autoretarded oxidation in the second stage. This is supported by the two-stage decomposition process observed[54] for hydroperoxides: an initial induction period involving no (or very little) peroxide decomposition during which the disulphide is oxidised to more powerful products which are responsible for the second rapid catalytic stage (Fig. 1, inset). This two-stage behaviour has also been observed by other workers.[43,57]

The length of the induction period of peroxide decomposition was shown[54] to decrease with increasing disulphide concentration. Therefore, at the stoichiometric ratio the induction period is not observed but is replaced by a rapid and effective decomposition from the beginning of the reaction. This accounts for the observed initial low extent of oxidation (less initial pro-oxidant effect) of decalin (in presence of CHP) at high disulphide concentrations [Fig. 1, curve (c)]. Examination of the product distribution of the peroxide decomposition reaction at high disulphide concentration (i.e. high molar ratio of [CHP]/[DRDS], e.g. 5)[54] revealed that at the beginning of the reaction the decomposition occurs mainly via a homolytic process (which accounts for the pro-oxidant stage of decalin oxidation) while the ionic products build up at later stages of the reaction. Furthermore, we have shown previously[34] that although thiophosphoryl disulphide decomposes hydroperoxides mainly by an ionic process at all ratios, there is a small contribution of the homolytic process at both stoichiometric and catalytic ratios. This suggests that the further oxidation products which are responsible for the ionic decomposition must be produced during the initial homolytic process involving the disulphide and the hydroperoxide.

To clarify the question of whether thiophosphoryl disulphide itself, in addition to its transformation products, plays a direct role in the decomposition of hydroperoxides, ^{31}P-NMR was used to follow the fate of the disulphide and the formation of transformation products during its reactions with hydroperoxides (e.g. tertiary butyl hydroperoxide (TBH) or CHP at 100°C and 110°C, respectively) and at high molar ratios of hydroperoxide to sulphur compound (e.g. at ratios of 1:5).[53] At high disulphide concentrations (e.g. 1×10^{-1} M), the disulphide itself is responsible for the initial (80–90%) peroxide decomposition (see Fig. 2 and Table 1) since almost no transformation products were observed during this period during which the concentration of the disulphide has decreased only very slightly. It is important to point out, that under these conditions, the first slow peroxide decomposition step (see inset of Fig. 1) is now replaced by an initial rapid decomposition stage (see Fig. 2). After this initial period of almost constant concentration of disulphide, there is a sharp fall in the disulphide concentration and this paralleled by a rise in the concentration of the oxidation products during a second stage of slow peroxide decomposition (Fig. 2). The direct contribution of the disulphide to peroxide decomposition, under these conditions, is further supported by the fact that, in the absence of added peroxides, the disulphide became effective as an antioxidant only after an initial period of inactivity during the autoxidation of decalin or white oil at 130°C [Fig. 1, curve (d)]. The oxidation products are, therefore, ultimately responsible for the final decomposition of the peroxide.

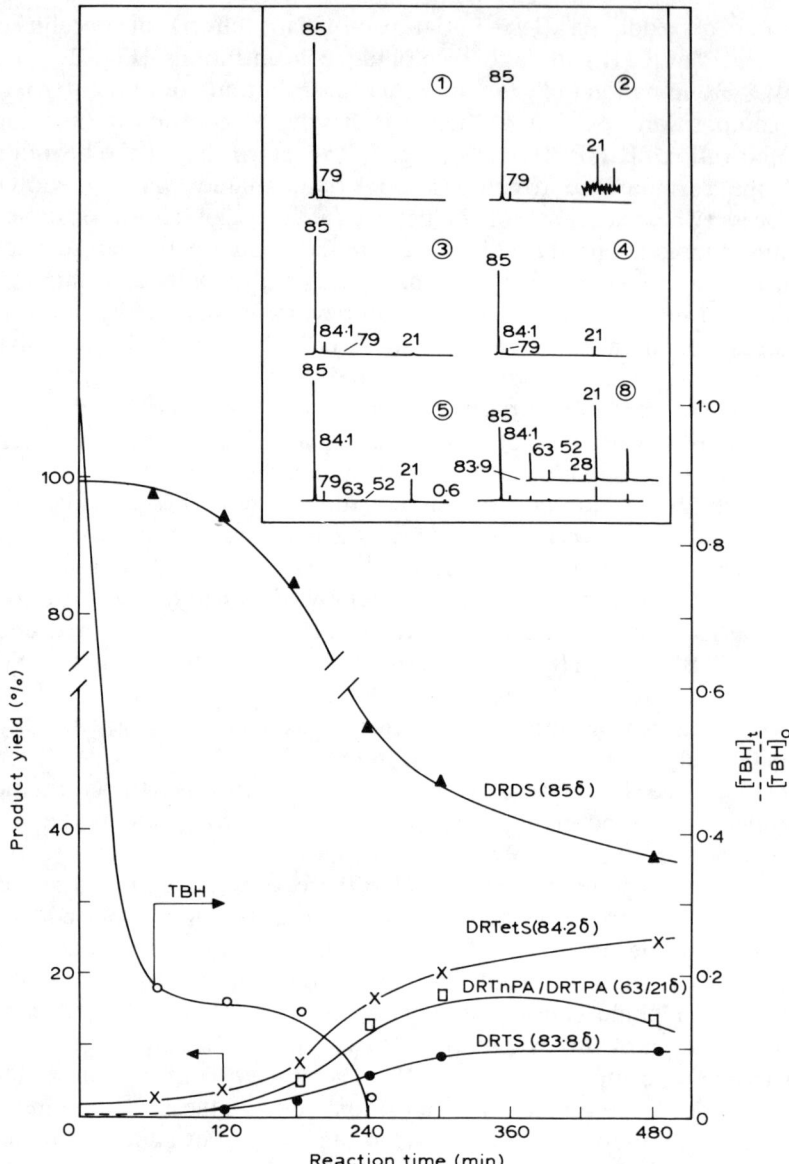

Fig. 2. kinetics of product formation during the reaction of DRDS, R = n-butyl (0·5 M) with TBH (1 M) in chlorobenzene at 100°C. Inset shows ^{31}P-NMR spectra of this reaction. Encircled numbers are reaction times in hours.

Table 1
Products of oxidation of DRDS $(1 \times 10^{-1}$ M) during its reaction with CHP in chlorobenzene at 110°C
The [DRDPA]/[CHP] molar ratio was 1:5.

Reaction time (min)	Phosphorus yield (%)			CHP decomposed (%)
	DRDS (85·2)[a]	DRTetS (84·2)[a]	DRTPA (21)[a]	
0	100	0	0	0
2	99	1	0	75
10	92	6	3	90
30	91	6	2	95

[a] ^{31}P shift, δ ppm.

The main oxidation products are the tetrasulphides (DRTetS, **XII**) and trisulphides (DRTS, **XIII**), in addition to thiophosphoric (DRTPA, **XIV**) and thionophosphoric (DRTnPA, **XV**) acids; see Fig. 2. The concentration of these acids was found[53] to increase in the presence of excess hydroperoxides. Thionophosphoric acid (DRTnPA) has been identified[54,58] as one of the final transformation products of nickel and zinc dithiophosphates, both of which give the disulphide as the initial transformation product (see later sections). The mechanism of antioxidant action of the disulphide is outlined in Scheme 2. It is worth mentioning here, that in the case of polymer stabilisation the concentration of stabilisers (e.g. dithiophosphates) is normally quite low (of the order of 10^{-4} M) and under these conditions, oxidation products of the disulphides, rather than the disulphides themselves, must play the major role in the stabilisation mechanism.

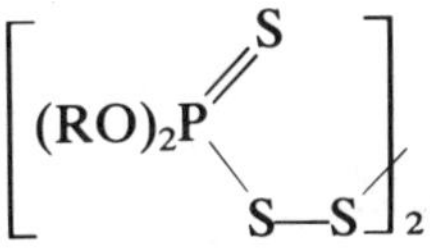

$$\left[(RO)_2P \underset{S-S}{\overset{\displaystyle S}{\diagup}} \right]_2 \qquad \left[(RO)_2P \underset{S}{\overset{\displaystyle S}{\diagup}} \right]_2 S$$

DRTetS, **XII** 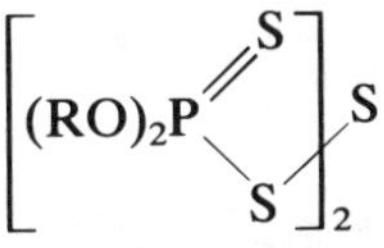   DRTS, **XIII**

$$(RO)_2P \underset{SH}{\overset{\displaystyle O}{\diagup}} \qquad (RO)_2P \underset{OH}{\overset{\displaystyle S}{\diagup}}$$

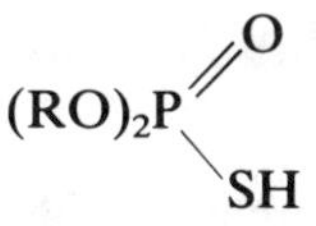

DRTPA, **XIV** 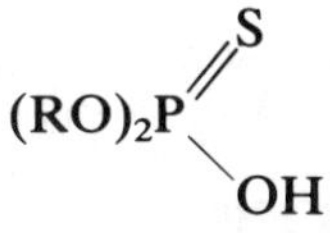   DRTnPA, **XV**

5.2 Dithio (DRDPA) and Thio (DRTPA) Phosphoric Acids

The role of dialkyl dithiophosphoric acid (DRDPA, **I**) as a catalyst responsible for the ionic decomposition of hydroperoxides (e.g. CHP) in the presence of metal dithiophosphates is still controversial.[36,59,60] However DRDPA was shown to be a good peroxide decomposer[62] and a radical trap.[35] Figure 3a shows that dithiophosphoric acid is a very effective inhibitor for the CHP-initiated oxidation of decalin at 130°C.[54] The absence of a dramatic initial oxidation of the substrate, when compared with that of the disulphide [cf. Fig. 3(a) and Fig. 1], suggests that the acid itself is a much more effective peroxide

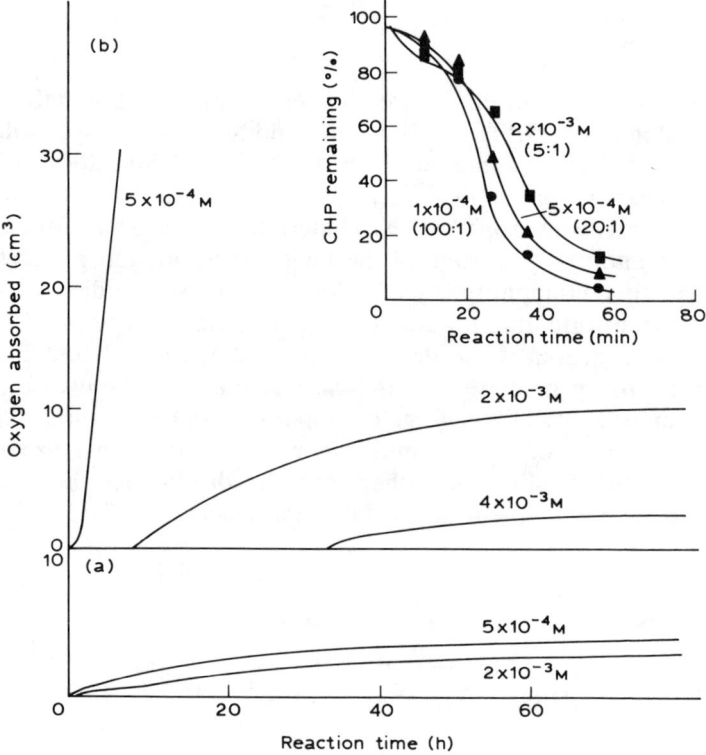

Fig. 3. Effect of dithiophosphoric acid. DRDPA, R = hexyl, on the oxidation of decalin at 130°C in the absence (b) and presence (a) of 1×10^{-2} M CHP. Inset shows the decomposition of CHP (1×10^{-2} M) by DRDPA in chlorobenzene at 110°C. The molar concentrations of the acid are given on the curves.

Table 2
Products of oxidation of DRDPA (2×10^{-1} M) during its reaction with CHP in chlorobenzene at 110°C
The [DRDPA]/[CHP] molar ratio was 1:2·5.

Reaction time (min)	Phosphorus yield (%)					CHP decomposed
	DRDPA (85·6)[a]	DRDS (85·2)[a]	DRTetS (84·2)[a]	$(RO)_2P(O)SR$ (24·8)[a]	DRTPA (21)[a]	CHP(%)
0	100	0	0	0	0	10
2	0	90	5	0	5	90
12	0	80	15	0	5	95
30	0	73	18	4	5	98

[a] ^{31}P shift, δ ppm.

decomposer, even at low concentrations, when compared with the disulphide. Further evidence which supports the role of the acid is the occurrence of an induction period when it is used to inhibit the oxidation of decalin in the absence of added peroxide at 130°C; see Fig. 3(b). The overall autoretarded inhibition shown in Fig. 3 does indicate, however, that the acid must be subsequently oxidised to more effective catalysts. Examination of the kinetics of the reaction of DRDPA with CHP (at 110°C) and the nature of its transformation products reveal two important points,

1. At all acid concentrations used (covering molar ratios of CHP/DRDPA = 2–100), there is a rapid one-stage CHP decomposition; see Fig. 3, inset.
2. The acid is fully transformed, very early in the reaction (e.g. after 2 min, at a molar ratio of 2·5, to further oxidation products: the corresponding disulphide, DRDS (90% yield), tetrasulphide, DRTetDS, and thiophosphoric acid, DRTPA (Table 2). The latter two products may be derived, in part, from the disulphide (see Table 1).

Since the yield of the disulphide formed from oxidation of the acid was so high, it is most likely to be present in the system (with CHP) at near to a stoichiometric ratio, conditions under which the disulphide was shown (see Section 5.1) to be able to decompose the hydroperoxide, to a large extent, without itself undergoing further oxidation (see Figs 1 and 2). The presence of the disulphide at such high concentration, in addition to the other powerful transformation products, e.g. thiophosphoric acid (DRTPA), is responsible for the effective inhibition of CHP-initiated oxidation of hydrocarbons (e.g. decalin). The formation of the disulphide from oxidation of dithiophosphoric acid has also been suggested by others.[36,61] Dithiophosphoric acid, DRDPA, was proposed by some workers[59,61] to be the catalyst responsible for the ionic decomposition of CHP in the presence of ZnDRP, although it was not observed as a reaction product. The rapid destruction of the acid by hydroperoxides (at all molar ratios) suggests that even if it is formed during the reaction of metal dithiophosphates with excess hydroperoxides, it will be only a transitory intermediate which is transformed mainly to the corresponding disulphide.

Very little is known about the antioxidant activity (e.g. peroxide decomposition and radical trapping) of thiophosphoric acids (DRTPA) and thionophosphoric acids (DRTnPA). Both were shown to be produced from nickel and zinc dithiophosphates during their reaction with hydroperoxides at different temperatures.[14,54,58] We have recently studied the effectiveness of thionophosphoric acids as peroxide decomposers and the nature of their transformation products under similar conditions to those used for other derivatives of dithiophosphoric acid.[53,54] Figure 4[54] shows that the oxidation of decalin (at 130°C) in the presence of thionophosphoric acid (DRTnPA) is autoretarded from the beginning of the reaction (no induction period) when the concentration of the acid is high, but that oxidation is rapid at low concentrations (e.g. 5×10^{-4} M). This suggests that DRTnPA itself is not a very effective antioxidant but when present at high enough concentrations it oxidises to a very powerful antioxidant which is responsible for the effective inhibition. This accounts for the observation that the same small concentration of the acid (5×10^{-4} M) autoretards the CHP-initiated oxidation of decalin in the same way as it does when the concentration of the acid is high (Fig. 4, inset). ^{31}P-NMR studies[53,54] have shown that DRTnPA is quantitatively transformed by hydroperoxides at high temperatures to thiophosphoric acid (DRTPA) during the early stages of the reaction (after 3 min; see Table 3); DRTPA is the real catalyst for the ionic decomposition of hydroperoxide. Furthermore, thionophosphoric acid is so readily oxidised in the presence of hydroperoxides that even at room temperature it is quantitatively converted to thiophosphoric acid;[53] see Scheme 2(g). Table 3 suggests that thiophosphoric acid is a stable end-product although a small amount (<10%) is oxidised

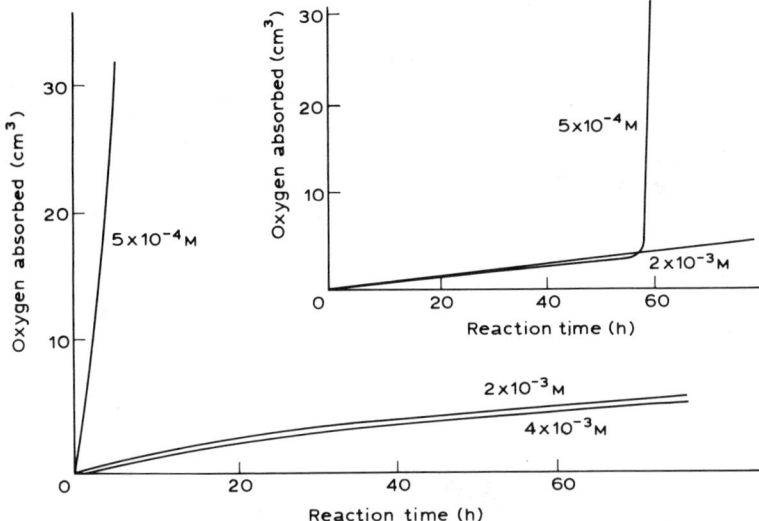

Fig. 4. Effect of thionophosphoric acid, DRTnPA, R = isobutyl, on the oxidation of decalin at 130°C in the absence and presence (inset) of 1×10^{-2} M CHP. The molar concentrations of the acid are given on the curves.

further to phosphoric acid (DRPA, **XVI**) which has been shown[62] to be an ineffective peroxide-decomposer [see Scheme 2 (i), (j)]. It is also possible that thiophosphoric acids may oxidise further to the corresponding unstable disulphides (DRDOS, **XVII** which have the same ^{31}P chemical shift as the acids). The intermediate formation of

Table 3
Products of oxidation of DRTnPA (2×10^{-1} M) during its reaction with CHP in chlorobenzene at 110°C
The [DRTnPA]/[CHP] molar ratio was 1:2·5

Reaction time (min)	Phosphorus yield (%)			
	DRTnPA (63)[a]	DRTPA (21)[a]	$(RO)_3P{=}O$ $(-0·1)$[a]	$(RO)_3P{=}S$ $(64·8)$[a]
0	100	0	0	0
3	0	99	1	0
20	0	91	9	0
45	0	88	9	3

[a] ^{31}P shift, δ ppm.

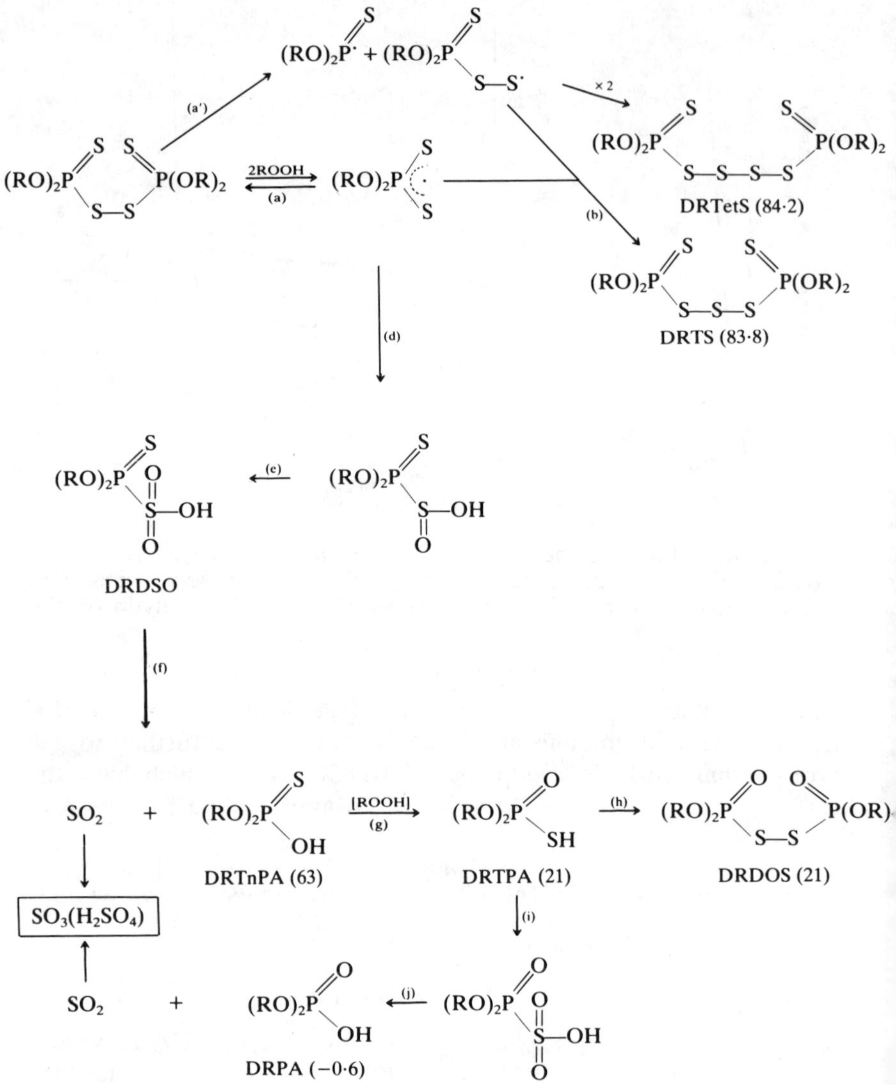

Scheme 2. Oxidation of thiophosphoryl disulphide in the presence of hydroperoxides. Numbers in parentheses are ^{31}P-NMR chemical shifts.

DRDOS during the oxidation of ZnDRP during its antioxidant action has been suggested by some workers (see Section 6.1) to occur by different routes (see Scheme 3).

$$(RO)_2P\underset{OH}{\overset{O}{\diagup}}$$

DRPA, **XVI**

$$(RO)_2P\underset{S-S}{\overset{O\ \ \ O}{\diagup\ \ \ \diagdown}}P(RO)_2$$

DRDOS, **XVII**

6 THE ANTIOXIDANT ACTION OF METAL COMPLEXES OF DITHIOPHOSPHORIC ACIDS

6.1 Zinc Dithiophosphates (ZnDRP)

Model studies of reactions of ZnDRP (**III**) with hydroperoxides in both oxidisable and non-oxidisable substrates have given an insight to the detailed mechanism of its antioxidant action. At low temperatures (typically 70°C), the reaction of ZnDRP with CHP in chlorobenzene (non-oxidisable substrate) was found to exhibit three stages.[43,55,57] A rapid initial stage favoured by a low molar ratio of the peroxide to the complex which gives rise to homolytic products (e.g. α-cumyl alcohol) is followed by a period of slow reaction (induction period) leading to a third fast stage during which all the remaining peroxides are destroyed. This final stage assumes greater significance at higher molar ratios of peroxide to complex and gives rise to an ionic decomposition of hydroperoxide. The pro-oxidant effect observed during the oxidation of hydrocarbons in the presence of ZnDRP and CHP[40,57] constitutes the initial stage of decomposition of hydroperoxide.[43] The consumption of the zinc complex during the first stage suggests that the catalyst for the third ionic stage is directly formed from the reaction of the zinc complex with the peroxide.[43]

The identity of the ionic catalysts produced from reactions of ZnDRP with hydroperoxides has not yet been fully resolved. Dialkyl thiophosphoryl disulphide, DRDS, for example, was isolated[43] as a major product of the first-stage reaction but its role as a catalyst has been questioned.[43,46,55,63] A number of other sulphur-containing compounds were identified in different laboratories as products of the oxidation of ZnDRP by CHP. At low molar ratios of peroxide to complex, basic zinc dithiophosphate (b-ZnDRP; **XVIII**) and the disulphide, DRDS, were isolated[27] (see Scheme 3). Further oxidation

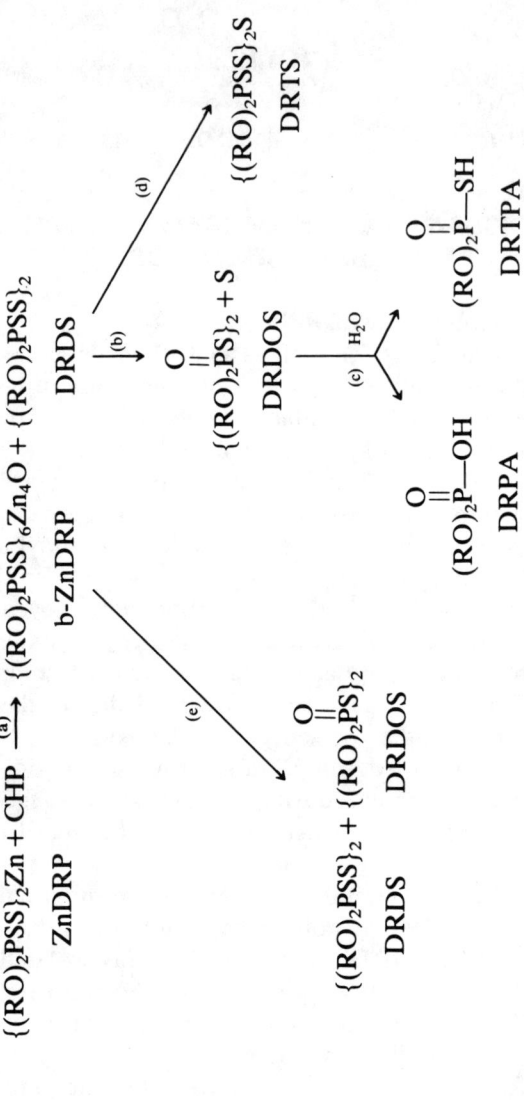

Scheme 3. Oxidation of ZnDRP by CHP as suggested by Sanin et al.[27], path (b)–(d), and by Rossi et al.[57], path (e).

of the disulphide was found to produce dialkylphosphoryl disulphide (DRDOS) which on hydrolysis gave dialkylthiophosphoric acid (DRTPA) and dialkylphosphoric acid (DRPA); see Scheme 3(b), (c).[27] Dialkylthiophosphoryl trisulphide (DRTS) was also identified as a product of the prolonged oxidation of ZnDRP by CHP [Scheme 3(d)].[27] The formation of the phosphoryl disulphide (DRDOS) was proposed via oxidation of the basic zinc complex which is formed in the initial reaction of zinc dithiophosphate with hydroperoxide [Scheme 3(e)].[57] Several other phosphorus-containing species have been reported from reactions of ZnDRP with hydroperoxides (at low temperatures) such as esters of dithiophosphoric (**XIX**) and phosphoric (**XX**) acids[64] as well as zinc dialkylthiophosphate (ZnDRT; **XXI**).[65]

$$(RO)_2P\underset{\underset{S}{\|}}{-}SR \qquad (RO)_3P{=}O \qquad \{(RO)_2PSO\}_2Zn$$

(TRDTP, **XIX**)　　　(O,O,O-TRPP, **XX**)　　　(ZnDRT, **XXI**)

Much of the mechanistic work described in the literature was conducted at relatively low temperatures. We have conducted extensive studies on the behaviour of ZnDRP, and some of its transformation products, as antioxidants in both oxidisable and non-oxidisable substrates, in the presence and absence of CHP, and at high temperatures. The conditions used for studying the antioxidant role of ZnDRP were similar to those used for other derivatives of dithiophosphoric acids. At 110°C, the reaction of ZnDRP with CHP in a non-oxidisable substrate (chlorobenzene) was found to proceed in two stages (see Fig. 5),[54] compared with three stages at lower temperatures (70°C).[43,55,57] The initial fast decomposition stage which corresponds to the first stage reported at lower temperatures gives rise to homolytic products, e.g. α-cumyl alcohol (αCA) and acetophenone (AC) (Fig. 5, inset). The disappearance of α-cumyl alcohol and the build-up of α-methylstyrene (αMS) and phenol (PH) during the slower second stage suggest an acid-catalysed reaction—hence the ionic mechanism (Scheme 4). The initial stage of homolytic decomposition of hydroperoxide which leads to the generation of free radicals was found to be responsible for the initial pro-oxidant stage observed during the inhibited oxidation of decaline by ZnDRP in the presence of CHP at high temperature, 130°C (see Fig. 6).[54] The subsequent effective

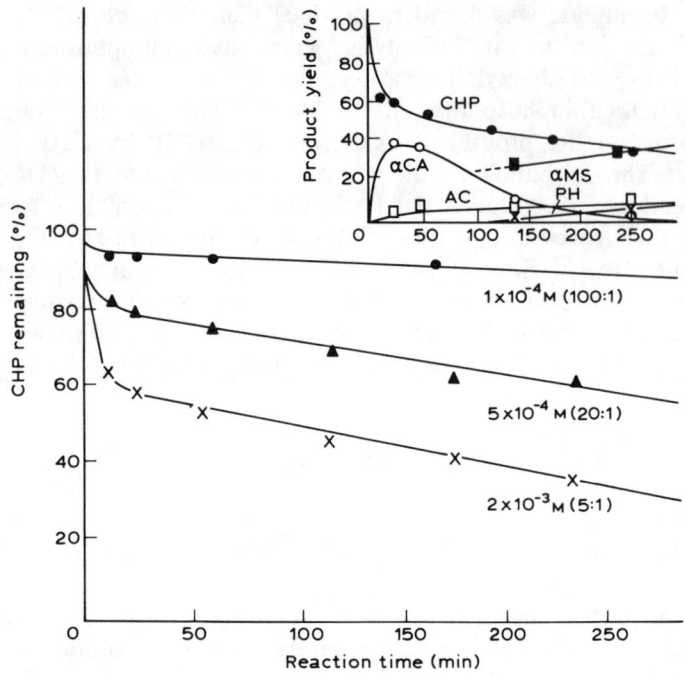

Fig. 5. Decomposition of CHP (1×10^{-2} M) by zinc dithiophosphate, ZnDRP, R = isobutyl in chlorobenzene at 110°C. The molar concentrations of ZnDiBP are given on the curves. Inset shows products formed from the decomposition of CHP (1×10^{-2} M) by ZnDiBP (2×10^{-3} M) in chlorobenzene at 110°C.

autoretarding inhibition of decalin is a consequence of the second-stage ionic decomposition of hydroperoxide (see Fig. 5). The basic zinc complex (b-ZnDRP) shows very similar behaviour when examined under the same experimental conditions[54] except that a much greater amount of CHP is needed to break down the b-ZnDRP than is required to destroy the same molar concentration of the ZnDRP due to the larger number of thiophosphoryl ligands present in the former case.

Examination (using ^{31}P-NMR) of the nature of the transformation products which are formed from the zinc dithiophosphate complex during its reaction with CHP at 110°C, Table 4,[54] shows that the nature and the concentration of the products formed depend on the

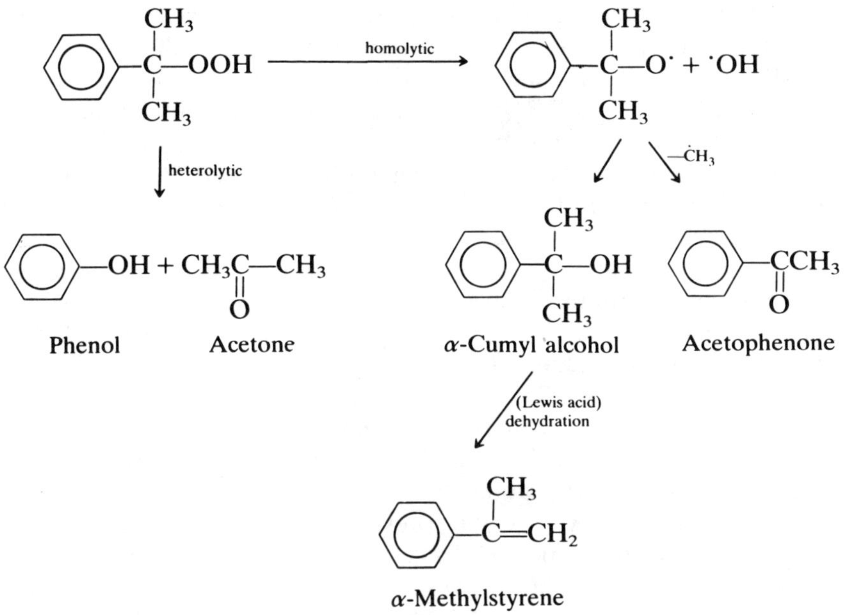

DITHIOPHOSPHORIC ACID DERIVATIVES

Scheme 4. Mechanisms of CHP decomposition via catalysed homolytic and heterolytic pathways.

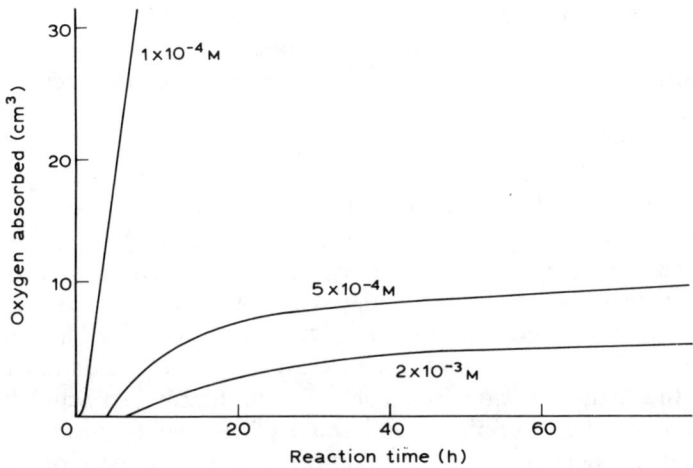

Fig. 6. Effect of ZnDRP, R = isobutyl, on the oxidation of decalin at 130°C in the presence of 1×10^{-2} M CHP.

initial ratio of metal complex to peroxide. It can be seen from Table 4 that the major products from a stoichiometric reaction of ZnDRP with CHP (at 110°C) are the basic zinc complex and the disulphide (b-ZnDRP, $\delta = 103$; DRDS, $\delta = 85\cdot2$). At higher molar ratios, e.g. [CHP]/[ZnDRP] > 5, the basic zinc complex was not detected. The original zinc dithiophosphate is completely consumed during the early stages of the reaction (<4 min); the small amount found at the end of a stoichiometric reaction must arise from subsequent decomposition of the basic zinc complex itself (compare Tables 4 and 5). The major transformation products observed at these high molar ratios are polysulphides, including di-, tri- and tetra-sulphides (DRDS, DRTS and DRTetS) in addition to monosulphide (DRMS) and thiophosphoric acid (DRTPA). The unstable nature of zinc dialkylthiophosphate (ZnDRT) is indicated from its initial build-up and subsequent consumption when the molar ratio of peroxide to complex is greater than 5[54]; a similar observation was made by Sher *et al.*[65]

It has been shown in the previous section that oxidation of thiophosphoryl disulphide (DRDS) leads mainly to tetrasulphides, thio- and thiono-phosphoric acids, in addition to trisulphide (see Fig. 2). Subsequent oxidation of thiophosphoric acid (DRTPA) may lead to the corresponding phosphoryl disulphide (DRDOS); see Scheme 2(h). However, although in our studies no further attempt was made to separate these two products (which are expected to give a similar NMR chemical shift), it is highly unlikely that DRDOS is formed directly from DRDS under these conditions, and is subsequently hydrolysed to the acid, DRTPA, as has been suggested previously [see Scheme 3(b), (c)].[27,57]

Reactions of the basic zinc complex with CHP at 110°C were shown to give predominantly disulphides, and the zinc thiophosphate complex (ZnDRT) (Table 5).[54] The formation of the latter from the basic zinc complex (Table 5) suggests that this species, which was found to build up during the early stages of oxidation of ZnDRP by CHP (see Table 4), may be derived from the basic zinc complex. Although the basic zinc complex was not detected at higher molar ratios of peroxide to complex, because of its fast oxidation reaction with the hydroperoxide (see Table 5), its intermediacy and transformation to the disulphide and ZnDRT is highly probable. It is suggested, therefore, that in addition to ZnDRT, part of the disulphide which is formed at all molar ratios of CHP to ZnDRP must originate from the basic zinc complex; the remaining amounts are

Table 4

Products of oxidation of ZnDRP by CHP in chlorobenzene at 110°C where TP is $(RO)_3P=S$

[ZnDRP]/[CHP] molar ratio	Reaction time (min)	b-ZnDRP (103)[a]	ZnDRP (98.2)[a]	DRDS (85.2)[a]	DRTetS (84.2)[a]	DRMS (79)[a]	ZnDRT (48.3)[a]	? (46.3)[a]	DRTS (83.8)[a]	DRTPA (21)[a]	DRPA (−0.6)[a]	TP (64.2)[a]	TIP (24.5)[a]
1:1	0	<1	99	0	0	<1	0	<1	—	—	—	—	—
1:1	30	33	9	39	6	4	7	2	—	—	—	—	—
5:1	0	<1	99	0	0	<1	0	<1	—	—	—	—	—
4:1	4	0	0	48	16	8	18	3	—	7	—	—	—
4:1	10	0	0	45	17	9	16	0	—	12	—	—	—
4:1	60	0	0	44	27	12	0	0	—	16	—	—	—
10:1	0	<1	99	0	0	<1	—	—	0	0	0	—	—
10:1	5	0	0	36	15	13	—	—	11	17	0	—	—
10:1	60	0	0	34	18	10	—	—	6	9	13	—	—
50:1	0	<1	99	0	0	<1	—	—	0	0	0	—	0
50:1	10	0	0	23	11	12	—	—	14	31	0	0	0
50:1	180	0	0	14	15	12	—	—	0	0	8	16	12

[a] ^{31}P shift, δ ppm.

Table 5
Products of oxidation of b-ZnDRP during its reaction with CHP in chlorobenzene at 110°C

[b-ZnDRP]/[CHP] molar ratio	Reaction time (min)	Phosphorus yield (%)					
		b-ZnDRP (104)[a]	ZnDRP (98·2)[a]	DRS (85·2)[a]	DRTetS (84·2)[a]	DRMS (79)[a]	ZnDRT (48·3)[a]	? (46·3)[a]
1:1	0	90	6	0	0	0	3	0
1:1	30	62	9	11	2	3	11	2
1:3	0	90	6	0	0	0	3	0
1:3	2	10	0	50	13	10	13	3
1:3	30	10	0	49	11	10	17	1

[a] ^{31}P shift, δ ppm.

derived directly from the zinc dithiophosphate itself. Reactions of disulphide with hydroperoxides at high temperature have been shown to give thiophosphoric acid as one of the major products, especially at later stages of the reaction; see Fig. 2.[53,54] Zinc thiophosphate, ZnDRT, must also contribute, in part, to the formation of this acid as the disappearance of the latter at the end of the reaction of CHP with ZnDRP (e.g. at a molar ratio of 5) is paralleled by the appearance of the acid DRTPA; see Table 4. Thiophosphoric acid must therefore be formed from both the disulphide and ZnDRT. This is in agreement with the finding of Rossi & Imperato[57] that transformation of ZnDRT leads to phosphoryl disulphide (DRDOS), and this was shown here to be the oxidation product of the acid DRTPA. At higher molar ratios of peroxide to zinc dithiophosphate (e.g. a ratio of 50) the formation of thiophosphoric acid predominates (see Table 4), and this leads to the main catalysts responsible for ionic decomposition of hydroperoxides. Similarly, the formation of the acid was found to dominate at higher molar ratios of peroxide to thiophosphoryl disulphide. Thiophosphoric acid was also found to be one of the major products formed during reactions of nickel dithiophosphate and thiophosphoryl disulphide with hydroperoxides at high molar ratios.

In the presence of excess hydroperoxide, the thiophosphoric acid (DRTPA) may oxidise further to the corresponding dialkylphosphoryl disulphide (DRODS), and both can subsequently give rise to dialkylphosphoric acid (DRPA, $\delta = -0\cdot 9$) and SO_2/SO_3. Phosphoric acid was shown[27] to be unreactive towards hydroperoxides and is, therefore, a stable reaction product. Oxidation of thiophosphoric acid gives, in addition to DRPA, esters of dithiophosphoric acids such as TP (**XXII**) and TIP (**XXIII**). These esters were also observed by other workers as products of the reaction of ZnDRP with CHP at low temperatures.[64] DRODS has been suggested by others from studies at lower temperatures although different routes to its formation were proposed (see Scheme 3). Scheme 5 summarises the mechanism of inhibition by ZnDRP based on the above finding.

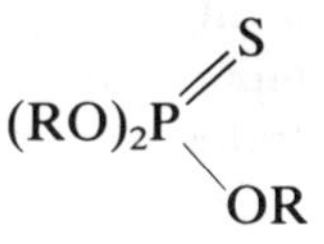

$$(RO)_2P\begin{smallmatrix}\nearrow S \\ \searrow OR\end{smallmatrix} \qquad (RO)_2P\begin{smallmatrix}\nearrow O \\ \searrow SR\end{smallmatrix}$$

TP, **XXII** 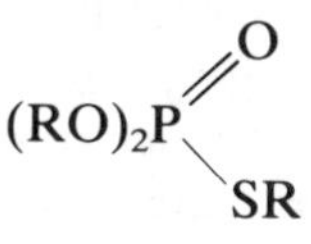   TIP, **XXIII**

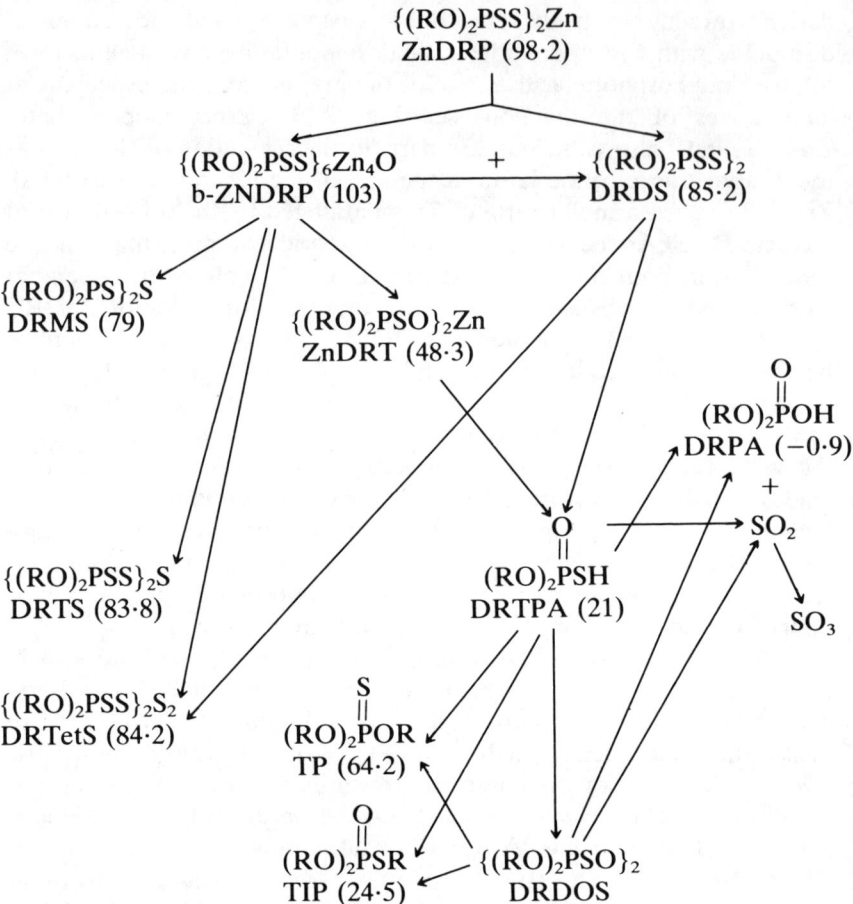

Scheme 5. Transformation products of ZnDRP during its reaction with CHP at high temperature (110°C). Numbers in the parentheses are ^{31}P chemical shifts.[54]

6.2 Nickel Dithiophosphates (NiDRP)

The behaviour of nickel dithiophosphates during their reactions with hydroperoxides (e.g. CHP) at different temperatures (30–150°C) in the presence and absence of oxidisable substrates was shown[34,56] to be similar to that of other nickel dithiolate (e.g. dithiocarbamate and xanthate) complexes. Nickel dithiolates decompose hydroperoxides by both homolytic (radical) and heterolytic (ionic) processes: the relative

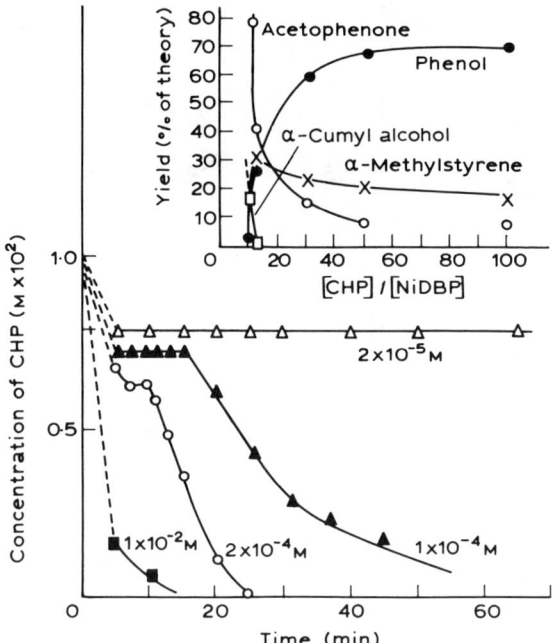

Fig. 7. Decomposition of CHP (1 × 10^{-2} M) in chlorobenzene at 110°C in the presence of NiDRP, R = n-butyl, at the molar concentrations shown. Inset shows product distribution of reactions of NiDBP with CHP at 110°C in chlorobenzene at various molar ratios [CHP]/[NiDBP].

contribution of each process is determined by the molar ratio of the two components. Nickel dithiophosphate was shown[34] to decompose CHP at 110°C typically in three stages (Fig. 7). An initial rapid catalytic stage (favoured by a low molar ratio of hydroperoxide to complex) gives rise to homolytic products (Fig. 8) followed by an induction period, during which oxidation products are formed which are the precursors of the second-stage catalysts (see Fig. 8, inset). The induction period leads to the slower catalytic stage which assumes greater significance at higher molar ratios of hydroperoxide to complex.

Figure 8 shows that for [CHP]/[NiDRP] = 100, at 110°C a clear change-over is obtained from an essentially homolytic to an essentially heterolytic reaction which dominates the later stages of the reaction. The first catalytic free-radical pathway, on the other hand, is more

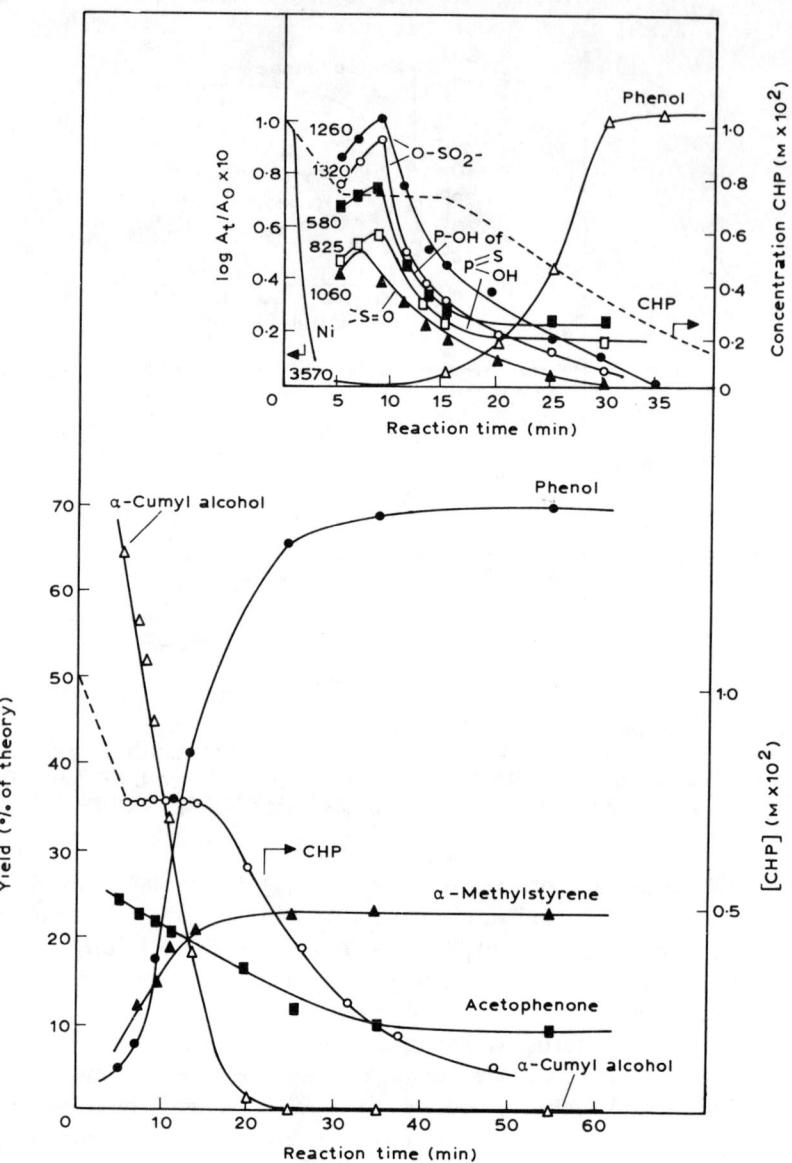

Fig. 8. Kinetics of ionic and radical products formation of the reaction of CHP (1×10^{-2} M) with NiDRP, R = n-butyl (1×10^{-4} M) in chlorobenzene at 110°C. The CHP decomposition curve for the same reaction is also shown. Inset shows formation and decay of products in the above reaction.

important at low molar ratios: α-cumyl alcohol and acetophenone (the homolytic products of CHP) are exclusively formed at molar ratios of [CHP]/[NiDRP] < 10 (see Fig. 7, inset). The relative contribution of the initial catalytic process to the overall stabilising effect of NiDRP was found to decrease with increasing complex concentration.[17] A linear relationship was found between the amount of hydroperoxide decomposed during the first catalytic stage and the concentration of the nickel complex.[17] In contrast, an inverse relationship was found between the length of the secondary induction period and the concentration of the nickel complex at constant peroxide concentration (see Fig. 7).[34] The length of this apparent period of inactivity was directly related to the time required to procure a critical threshold concentration of the active catalyst from the nickel complex. An opposite relationship was reported by Burn et al.[43] for reactions of ZnDRP with CHP (at a lower temperature, 70°C), where the length of the induction period was found to increase with increasing the initial zinc complex concentration. This observation had led Burn to discount the formation of active catalyst from the metal complex under these conditions.

It is now well established[17,42,46] that metal dithiolates are not themselves responsible for the heterolytic decomposition of hydroperoxides but, rather, their transformation products, which are formed by oxidation in the presence of hydroperoxides. At all molar ratios of hydroperoxide to nickel dithiophosphate, the nickel complex was shown to be completely destroyed before the onset of the secondary catalytic stage (see Fig. 8, inset). This confirms that the metal complex is a precursor for the effective catalytic peroxide decomposers. The nature of the ionic catalyst(s) for peroxide decomposition formed from reactions of NiDRP with hydroperoxides at different temperatures (25–150°C) was investigated.[14,58,66] Thiophosphoryl disulphide (DRDS) was found to be a primary intermediate of the oxidation of nickel dithiophosphate at high molar ratios of peroxide to complex and at elevated temperatures (e.g. 110°C).[34,58] The formation of DRDS is associated with the first rapid homolytic peroxide decomposition stage. The disulphide is further oxidised to unstable intermediates which break down to give the catalysts responsible for the effective hydroperoxide-decomposing activity of the nickel complex. Figure 8 (inset) shows that thiosulphonic acid (DRTSA) and thionophosphoric acid (DRTnPA) are formed slowly during the induction period; the former loses SO_2 to give the latter [see Scheme 6(f)] which can be seen to be a relatively stable product. Figure 8 (inset) also shows that although CHP is decomposed at the beginning of the reaction, no phenol (the ionic decomposition product) was found initially during the build-up of the disulphide-derived oxidation products. Instead the formation of phenol was found to coincide with the onset of decomposition of oxidation intermediates. This observation clearly identifies the antioxidant activity with the sulphur acids formed from NiDRP by hydroperoxide oxidation.

Examination of products formed from reactions of NiDRP with TBH at room temperature using ^{31}P-NMR shows that the nature and proportion of the transformation products is highly dependent on the ratio of the peroxide to complex.[14] Monosulphide, disulphide, thiono- and thio-phosphoric acids are the major transformation products observed during oxidation reactions of NiDRP with CHP (or TBH) at different molar ratios (see Fig. 9). A small amount (about 10% of the total products) of other phosphorus-containing species, mainly esters of dithiophosphoric (DTP) and thiophosphoric (TP) acid [Scheme 6(g), (j)] are also formed. Thiophosphoric acid (DRTPA), which becomes more evident at higher molar [TBH]/[NiDRP] ratios (Fig. 9), is most likely formed via reaction of thiophosphoryl radical with the hydroperoxide [Scheme 6, reactions (d)–(f), (k)]. Under the low-temperature conditions used in these studies, the formation of thionophosphoric acid from the disulphide [route (l)–(f)] seems to be unlikely in the light of the observation[67] that disulphides of dithioic acids are unable to effect thermal decomposition of TBH at temperatures below 70°C. Furthermore, the concentration of the thio- and thiono-acids in the system was found to be a function of the amount of available hydroperoxide.[14,53] Scheme 6 summarises the reactions involved in the antioxidant function of nickel dithiophosphates.

6.3 Iron Dithiophosphate (FeDRP)

Much less work has been done on reactions of FeDRP with hydroperoxides and its role as an antioxidant, when compared with other metal dithiophosphates, e.g. nickel and zinc analogues. The lower interest in FeDRP must be attributed to earlier reports of the instability of iron complexes of dithiophosphoric acids in the presence of traces of hydrogen chloride as a result of using iron(III) chloride for their preparation.[68,69] Stable iron(III) di-isobutyl dithiophosphate was prepared for the present investigation using a recent preparative method[70] and was used for reactions with hydroperoxides at high

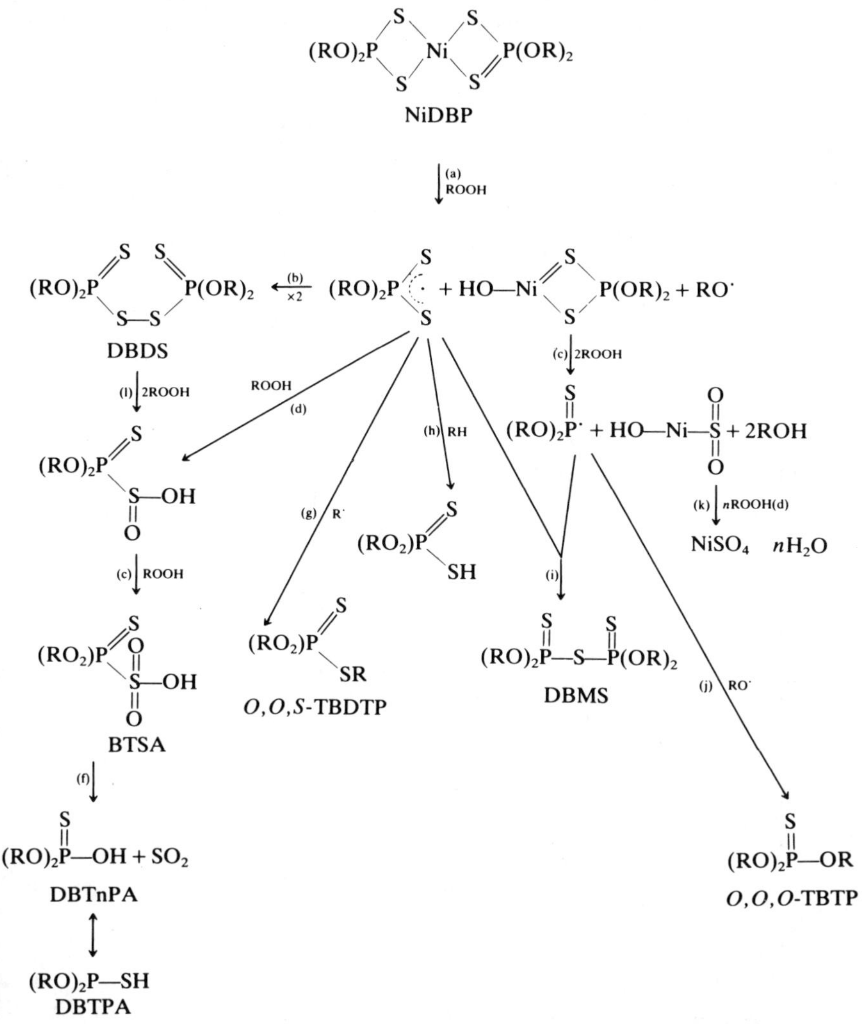

DITHIOPHOSPHORIC ACID DERIVATIVES

$$(RO)_2P\underset{S}{\overset{S}{\diagup}}\underset{S}{\overset{S}{Ni}}\underset{S}{\overset{S}{\diagdown}}P(OR)_2$$

NiDBP

↓ (a) ROOH

$(RO)_2P\underset{S-S}{\overset{S\ \ \ S}{\diagup\ \ \diagdown}}P(OR)_2$ ⇐(b)/×2 $(RO)_2P\overset{S}{\underset{S}{\cdot}}$ + HO—Ni$\underset{S}{\overset{S}{\diagdown}}$P(OR)$_2$ + RO·

DBDS   (c)│2ROOH

(l)│2ROOH   ROOH
                 (d)

$(RO)_2P\overset{S}{\underset{O}{\parallel}}\!\!\!-\!\!\text{S—OH}$ (h)│RH $(RO)_2P\cdot + HO\!-\!Ni\!-\!\overset{O}{\underset{O}{\overset{\parallel}{\underset{\parallel}{S}}}}\! + 2ROH$

                              (g)│R·   (k)│nROOH(d)
                                    $(RO_2)P\overset{S}{\underset{SH}{\diagdown}}$ NiSO$_4$ nH$_2$O
(c)│ROOH   (i)

$(RO_2)P\overset{S}{\underset{O}{\overset{\parallel}{\underset{\parallel}{S-OH}}}}$ $(RO_2)P\overset{S}{\underset{SR}{\diagdown}}$ $(RO)_2\overset{S}{\overset{\parallel}{P}}\!-\!S\!-\!\overset{S}{\overset{\parallel}{P}}(OR)_2$ (j)│RO·

BTSA   O,O,S-TBDTP   DBMS

(f)│

$(RO)_2\overset{S}{\overset{\parallel}{P}}\!-\!OH + SO_2$ $(RO)_2\overset{S}{\overset{\parallel}{P}}\!-\!OR$

DBTnPA   O,O,O-TBTP

↓

$(RO)_2P\!-\!SH$
DBTPA

Scheme 6. Antioxidant mechanisms of nickel dibutyl dithiophosphate (NiDBP).

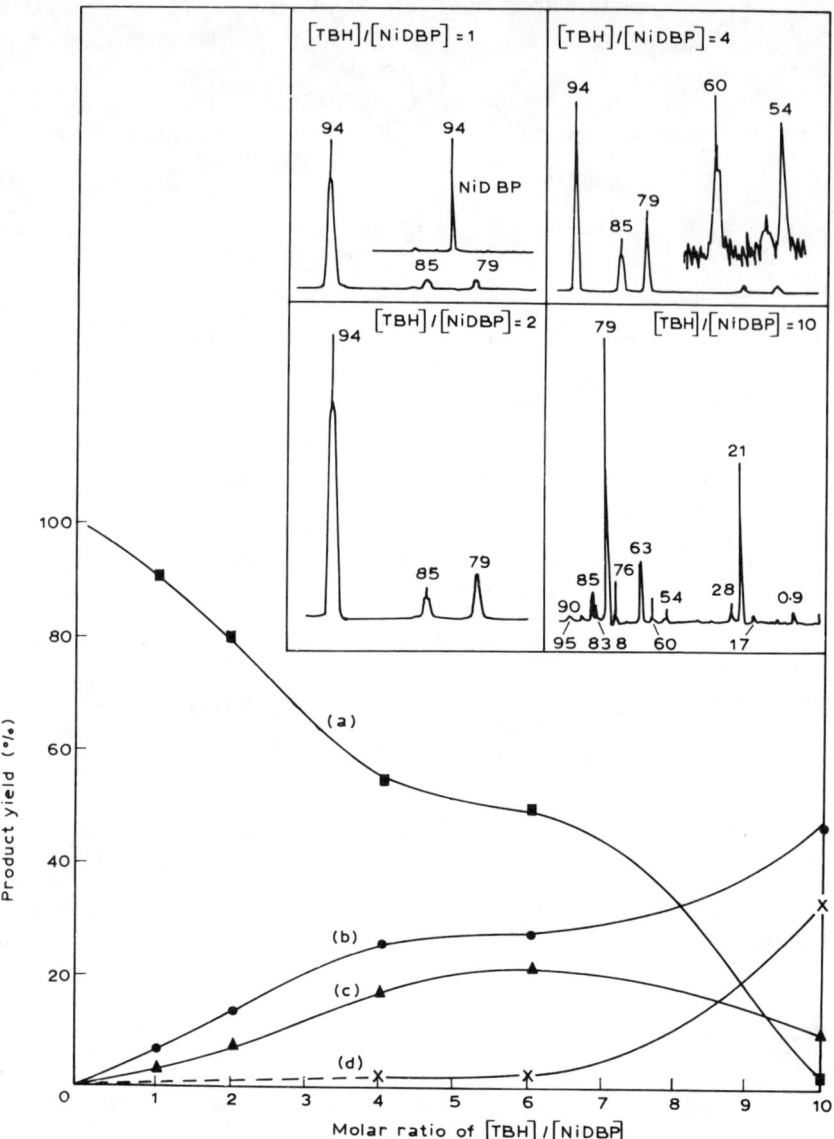

Fig. 9. Major product yield after complete reaction of NiDRP, R = n-butyl with TBH in cyclohexane at 25°C at different molar ratios [TBH]/[NiDBP]: (a) NiDBP (δ = 94 ppm); (b) DBMS (δ = 79 ppm); (c) DBDS (δ = 85 ppm); (d) DBTnPA (δ = 63 ppm), DBTPA (δ = 21 ppm) and $(BuO)_2PSH$ (δ = 60 ppm). Inset shows ^{31}P-NMR spectra of products formed at the end of reactions of NiDBP (0·3 M) and TBH at different molar ratios in cyclohexane at 25°C. Numbers on signals are chemical shifts in ppm.

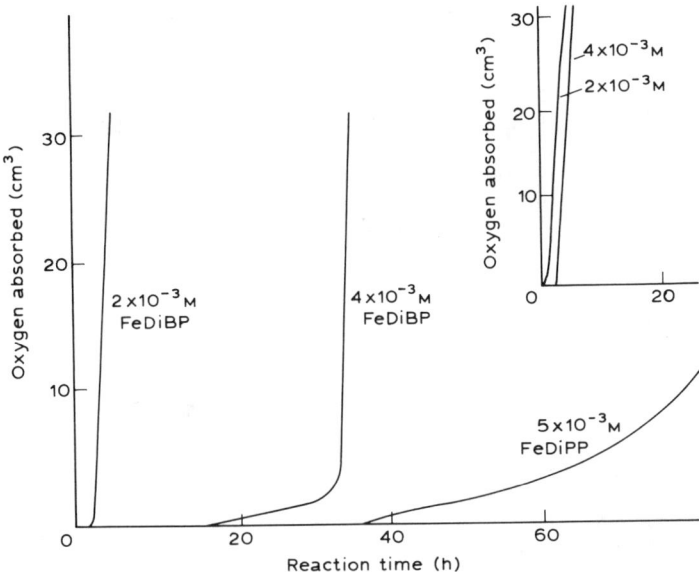

Fig. 10. Effect of FeDRP, R = isobutyl, on the oxidation of decalin at 130°C in the absence and presence (inset) of 1×10^{-2} M CHP. The molar concentrations of FeDiBP are given on the curves.

temperatures to evaluate its role as antioxidant under conditions similar to those used for the other derivatives of dithiophosphoric acid.

Iron di-isobutyl dithiophosphate was found to be a reasonably good inhibitor of oxidation of hydrocarbons (e.g. decalin at 130°C) in the absence of peroxides but is totally ineffective when excess hydroperoxide is present.[54] The behaviour of the iron complex when used as an inhibitor for the non-catalysed oxidation of decalin at 130°C (Fig. 10) was shown to be analogous to that of the corresponding zinc complex (examined under the same conditions) in that the length of the induction period increases with increasing the complex concentration but, on the whole, the iron complex is less effective. In studies on the effect of MRDP on AIBN-catalysed oxidation of cumene at 60°C, Burn found that the iron complex is a good trapping agent for peroxyl radicals.[30,31]

The behaviour of FeDRP in the presence of CHP (at 110°C) was found to be similar to that shown by the metal complexes (e.g. nickel

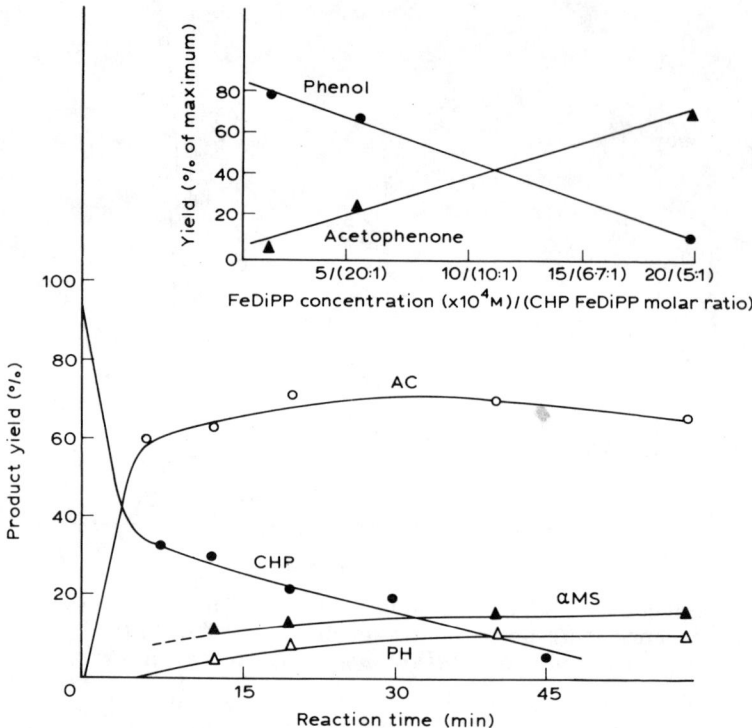

Fig. 11. Kinetics of products formed during reaction of CHP (1×10^{-2} M) with FeDBP, R = isobutyl (20×10^{-4} M) in chlorobenzene at 110°C. The CHP decomposition curve for the same reaction is also shown. Inset shows changes in concentrations of phenol and acetophenone (ionic and radical products of CHP) formed at the end of reactions of CHP with FeDiBP (chlorobenzene, 110°C) at different molar ratios.

and zinc); in the two-stage decomposition process, the relative importance of each stage depends on the initial [CHP]/[FeDRP] molar ratio. The first stage (Fig. 11) is a homolytic process with acetophenone as the major product; this stage predominates at high concentrations of the iron complex (see inset of Fig. 11). The predominance of the free-radical process at lower molar ratios (<5) must be responsible for the inability of FeDRP to prevent the CHP-initiated oxidation of decalin (inset, Fig. 10). Burn has also found that the effectiveness of FeDRP as an antioxidant for cumene (at 60°C) was seriously affected by the presence of CFP.[31] The first stage of the FeDRP–hydroperoxide reactions is accompanied by a colour change—the initial dark green changes to orange before yielding an orange/brown inorganic precipitate (most likely Fe_2O_3 or its hydrated form) with a clear colourless supernatant liquid. This suggests that all the iron is destroyed (oxidised) by the end of the first stage. The corresponding thiophosphoryl disulphide (DRDS) was identified by ^{31}P-NMR as the major phosphorus-containing product present in the supernatant liquid.[54] The second stage of CHP decomposition is an ionic process with phenol as the major product (see inset, Fig. 11); this stage predominates at low concentrations of FeDRP (i.e. higher molar ratios). The ionic decomposition associated with this stage must be caused by sulphur acids; these are the major oxidation products of disulphides which are released during the first stage of CHP-initiated decomposition of the iron complex (see Scheme 2).

7 THE ROLE OF DITHIOPHOSPHATES AS ANTIOXIDANTS FOR POLYMERS

It is now well known that one of the notable features of sulphur-containing antioxidants is that they exhibit a dual role as pro- and anti-oxidants in the same substrate depending on the prevailing conditions.[47,71] Mechanistic studies in hydrocarbon model systems have illustrated clearly that these compounds (see the previous sections for examples) undergo a complex series of oxidation reactions involving intermediate free radicals (pro-oxidants) to give sulphur acids which are the real catalysts for the decomposition of the hydroperoxides. Sulphur compounds which give an initial severe pro-oxidant effect are not useful on their own and are usually used in combination with chain-breaking (CB) antioxidants which either reduce the extent of, or eliminate, the initial radical-generating pro-oxidant step.[72] In the case of metal dithiolates, however, it was shown[17] that the pro-oxidant stage is not observed during both photo-oxidation and thermal oxidation of polymers; the radical-scavenging ability of metal dithiolates may be responsible for the absence of this stage. Dithiolates have, therefore, been used alone[7-11] for stabilising polymers (melt, thermal and photo-stabilisation) without the need for additional CB antioxidants due to their broad spectrum of activities.[17]

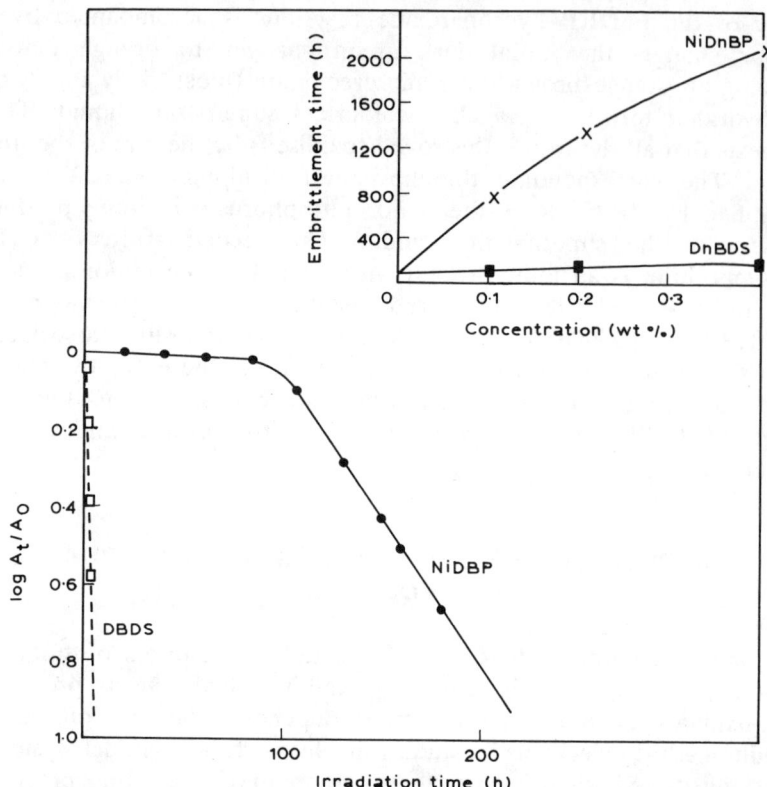

Fig. 12. Kinetics of disappearance of the 316 nm band of NiDBP and the 265 nm band of DBDS during UV exposure (at 30°C) in hexane solution (at 0.8×10^{-4} M). Inset shows effect of concentration of NiDBP and DBDS on the UV stability of PP.

A normal requirement for an effective peroxidolytic photo-antioxidant is long-term photostability where the parent compound is neither destroyed by light nor will give rise to sensitising products. Metal dithiophosphates have much higher UV stability (and higher extinction coefficients) than thiophosphoryl sulphides (e.g. disulphide; see Fig. 12) and simple sulphides. The metal ion plays a crucial part in their overall effectiveness; for example, transition metal complexes such as NiDRP are more stable towards UV light and are, therefore, better UV stabilisers than Group II metal complexes (e.g. ZnDRP). Another difference between the nickel complex and the disulphide is

Table 6
Comparison of the effectiveness of NiDRP (R = n-Bu) and Cyasorb UV 531 as UV screens in unstabilised polypropylene processed at 190°c for 10 min in a closed internal mixer

Additive	Concentration (mol/100g)	Embrittlement time (h)		Contribution of screening mechanism (%)
		Additive	Screen	
UV 531	3×10^{-4}	310	270	87
UV 531	6×10^{-4}	520	450	87
NiDRP	3×10^{-4}	950	320	33·6
NiDRP	6×10^{-4}	1 500	480	32

that, unlike the latter, the UV stabilising action of the former is strongly concentration-dependent (Fig. 12, inset). In addition to the major role of dithiophosphates as peroxidolytic and, to a lesser extend, radical-scavenging agents, they also act by other mechanisms as UV screens and quenchers. Table 6 shows that although the UV-screening mechanism contributes to the overall UV-stabilising effectiveness of NiDRP, this mode of action is far less significant than for typical commercial UV absorbers such as substituted hydroxybenzophenones, e.g. Cyasorb UV 531. In contrast to the metal complexes which are quite stable to UV light in the harmful region of sunlight, the corresponding thiophosphoryl disulphides are not very stable to UV light and hence are not good UV screens.

The effect of processing severity on the photoantioxidant efficiency of NiDRTP and DRDS is also different, reflecting the above differences in the antioxidant behaviour of these compounds. Severe processing drastically affects the persistence of the nickel complex in the polymer, giving rise to a shorter photo-oxidative induction period (the induction period was shown[8] to be directly proportional to the amount of NiDRP which survives the processing operation) and hence reduced performance (Fig. 13). The nickel complexes, therefore, act as light-stable reservoirs for the controlled release of active antioxidant species during UV exposure. On the other hand, the effect of processing on the stabilising action of the disulphide, which is the primary oxidation product of NiDRP, is opposite to that of the parent nickel complex: the performance of the disulphide as a UV stabiliser improves when it is subjected to the same severe processing operation (Fig. 13).

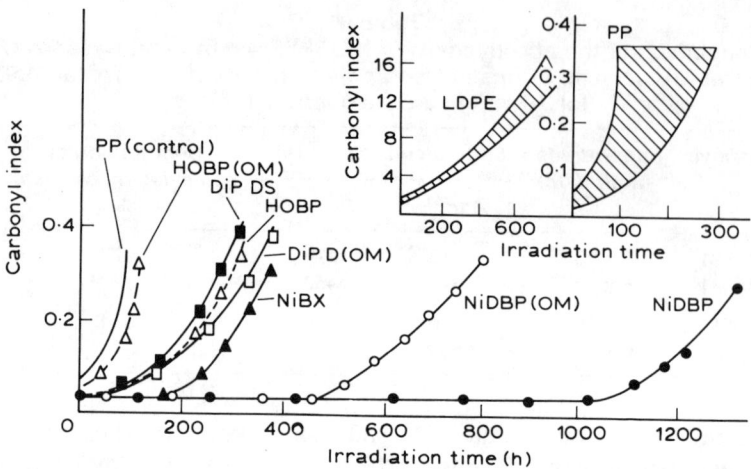

Fig. 13. Effect of mixing procedures (processing in a closed mixer, CM, and openmixer, OM, in an internal mixer) on photo-oxidative stability of PP-containing additives (2.5×10^{-4} mol/100g). Inset compares photo-stabilising effectiveness of mildly processed (CM) DRDS, R = iso-PR in LDPE and PP.

The behaviour of thiophosphoryl disulphides during processing has been exploited[10,16] for the development of very effective stabiliser systems. The stabiliser system is produced by a controlled oxidation of a sulphur-based antioxidant during processing of the polymer. Figure 14 compares the UV-stabilising effectiveness of an oxidatively processed thiophosphoryl disulphide containing NiDRP and Cyasorb UV531 as synergists (here called COPS) with that of a number of commercially based synergistic systems and, also, with the same system under normal (mild) processing conditions (here called CMPS). It is important to mention here that oxidative processing is not normally used by polymer manufacturers since it affects adversely the UV stability of the product due to the formation of sensitising groups, especially hydroperoxides. In a controlled oxidatively processed system, the active antioxidant is generated *in situ* through peroxide-induced transformations of antioxidants or their precursors, during processing. This is a new approach to the current polymer stabilisation practice and since the oxidation process is carried out separately in the form of masterbatch concentrates, it should be acceptable to industry.

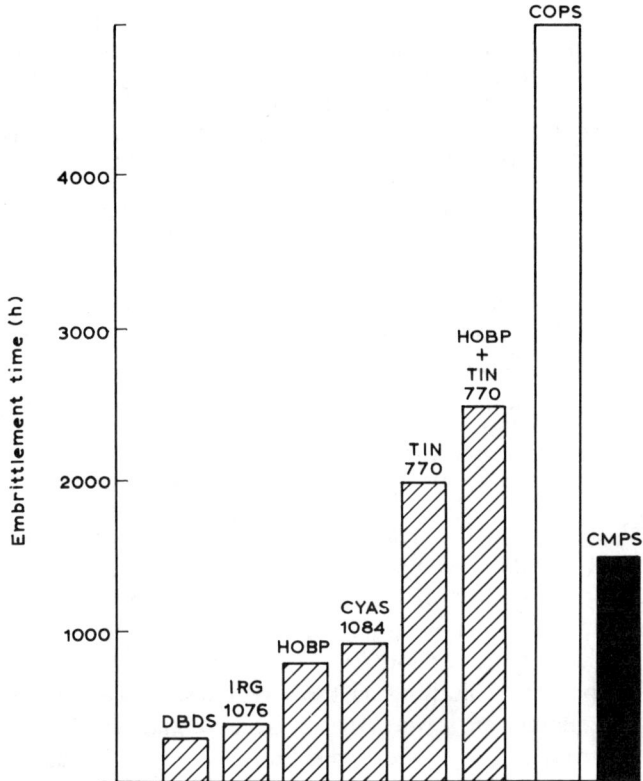

Fig. 14. Comparison of the effectiveness of different commercial UV stabilisers with oxidatively processed, under controlled conditions, polymer containing sulphur antioxidants in addition to a small concentration of a UV absorber, hydroxy octyloxy benzophenone, HOBP, (DBDS + NiDBP + HOBP) referred to as COPS, and with the same system but processed mildly in a closed mixer (referred to as CMPS). Total concentration of the stabiliser in each case is 0·4%.

It is well known that for antioxidants to be effective, they must be retained in the polymer without physical loss. The equilibrium solubility of an additive in a polymer is a very important property of the system since it determines how much additive can exist in the polymer in homogeneous solution. In general, solubility of stabilisers in polymer substrates can be improved by introducing longer-chain substituents in stabilisers which have similar character (e.g. similar

Table 7
Extent of binding of dithiophosphoric acid (DRDP) and its ammonium salt (ARDP) to natural rubber (NR) latex during its reaction with TBH in the presence and absence of activators
The [TBH]/[Dithiophosphate] molar ratio was 3:1. Reaction time = 16h; reaction temperature = 60°C.

Dithiophosphate	Activator	Concentration (g/100g latex)	Percentage bound (g/100g additive)
DRDP	None	—	44
	TEPA^a	0·2	24
	Na$_2$WO$_4$	0·2	50
	(NH$_4$)$_2$Ce(SO$_4$)$_3$	0·2	61
ARDP	None	0·2	44

^{a} TEPA, tetraethylene pentamine.

functional groups) to the host polymer. It was shown that increasing the length of the alkyl groups of metal dithiophosphates (e.g. NiDRP) increases the solubility of these stabilisers in the polymer during processing, leading to substantial improvements in the subsequent UV stability of the polymer (Table 7).[12] However, it is clear from Table 6 that this approach is limiting as the usefulness of the stabiliser may actually be reduced on a weight basis because the weight ratio of the functional group to the inert residue is steadily reduced in the antioxidant.[12]

A potential solution to the problem of physical loss of antioxidants from polymers involves the use of reactive antioxidants that are capable of reacting with different polymers (e.g. rubbers and unsaturated thermoplastics such as ABS) in a reactive processing procedure to produce polymer-bound functions which can be removed only by breaking chemical bonds.[73] For example, using simple thiols and sulphide antioxidants, Kharasch-type addition[74] of the sulphur function to the double bond has been achieved in polymer melts initiated by shear, and this method was used to make masterbatch concentrates of bound antioxidants.[75] We have shown that dialkyl dithiophosphoric acids (DRDPA; I) and their salts (e.g. ammonium) can readily undergo Kharasch addition to double bonds in rubbers in the presence of free radicals formed either by redox reactions with hydroperoxides (Table 7), or by mechanical shear of the macromolecule (Table 8) to give a phosphoryl-bound moiety.[76] It was further shown[76] that the use

Table 8
Extent of binding of dithiophosphoric acid (DRDPA), the disulphide (DRDS) and the ammonium salt (ADRP) to NR during processing in a shearing mixer
Reaction time = 10 min; 10g additive/100g rubber.

Dithiophosphate	Processing temperature (°C)	Extent of binding (g/100g additive)			
		Open	Closed	Closed + TBH	Closed + DCP
DRDP	140	13	22	20	—
	160	d	d	17	—
DRDS	140	9	20	—	26
	160	d	d	—	22
ARDP	140	12	23	—	—
	160	d	12	—	—

of reducing agents in combination with hydroperoxides has no beneficial effect on the above addition reactions. This is not surprising since it was mentioned earlier that the dithiophosphoric acids themselves are good reducing agents [reaction (5)], which, in this case, will be followed by reactions (6) and (7). Reaction (5) can be facilitated by using transition metal ions as activators (see Table 7) but the adduct reaction [reaction (7)] becomes reversible at high temperatures (e.g. >140°C) and high yields cannot be achieved by this procedure.[76] Examination of the effect of vulcanisation (e.g. conventional sulphur vulcanisation) on the polymer adduct revealed that about half of the thiophosphoryl adduct is eliminated from the polymer chain during vulcanisation [reaction (8)], and this supports the above reversibility of the reaction.

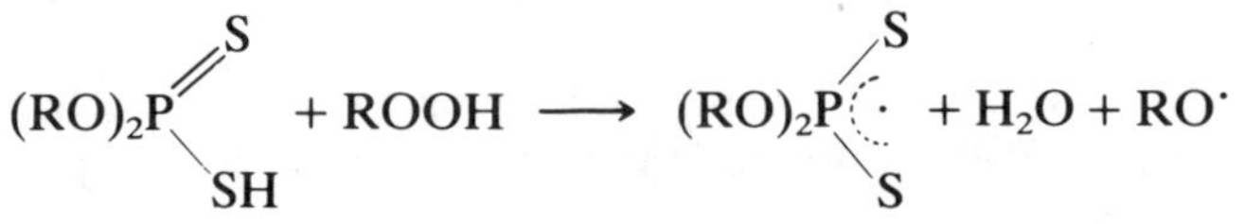

$$(RO)_2P\begin{matrix}S\\SH\end{matrix} + ROOH \longrightarrow (RO)_2P\begin{matrix}S\\S\end{matrix} + H_2O + RO^\cdot \quad (5)$$

$$(RO)_2P\begin{matrix}S\\S\end{matrix} + -CH_2\overset{CH_3}{\underset{}{C}}=CHCH_2- \longrightarrow -CH_2\overset{CH_3}{\underset{S-P(OR)_2}{C}}-CHCH_2- \quad (6)$$
$$\underset{\|}{}$$
$$S$$

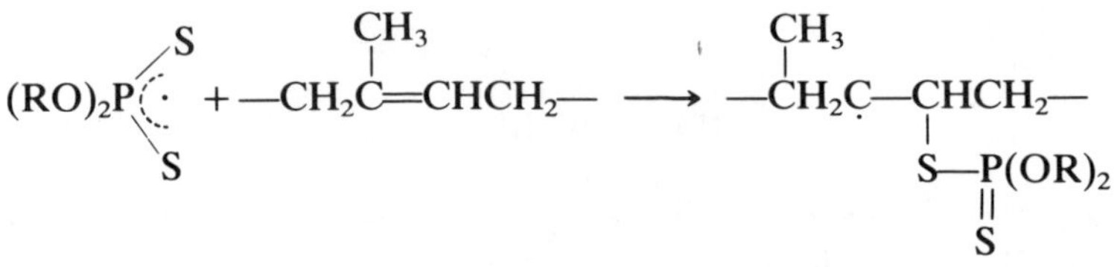

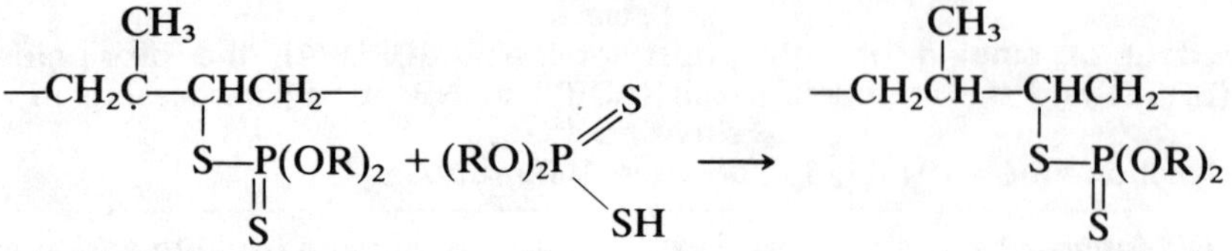

$$\begin{array}{c}\text{CH}_3\\|\\-\text{CH}_2\text{C}-\text{CHCH}_2-\\|\\\text{S}-\text{P(OR)}_2\\\|\\\text{S}\end{array} + (\text{RO})_2\text{P}\begin{array}{c}\nearrow\text{S}\\\searrow\text{SH}\end{array} \longrightarrow \begin{array}{c}\text{CH}_3\\|\\-\text{CH}_2\text{CH}-\text{CHCH}_2-\\|\\\text{S}-\text{P(OR)}_2\\\|\\\text{S}\end{array}$$

(7)

$$\begin{array}{c}\text{CH}_3\\|\\-\text{CH}_2\text{CH}-\text{CHCH}_2-\\|\\\text{S}-\text{P(OR)}_2\\\nearrow\\\text{S}\end{array} \longrightarrow -\text{CH}_2\text{C}=\text{CHCH}_2- + (\text{RO})_2\text{P}\begin{array}{c}\nearrow\text{S}\\\searrow\text{SH}\end{array}$$

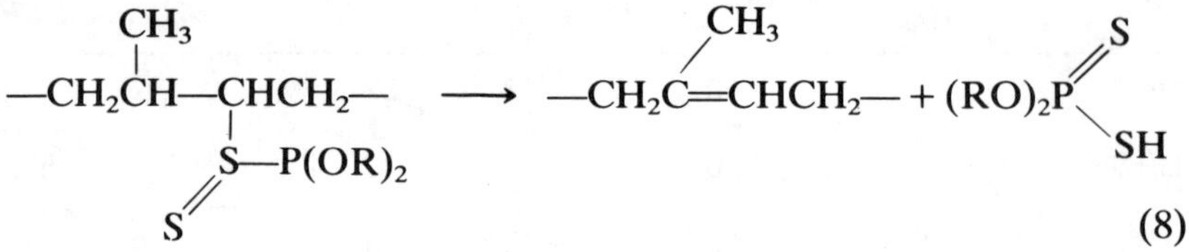

(8)

To achieve optimal binding in the case of dialkyl dithiophosphoric acids and their derivatives, at least a three-fold excess of hydroperoxide is needed[76] (unlike the case of alkyl thiols[77,78] which normally require a molar ratio of 1) and the yield of the adduct is never greater than 50%, suggesting that competing reactions occur, leading to non-desired products. This chapter has shown clearly that dithiophosphates are the precursors of very powerful peroxidolytic agents which form as a result of further oxidation of the starting compounds by hydroperoxides and this is how they function as antioxidants in technological media. The peroxidolytic products will, therefore, rapidly destroy hydroperoxides by a non-radical mechanism once they are formed and the adduct process becomes self-limiting.

ACKNOWLEDGEMENTS

I acknowledge with gratitude my colleagues Professor G. Scott and Drs M. Coker and P. Smith for their contributions to the previously unpublished results.

REFERENCES

1. Howard, J. A., Ohkatsu, Y., Chenier, J. H. B. & Ingold, K. U., *Can. J. Chem.*, **51** (1973) 1543.

2. Reid, E. E., in *Organic Chemistry of Bivalent Sulphur*, Vol. 1. Chemical Publishing, New York, 1958, p. 305.
3. Pimblott, J. G., Scott, G. & Stuckey, J. E., *J. Appl. Polym. Sci.*, **19** (1975) 865; **23** (1979) 3621.
4. Chadwick, D. H. & Watt, R. S., In *Phosphorus and its Compounds*, ed. J. R. V. Wazer, Vol. II. Interscience Publishers, New York, 1961, p. 1264.
5. Scott, G., In *Atmospheric Oxidation and Antioxidants*. Elsevier, London, 1965, p. 271.
6. US Patent 2 884 405.
7. Al-Malaika, S. & Scott, G., *Europ. Polym. J.*, **16** (1980) 709.
8. Al-Malaika, S. & Scott, G., *Polym. Degrad. Stab.*, **5** (1983) 415.
9. Al-Malaika, S. & Scott, G., *Europ. Polym. J.*, **19** (1983) 241.
10. Al-Malaika, S., *Brit. Polym. J.*, **16** (1984) 301.
11. Al-Malaika, S., Moraue, B. & Scott, G., *Plast. Rubb. Proc. Appl.*, **4** (1984) 365.
12. Al-Malaika, S., Desai, P. & Scott, G., *Plast. Rubb. Proc. Appl.*, **5** (1985) 15.
13. Al-Malaika, S., Coker, M. & Scott, G., *Polym. Degrad. Stab.*, **10** (1985) 173.
14. Al-Malaika, S., Coker, M. & Scott, G., *Polym. Degrad. Stab.*, **22** (1988) 147.
15. Hercules Powder, French Patent 1 314 410 (1961).
16. Al-Malaika, S. & Scott, G., British Patent 2 117 779A (1983).
17. Al-Malaika, S., Chakraborty, K. B. & Scott, G. In *Developments in Polymer Stabilisation—6*, ed. G. Scott. Applied Science Publishers, London, 1983, Chapter 3.
18. Ivanov, S. K. In *Developments in Polymer Stabilisation—3*, ed. G. Scott. Applied Science Publishers, London, Chapter 3.
19. Carius, Z., *Ann.*, **119** (1861) 289.
20. Kosolapoff, G. M. In *Organophosphorous Compounds*, ed. G. M. Kosolapoff. Wiley, New York, 1985, p. 236.
21. Corbridge, D. E. C. In *Phosphorus, an outline of its Chemistry, Biochemistry and Technology*, Elsevier, Amsterdam, 1978, Chapter 7.
22. Bacon, W. E. & LeSuer, W. M., *J. Am. Chem. Soc.*, **76** (1954) 670.
23. Ailman, D. E. & Magee, R. J. In *Organophosphorous Compounds*, Vol. 7, ed. G. M.Kosolapoff. Wiley, New York, 1985, Chapter 19, p. 523.
24. Oae, S., Nakanishi, A. & Tsujimoto, N., *Tetrahedron*, **28** (1972) 2981.
25. Jorgenson, C. K. In *Inorganic Complexes*. Academic Press, New York, 1963, Chapter 7, p. 138.
26. Coucouvanis, D., *Prog. Inorg. Chem.*, **11** (1970) 233.
27. Sanin, P., Blagovidov, I., Vipper, A., Kuliev, A., Krein, S., Ramaya, A. K., Shor, G., Sher, V. & Zaslavsky, Y., *Proc. 8th World Petroleum Congress*, **5** (1971) 91.
28. Colclough, T. & Cunneen, J. I., *J. Chem. Soc.* (1964) 4790.
29. Ivanov, S. K. & Shopov, D., *Compt. Rend. Acad. Bulg. Sci.*, **18** (1965) 845.
30. Burn, A. J., *Tetrahedron*, **22** (1966) 2153.

31. Burn, A. J., *International Oxidation Symposium*, **1** (1967) 323.
32. Willermet, P. & Kandah, S., *ASLE Transactions*, **27** (1983) 67.
33. Howard, J. & Tong, S., *Can. J. Chem.*, **58** (1979) 92.
34. Al-Malaika, S. & Scott, G. *Europ. Polym. J.*, **16** (1980) 503.
35. Korcek, S., Mahoney, L., Johnson, M. & Siegl, W., Society of Automotive Engineers, Technical Paper 810014 (1981).
36. Johnson, M., Korcek, S. & Zinbo, M., Society of Automotive Engineers, Technical Paper SP-558 (1983) p. 71.
37. Burn, A. J. In *Oxidation of Organic Compounds*, American Chemical Society, Washington, 1968.
38. Mahoney, L. R., Korcek, S., Hoffman, S. & Willermet, P., *Ind. Eng. Chem. Prod. Res. Dev.*, **17** (1978) 250.
39. Howard, J. A. In *Frontiers of Free Radical Chemistry*, ed. W. A. Pryor. Academic Press, New York, 1980, p. 237 et seq.
40. Kennerley, G. & Patterson, W., *Ind. Eng. Chem.*, **48** (1956) 1917.
41. Hock, H. & Lang, S., *Ber.*, **77B** (1944) 257.
42. Holdsworth, J. D., Scott, G. & Williams, D., *J. Chem. Soc.* (1964) 4962.
43. Burn, A. J., Cecil, R. & Young, V., *J. Inst. Petr.*, **57** (1971) 319.
44. Sexton, M., *J. Chem. Soc., Perkin Trans.* 2 (1984) 1771.
45. Husbands, M. & Scott, G., *Europ. Polym. J.*, **15** (1979) 249.
46. Larson, R., *Sci. Lubrication*, **10** (1958) 12.
47. Ivanov, S. K., Yurittsin, V. S. & Scopov, D., *Coll. Czech Chem. Commun.*, **37** (1972) 3284.
48. Laver, H. S. In *Developments in Polymer Stabilistation—1*, ed. G. Scott. Applied Science Publishers London, 1979 Chapter 5.
49. Carlsson, D. J. & Wiles, D. M., *J. Macromol. Sci., Macromol. Chem.*, **C14**(65) (1976) 155.
50. Al-Malaika, S., Marogi, A. & Scott, G., *J. Appl. Polym. Sci.*, **30** (1985) 789.
51. Al-Malaika, S., Marogi, A. & Scott, G., *Polym. Degrad. Stab.*, **10** (1985) 237.
52. Vassil'ev, R. F. & Vichutinskii, A. A., *Nature (London)*, **194** (1962) 1276.
53. Al-Malaika, S., Coker, M. & Scott, G., unpublished work.
54. Al-Malaika, S., Smith, P. & Scott, G., unpublished work.
55. Ohkatsu, Y., Kikkawa, K. & Osa, T., *Bull. Chem. Soc. Japan*, **51** (1978) 3606.
56. Al-Malaika, S., Marogi, A. & Scott, G. *J. Appl. Polym. Sci.*, **33** (1987) 1455.
57. Rossi, E. & Imperato, L., *Chimica e Industria*, **53** (1971) 838.
58. Al-Malaika, S. & Scott, G. *Polym. Commun.*, **23** (1982) 1711.
59. Bridgewater, A., Dever, J. & Sexton, M., *J. Chem. Soc., Perkin Trans.*, **11** (1980) 1006.
60. Cherkasova, O., Mukmeneva, N., Chebotareva, E., Ovchinnikov, V., Pobedimskii, D. & Kirpichnikov, P., *Neftekhimiya*, **24** (1984) 76; *Chem. Abstr.*, **101** 7303p.
61. Grishina, O. & Bashinova, V., *Neftekhimiya*, **14** (1974) 142; *Chem. Abstr.*, **80**, 145654c.

62. Ivanov, S. & Kateva, I., *Neftekhimiya*, **18** (1978) 417; *Chem. Abstr.*, **89** 146540k.
63. Shkhiyants, I., Voevoda, N., Komissarova, N., Chernyav, S., Kaya, L., Sher, V. & Sanin, P. *Neftekhimiya*, **14** (1974) 312; *Chem. Abstr.* **81**, 13216x.
64. Kozak, P. & Rabel, V., *Scientific Papers of the Prague Institute of Chemical Technology*, **D39** (1978) 141.
65. Sher, V., Markova, E., Khanakova, L., Kuzminav, G. & Sanin, P., *Neftekhimiya*, **13** (1973) 876; *Chem. Abstr.*, **80** 95168z.
66. Al-Malaika, S. & Scott, G., *Europ. Polym. J.*, **19** (1983) 235.
67. Al-Malaika, S. & Scott, G., *Europ. Polym. J.*, **24** (1983) 25.
68. Jorgenson, C. K., *J. Inorg. Nucl. Chem.*, **24** (1962) 1571.
69. Lebedda, J. & Palmer, R., *Spectrochim. Acta, Part A*, **29** (1973) 1371.
70. Hopkins, W. & Mitchell, P., Reading University, unpublished work.
71. Shelton, J. R. In *Developments in Polymer Stabilisation—4*, ed. G. Scott. Applied Science Publishers, London, 1981, Chapter 2.
72. Scott, G. In *Developments in Polymer Stabilisation—4*, ed. G. Scott. Applied Science Publishers, London, 1981, Chapter 1.
73. Al-Malaika, S., *Polymer Prepr.*, **29**(1) (1988) 555.
74. Karasch, M. S., Nudenberg, W. & Mantell, G. T., *J. Org. Chem.*, **16** (1951) 524.
75. Scott, G. In *Developments in Polymer Stabilisation—4*, ed. G. Scott. Applied Science Publishers, London, 1981, Chapter 6.
76. Al-Malaika, S., Honggokusumo, S. & Scott, G., *Polym. Degrad. Stab.*, **16** (1986) 25.
77. Scott, G. & Suharto, R., *Europ. Polym. J.*, **20** (1984) 139.
78. Scott, G. & Suharto, R., *Europ. Polym. J.*, **21** (1985) 765.

Chapter 4

Polymers and High-Energy Irradiation: Degradation and Stabilization†

D. J. CARLSSON & S. CHMELA‡

*Division of Chemistry, National Research Council of Canada,
Ottawa, Canada*

ABSTRACT

The primary and secondary events which occur in polymers exposed to high-energy radiation are briefly surveyed with emphasis on the role of oxygen in causing polymer degradation. In addition slow, post-irradiation reactions are discussed as they relate to deterioration during storage. Potential methods of polymer stabilization against high-energy radiation are then reviewed from the viewpoints of electron and ion scavenging, energy transfer processes, radical scavenging (both carbon-centred and peroxyl) and also acceleration of radical decay. Some recently developed additive combinations, based on hindered amines, are discussed in detail and shown to operate by suppressing the formation of hydroperoxides during the free-radical oxidative chain process which immediately follows exposure to high-energy irradiation. The absence of these thermally labile groups improves post-irradiation storage stability. Finally, the effects of the radiation and radiation-induced events on stabilizers themselves are considered.

† Issued as NRCC No. 31475.
‡ Permanent addresss: Polymer Insittute, Centre for Chemical Research, Slovak Academy of Sciences, Bratislava, Czechoslovakia.

1 INTRODUCTION

Plastics, synthetic fibres and composites are used in many areas where exposure to high-energy radiation can occur. Applications include nuclear power plants, radiation equipment (such as X-ray sources), particle-physics generators, sterilization systems (medical equipment, hospital clothing or foodstuffs packaging), etc. In addition, high-energy radiation is used to deliberately modify polymers so as to control molecular weight, crosslinking and long-chain branching, especially in high-resolution lithography. The types of radiation to which polymers may be exposed normally comprise γ-rays, X-rays and electron (e-) beams; heavy particles and neutrons are somewhat less common. Visible and UV radiation will not be discussed, although short-wave UV may produce quite similar chemistry.

The effects of high-energy radiation on polymers have been studied for many years. The pioneering work of Charlesby and Chapiro is particularly noteworthy, as are the more recent studies of Dole and co-workers, primarily on polyolefins.[1-4] The radiation chemistry of elastomers and polymers have been recently reviewed.[5,6] Before considering techniques for minimizing or controlling radiation effects on polymers, it is first necessary to review briefly the immediate effects of radiation on polymers as well as the post-irradiation reactions. The discussion will be confined to synthetic polymers and their radiation stabilization. There is copious literature on the radiation degradation of biopolymers and antirads to minimize their degradation.[7,8] However, biopolymers are outside the scope of this chapter.

2 PRIMARY RADIATION PROCESSES IN POLYMERS

Irradiation of polymers can produce crosslinking, backbone scission, hydrogen evolution, etc. The timescale for these events is shown in Table 1.[3-6] Chemical products result from the occurrence of a complex cascade of events such as reactions (1)–(6) (typical of γ-irradiation). Initial interaction of each γ-ray photon with the polymer yields some fast electrons (similar to those from direct e-beam irradiation), which in turn cause subsequent ejection of secondary electrons at some distance (several microns) from the primary event. At ambient temperatures, ion–electron recombination occurs quickly to give highly excited states (P*) and cations restructure.[9] At low

Table 1
Radiation-induced reaction sequence

Time (s)[a]	Process	Prevention/reduction
~10^{-18}	Ejection of primary, energetic electrons.	Radiation shielding.
10^{-17}–10^{-15}	Secondary electrons ejected, cations formed. Excited states formed.	
10^{-12}–10^{-3}	Ion–electron recombination. Excited states, H˙ and macroalkyl radicals formed. Chain scission.	e^-, ion scavenge. Excited state quenching. Radical scavenging.
10^{-10}–∞	Chemical events. 'Stable' radicals, trapped electrons formed. Ion diffusion occurs. O_2 involvement (O_3 reactions), oxidative chain scission. Initial product decomposition.	Radical scavenging. O_2 exclusion. Peroxide decomposition.

[a] Measured from the start of irradiation.

temperatures (≤ −100°C), ejected electrons may be trapped in the polymer matrix.[10] The excited states dissipate some of their excess energy by bond scission to give free radicals. The scission of C–H bonds is favoured over C–C backbone scission. This has been attributed to energy migration along the backbone by an exciton mechanism, minimizing energy localization in specific C–C backbone bonds, whereas energy deposited in C–H bonds cannot migrate.[11] An alternative suggestion is that in highly excited states, C–C bonds are in fact more stable than C–H bonds, contrary to the ground-state situation.[12] In addition, recombination of the two macroalkyl radicals

$$\text{Polymer (P)} \xrightarrow{\gamma} \text{Energy absorption} \quad (1)$$

$$\downarrow$$

$$e^-, P^+ \quad \text{Electron ejection} \quad (2)$$

$$e^- + nP \longrightarrow nP^+ + (n+1)e^- \quad \text{Secondary electron ejection} \quad (3)$$

$$e^- + P^+ \longrightarrow P^* \quad \text{Excited state formation} \quad (4)$$

$$P^* \longrightarrow P˙ + P˙ \quad \text{C–C scission} \quad (5)$$

$$\searrow P˙ + H˙ \quad \text{C–H scission} \quad (6)$$

resulting from backbone scission is facilitated by a solid matrix.[13] Loss of small side groups which can diffuse away is likely to be permanent.

For γ-, X-ray and e-beam exposure, reaction centres are quite widely separated, with only small clusters of products (two or four radicals) in close proximity. On the other hand, heavy particles lead to dense ionization along tracks with high local concentrations of products occurring in spurs.

3 SECONDARY REACTIONS IN IRRADIATED POLYMERS

The free radicals produced in reactions (5) and (6) lead to many of the chemical products associated with radiation effects. The combination of macro-alkyl radicals or their addition to unsaturated sites leads to chain branching and/or crosslinking. Hydrogen atoms mainly abstract from the polymer chain to give molecular hydrogen and fresh macroalkyl radicals [reaction (7)].

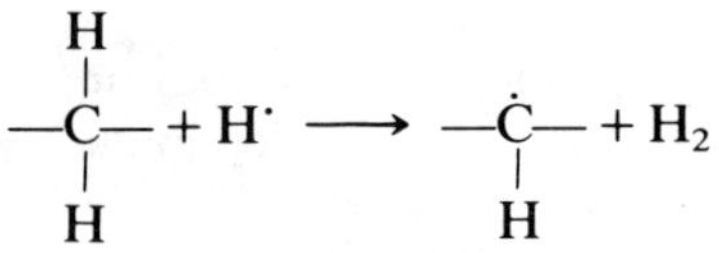

$$\begin{array}{c} \text{H} \\ | \\ -\text{C}- \\ | \\ \text{H} \end{array} + \text{H}^{\cdot} \longrightarrow \begin{array}{c} \\ | \\ -\dot{\text{C}}- \\ | \\ \text{H} \end{array} + \text{H}_2 \qquad (7)$$

In some polymers, main chain scission is followed by monomer elimination. Macroradical combination results in crosslink formation [reaction (8)].

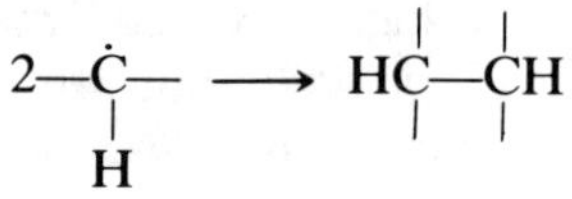

$$2 \begin{array}{c} \\ -\dot{\text{C}}- \\ | \\ \text{H} \end{array} \longrightarrow \begin{array}{c} | \quad | \\ \text{HC}-\text{CH} \\ | \quad | \end{array} \qquad (8)$$

Unsaturation is a major product from irradiated polyolefins and is believed to result from migration of radical sites by an inter- and intra-molecular hydrogen atom transfer until two sites come together.[4]

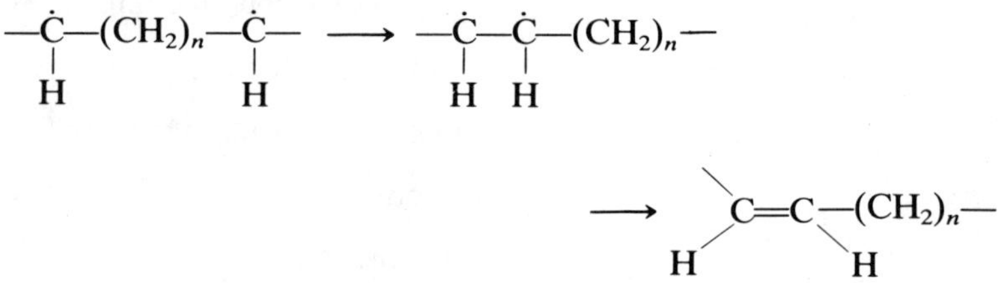

$$\begin{array}{c} -\dot{\text{C}}-(\text{CH}_2)_n-\dot{\text{C}}- \\ | \qquad \qquad | \\ \text{H} \qquad \qquad \text{H} \end{array} \longrightarrow \begin{array}{c} -\dot{\text{C}}-\dot{\text{C}}-(\text{CH}_2)_n- \\ | \quad | \\ \text{H} \ \text{H} \end{array}$$

$$\longrightarrow \begin{array}{c} \diagdown \\ \text{C}=\text{C}-(\text{CH}_2)_n- \\ \diagup \qquad \diagdown \\ \text{H} \qquad \text{H} \end{array} \qquad (9)$$

In the absence of oxygen, the net result of irradiation is the composite result of reactions (5)–(9) so that crosslinked gel or a degradation of molecular weight results.

In the absence of O_2, the behaviour of various polymers may be generalized into those which crosslink during irradiation [polyethylene, poly(methyl acrylate), poly(acrylic acid), polystyrene] and those which degrade [poly(methyl methacrylate), poly(methylacrylic acid), poly(α-methylstyrene), poly(butene-2)].[3] Polypropylene undergoes both scission and crosslinking. Crosslinking increases the stiffness of plastics and can render them inextensible. Poly(olefin sulphones) have been shown to be exceptionally sensitive to γ- or e-beam radiation and can be used as short-wavelength photoresists.[14]

Chain scission also leads to embrittlement, but the effect of direct, radiation-induced scission in commodity polymers is normally minor compared with oxidative chain scission (see below).

3.1 Influence of Oxygen

For polymers in equilibrium with air, O_2 pervades the amorphous domains of polymers and possibly also the defective crystalline-morphology produced in some commercial thermoplastic materials, as a result of rapid cooling from the melt. The O_2 solubility is $\sim 1 \times 10^{-3}$ mol kg^{-1} for semi-crystalline polyolefins in equilibrium with air, but is reduced by orientation of the polymer. Because of its biradical nature, O_2 reacts at close to the encounter frequency with carbon-centred radicals to give peroxyl radicals, by reaction (10)). A relatively slow hydrogen abstraction from the polymer matrix by the peroxyl radicals, reaction (II), completes a cycle of reactions which cause the progressive oxidation of the polymer.[15]

$$-\overset{|}{\underset{|}{C}}{}^{\cdot} + O_2 \longrightarrow -\overset{|}{\underset{|}{C}}-O_2^{\cdot} \qquad (10)$$

$$-\overset{|}{\underset{|}{C}}-O_2^{\cdot} + -\overset{|}{\underset{|}{C}}-H \longrightarrow -\overset{|}{\underset{|}{C}}-OOH + -\overset{|}{\underset{|}{C}}{}^{\cdot} \qquad (11)$$

The first molecular product, the hydroperoxide group ($-\overset{|}{\underset{|}{C}}-OOH$), is thermally unstable and cleaves readily at the O–O linkage to give a pair of radicals and so leads to a branching, thermal oxidation during

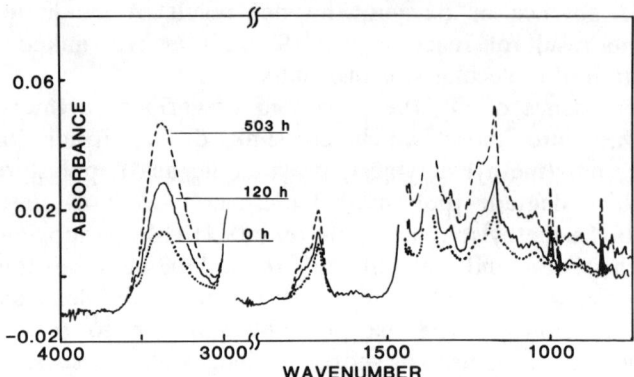

Fig. 1. Effects of γ-irradiation and subsequent storage on polypropylene films. Difference FTIR spectra. Times refer to storage at 23°C after exposure to 2·8 Mrad at 0·14 Mrad h^{-1}. Reproduced with permission from Ref. 16.

storage after irradiation.[16] Post-γ-oxidation is clearly shown in Fig. 1 for polypropylene film. This effect is of major concern for γ-sterilized medical equipment, implants, etc.

Post-γ-oxidation has been previously attributed to long-lived radicals, trapped in the crystalline phase in polyolefins.[17,18] Although long-lived peroxyl radicals can be detected after irradiation, they cannot be in the perfect (O_2-impermeable) crystal, but on fold surfaces or hindered in some other way. We have shown by ^{17}O labelling that these long-lived radicals are formed during irradiation, and not in the post-irradiation reactions.[19] However, they appear to contribute only slightly to the storage degradation, which is primarily controlled by the slow decomposition of the hydroperoxides at ambient temperature.[16]

3.2 Degradation of Physical Properties

Loss of physical properties in many polymers containing aliphatic backbone substituents results from the β-scission of alkoxyl radicals, reaction (12).[15] Alkoxyl radicals are formed by hydroperoxide

$$R'-\overset{\text{\}}{\underset{\text{\}}{C}}-O^{\cdot} \longrightarrow \mathord{\sim\sim}C\overset{O}{\underset{R}{\diagdown}}+\mathord{\sim\sim} \tag{12}$$

decomposition. They are also formed in the complex self-reaction of peroxyl radicals which may terminate the radicals. Elongation at break has been shown to be appreciably more sensitive to degradation than tensile strength.[20]

The relative ranking of polymers according to their radiation resistance is complicated by the marked effect of dose rate. For a given total radiation dose, low dose rate has been reported to be more damaging than high rates.[20-22] This effect is not fully understood, but has been discussed in detail by Gillen & Clough.[21] In addition, the study of dose-rate effects and the comparison of data from different authors is complicated by differences in exposure conditions. At very high dose rates, as are present during e-beam exposure, O_2 is completely consumed almost immediately and the bulk of the exposure is then anaerobic because O_2 cannot diffuse quickly enough to replenish the interior of even thin films and fibres. The oxidation of polypropylene resulting from a given exposure dose is appreciably less after e-beam exposure as compared with γ-irradiation.[22] Some data on relative radiation reactivity of aliphatic and aromatic polymers are collected in Tables 2 and 3 for widely differing irradiation conditions. Schönbacher has given an informative relative ranking of the radiation resistance of many rubbers, plastics and thermosets.[25] However, it must be stressed that the radiation sensitivity of any one polymer may be markedly affected by impurities, additives, dose rate, sample thickness and morphology. For example, the highly oriented, chain-

Table 2
Stability of polymers to γ-irradiation in air

Polymer	Dose to reduce elongation at break by 50% (Mrad)		
	At 100 rad $h^{-1\,a}$	At 5 krad $h^{-1\,b}$	At 0·1 Mrad $h^{-1\,b}$
Polytetrafluorethylene	1		
Polypropylene	3	1·3	3
Polyethylene (high density)	8	3·6	16
Polyamide-6	18	4·5	15
Poly(vinyl chloride), rigid	25		7
Polyoxymethylene	27		
Polyethylene (low density)	80	12	40
Poly(ethylene terephthalate)	130		180
Ethylene–propylene rubber	230		

[a] Extrapolated data from Ref. 20.
[b] From Ref. 23.

Table 3
Stability of aromatic polymers to e-beam irradiation[a]

Polymer	Dose to reduce elongation at break (Mrad)	
	By 50%	By 80%
Poly(ether imide)	1 000	7 000
Polyimide	2 700	6 400
Poly(ether ether ketone), 'amorphous'	2 000	5 000
Poly(1,4-phenylene isophthalamide)	1 200	2 500
Poly(bisphenol A–terephthalic acid), polyester	100	200
Polysulphone	50	200
Polyphenylene oxide/rubber-modified polystyrene blend	50	150

[a] Film samples, irradiated in air at 0·5 Mrad s^{-1} with 2·5 MeV electrons. Data abstracted from Ref. (24).

extended morphology in highly drawn PE fibres is much more γ-irradiation-resistant than the usual melt-quenched semicrystalline morphology.[26] This results both from restricted O_2 diffusion and effects on radical decay rates.

Contradictory information on the accumulation of oxidation products from the γ-irradiation of even the simplest polymer, PE, have appeared in the literature.[27–29] Our own data for additive-free, thin films shows a linear build-up of hydroperoxide (—OOH), alcohol (—OH), ketone (>C=O) and carboxylic acid [—C(=O)—OH] groups from the onset of irradiation (Fig. 2).[30] Previously reported induction periods and plateauing after longer exposures appear to result from the presence of unsuspected additives and sample thickness control of O_2 diffusion respectively. It must be emphasized that the chemistry leading to backbone scission (as indicated by the presence of carboxylic acid end groups) is as yet unclear, even for this simple polymer.

For aliphatic polymers, the rubbery polymers are most radiation-resistant, presumably because many scissions must occur to reduce significantly the integrity of the article. In highly crystalline polymers, only a relatively small number of scissions in the intercrystalline tie molecules is required to affect drastically the toughness of the material. Sasuga et al. have pointed out the extremely high radiation resistance of aromatic polymers.[24] Polyimides are particularly out-

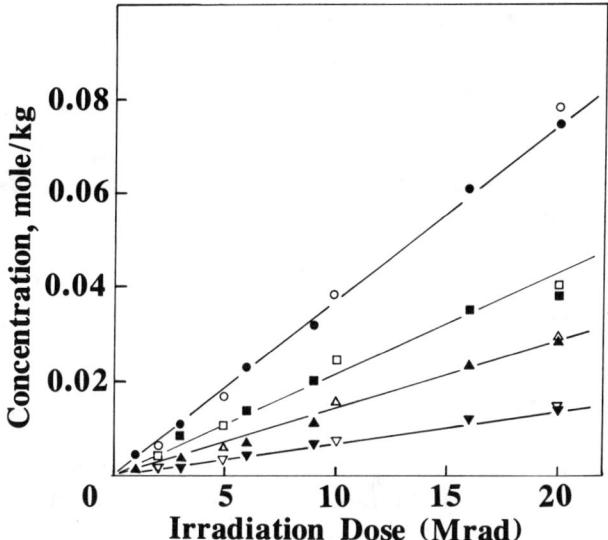

Fig. 2. Oxidation products from the γ-initiated oxidation of polyethylene films. Closed symbols HDPE (30 μm); open symbols LLDPE (120 μm): ●, ○, ketone (1718 cm^{-1}); ▲, △, carboxylic acid (from 1846 cm^{-1} after SF$_4$ exposure); ■, □, secondary hydroperoxide (as nitrate, after NO exposure); ▼, ▽, secondary alcohol (as nitrite, after NO exposure). Reproduced with permission from Ref. 30.

standing (Table 3). Sasuga *et al.* have ranked aromatic groups in order of increasing stability:

4 INHIBITION OF RADIATION DEGRADATION REACTIONS

The stabilization of polymers against high-energy radiation has been reviewed by several authors.[6,10,31,32] Protective reactions have been proposed to correspond to most of the radiation processes involved in the chemical and physical destruction of polymers. Preventative steps paralleling the destructive processes are listed in Table 1. Although radiation shielding will obviously reduce degradation, the tendency for some polymers to receive more damage from a given dose at a low dose rate, as compared with a higher rate, limits the effectiveness of this approach.

Because of the extremely short-lived nature of several of the intermediates produced by high-energy radiation, quenching and scavenging can be expected to be effective only with high concentrations (several weight %) of protectants. With longer-lived radicals and free-radical chain processes, protection can come from low levels of additives. Unambiguous evidence for stabilization by a specific mechanism is relatively rare, most work being confined to a measurement of the decrease in some practical property in the presence of different additives. In a few cases, evidence for the more important individual stabilization processes has been collected and will be reviewed.

4.1 Electron and Ion Scavenging

The early involvement of electrons and ions in γ-, X-ray and e-beam induced deterioration makes the scavenging of these species an attractive possibility. However, firm evidence to differentiate these processes from excited-state deactivation is often lacking. Cation deactivation by polynuclear aromatics, such as pyrene, aromatic amines and heterocyclics, such as phenothiazine, has been clearly demonstrated for irradiated alkane glasses and PMMA.[9,33,34] In particular, the antirad effectiveness has been found to increase in the sequence 1,3,5-trinitrobenzene < benzoic acid < benzophenone < naphthalene < diphenylamine < N,N' - tetramethylphenylenediamine.[10,33] This increase correlates with changes in the ionization potential of the additives. PMMA degradation is also suppressed by phenols and thiourea, with these and the above stabilizers becoming grafted during the irradiation process.[31]

The γ-irradiation of PVC leads to discolouration, evolution of HCl and loss in physical properties. The evolution of HCl is suppressed by aromatic compounds such as quinones. This has been attributed to electron scavenging.[10] However, more conventional PVC stabilizers also protect. Metal stearates retard colour development, but not loss in physical properties; organotin compounds do not slow discolouration, but do minimize embrittlement.[31] Combinations of organotin compounds with calcium and zinc compounds have been found to prevent both discolouration and loss of tensile properties up to and beyond 3-Mrad doses.[35]

4.2 Energy Transfer

The formation of highly excited molecular states as the result of cation–electron recombination makes energy transfer to deactivate these excited states an attractive possibility. Emission from blended acenaphthene or grafted acenaphthene in ethylene–propylene–diene rubbers has been observed after electron exposure in a pulse radiolysis system.[36] Very fast energy transfer was observed, but the grafted product was found to be most effective. Similarly, the drop in H_2 evolution (and so alkyl radical production) from irradiated ethylene–propylene rubbers containing propylfluoranthene has been attributed to energy transfer.[37,38] However a 5 wt% loading was required to give a ~50% reduction.

Incorporation of phenyl groups into polymers has been found to enhance radiation stability, possibly through an energy transfer process. In polyethylene (PE), hydrogen yields dropped when the polymer was blended with polystyrene, but grafted styrene was even more effective.[10] Similar effects were reported with PMMA, PAN and polybutadiene blends and copolymers.[10,31] Introduction of phenyl side groups into siloxanes enhances radiation stability, but possibly more by a simple dilution effect than by a proven energy transfer process.[10] Copolymerization of PMMA with styrene enhances radiation stability.[31] However, the presence of the phenyl group in the side chain, as in poly(phenyl methacrylate), did not protect.[10] In contrast, cellulose fibres can be stabilized against γ-degradation by benzoylation.[39] Even with spacings of up to 20 Å between benzoate groups along the cellulose backbone, marked protection against backbone scission was found.

The extreme radiation resistance of highly aromatic polymers (Table 3) must result from energy transfer, electron trapping and, above all, formation of excited states capable of harmlessly dissipating energy.

4.3 Radical Scavenging

Scavenging of hydrogen atoms or the initial macro-alkyl radicals before they can cause backbone scission or crosslinking reactions will minimize the radiation deterioration of many aliphatic polymers. Several studies of the effects of additives on H_2 yields have indicated radical scavenging by compounds such as mercaptans, thiourea, aminothiols, SO_2, iodine and benzophenone.[10] Phenothiazine has been found to react very rapidly with H˙ (but can also take part in electron transfer).[34]

Electron spin resonance spectroscopy allows the direct measurement of macroalkyl radical yields during irradiation. In polypropylene, the macroalkyl yield was reduced by both phenols and aromatic amines such as phenyl-β-naphthylamine.[31] Similar effects were found for propylfluoranthene in polypropylene and ethylene–propylene rubbers.[38] However, for polyamide-6, the aromatic amines were ineffective. Radicals from poly(dimethyl siloxane) can be trapped with 2,4,6-tri-*tert*-butylnitrosobenzene, but ionic species were concluded to be more important for degradation.[40]

Because changes in the tensile properties of polymers result from radiation-induced chain scission or crosslink formation, it may be possible to devise copolymers or blends of polymers where these two effects compensate.[6] Attempts at demonstrating this approach have been only partially successful with methyl methacrylate (PMMA undergoes chain scission upon irradiation) copolymerized with styrene–butadiene rubbers (SBR, crosslinks upon irradiation). An alternative approach to protect polymers which undergo backbone scission is to include a vinyl monomer in the polymer so that it may polymerize and bridge scission points as they are formed. Acrylic acid has been found to reduce the radiation degradation of polyamides and bromoacenaphthylene to reduce the degradation of ethylene–propylene copolymers possibly by this mechanism.[6,31] In air–saturated polymers, competition with O_2 for radical sites will severely limit the method, as well the copolymerization of O_2 to give labile peroxide links.

Carbon black at >5 wt% can dramatically improve the resistance of polytetrafluoroethylene to γ-initiated degradation in the absence of O_2 (Table 4).[41] Only certain types of carbon black may be effective. This may stem from the differing surface chemistries of the carbon blacks. Unfortunately, precise details of the carbons were not given, but the stabilizing carbon black belongs to a group with a high surface

Table 4
Stabilization of PTFE by carbon blacks[a]

Additive	(wt%)	Fractional tensile strength retained	
		2·5-Mrad dose	10-Mrad dose
None	—	0·8	0·56
Cabot Vulcan XC 72	10	0·56	0·38
Cabot Spheron	5	0·99	0·97
Cabot Spheron	10	1·02	1·05

[a] Dose rate 0·2 Mrad h^{-1} γ-irradiation, PTFE sheets, under vacuum, from Ref. 41.

oxidation level.[42] This surface oxidation includes both carboxylic acids and phenolic groups which may account for the stabilization. In addition, stable free radicals are also usually present in carbon blacks and could act as alkyl radical scavengers.

In polymer degradation at moderate dose rates (<5 Mrad/h), oxygen plays a dominant role in degradation reactions. This comes from both the oxidative chain reactions (10) and (11), and the formation of molecular products such as hydroperoxide groups. These groups decompose slowly but steadily even at ambient temperatures to re-initiate oxidation chain processes.[16,43] Above all, polymer backbone scission reactions result both from peroxyl radical self-reactions and from the decomposition of hydroperoxide species, reaction (12). Well-known scavengers of peroxyl radicals, such as phenols, piperidyl compounds, aromatic amines, and possibly some metal chelates, have been shown to reduce O_2 uptake and embrittlement for polyolefins and ethylene–propylene rubbers (Fig. 3).[16,37]

Wilski has compared a large number of stabilizers in polyolefins (Table 5).[20] Based on the criterion of retention of elongation, phenolic antioxidants are very effective at a low dose rate similar to that in nuclear power plants.

Polymers containing oxidizable aliphatic groups are particularly susceptible to O_2 effects after the actual irradiation because of a slow thermal degradation which continues during storage. Polypropylene medical equipment which has been γ-sterilized is particularly vulnerable to this insidious deterioration. Conventional radical scavengers (usually hindered phenols) can stop this oxidation. However, phenols

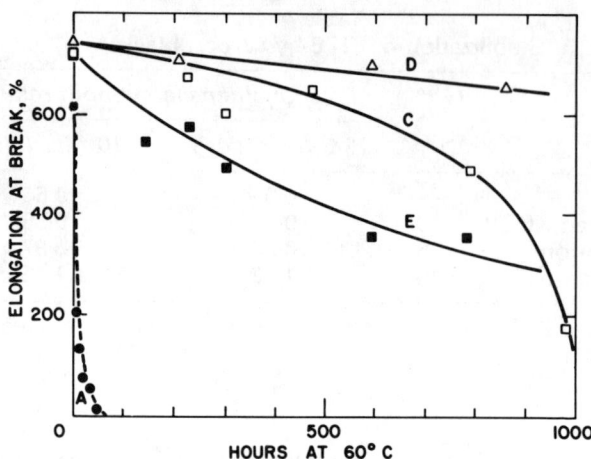

Fig. 3. Effects of stabilizers on the post-irradiation oxidation of polypropylene films.[16] Films γ-irradiated in air to 2·0 Mrad at 1·35 Mrad h^{-1}, then stored in air at 60°C. Initial elongation at break 1000%. ●, No additive; △, octadecyl β-(3,5-di-tert-butyl-4-hydroxyphenyl)propionate; □, 1,2,2,6,6-pentamethyl-4-piperidyl octadecanoate; ■, 2,2,6,6-tetramethyl-4-piperidyl-n-oxyl octadecanoate.

Table 5
γ-Stability of polyolefin formulations[a]

Additive	Dose to 50% loss of elongation (Mrad)	
	HDPE[b]	PP[c]
None	0·6	0·5
Trilauryl phosphite	0·6	0·6
1,3,5-Tris(3',5'-di-tert-butyl-4'-hydroxylbenzyl)-2,4,6-trimethylbenzene	0·8	1·7
2-Mercaptobenzothiazole	1·3	—
N,N'-Di-β-naphthyl-p-phenylenediamine	1·5	0·6
4,4'-n-Propylmethylene bis(2-tert-butyl-5-methylphenol)	2·4	1·2
Tetrabis[methylene-3-(3',5'-di-tert-butyl-4'-hydroxyphenyl)-propionate]methane	—	1·7

[a] Data from Ref. 20.
[b] Additive at 0·25 wt%.
[c] Additive at 0·50 wt%.

Table 6
Effect of γ-irradiation on stabilized PP moulded plaques[a]

Stabilizer[b]	As irradiated			After 6 months' storage at 22°C		
	Additive retained (%)	Flange bend[c] angle (deg.)	Colour[d]	Additive retained (%)	Flange bend[c] angle (deg.)	Colour[d]
Non-stabilized	—	70	5	—	≤20	5
Tetrakis[methylene-3-(3',5'-di-*tert*-butyl-4'-hydroxyphenyl)-propionate]methane	36	90	14	30	90	16
2,2,6,6-Tetramethyl-4-piperidyl sebacate	63	90	6	58	90	7
Tris(2,4-di-*tert*-butylphenyl)phosphite	21	90	10	0	50	13
Distearyl thiodipropionate	0	70	14	0	≤20	26

[a] 2·5 Mrad in air at 0·5 Mrad h^{-1}.[47]
[b] Stabilizer at ~0·5 wt%.
[c] 90° before irradiation.
[d] 5 ± 1 before irradiation by Gardner colorimeter.

are unacceptable medically because of the intense yellow discolouration which results from the formation of compounds such as stilbene quinones upon irradiation of the phenols.[44,45] These products may be chemically grafted to the polymer.[46] Much industrial effort has been devoted to the search for non-staining antioxidant systems, especially for disposable polypropylene syringes.

Horng & Klemchuk, in a well detailed study, have measured yellowing and embrittlement (as measured by bending a syringe flange to failure) for a series of stabilized PP formulations (Table 6).[47] In addition, the fraction of the additive which survived irradiation (and so is available to protect during storage) was also analysed. A substituted piperidine, developed originally for polymer stabilization against photo-initiated oxidation, was found to give good protection (little yellowing or embrittlement) against γ-irradiation, as well as post-irradiation storage under accelerated test conditions (60°C in air). In addition to the secondary piperidines, both tertiary amines and nitroxyl derived from this type of additive have been found to retard embrittlement during post-irradiation storage (Fig. 3).[16] Combinations of piperidines and thio-compounds have been found to give antagonistic effects during γ-irradiation.[48]

Large fractions of each initial additive were found to have been destroyed or chemically grafted during irradiation (Table 6). This is particularly surprising because of the low additive levels employed, and the statistical deposition of energy by γ-irradiation: only the polymer should be attacked. However, energy transfer and selective deposition in polar groups has been well documented.[49] Unfortunately, yields of radicals from the polymer immediately after irradiation with these additives were not reported, and it is possible that the additives were destroyed by processes which are unimportant components of the degradation of the polymer. Our own data indicate negligible loss of piperidyl groups during irradiation, with loss occurring in the post-irradiation period.[50]

Lyons & Lanza have stressed the difficulties of extrapolating to radiation stabilization from photostabilization, which usually involves very different energetic species, and from thermal antioxidants, which are designed to scavenge different radicals and are often optimized for higher temperatures.[31] Nevertheless, some interesting radiation stabilization effects have been found with compounds chosen for their intense absorption in the near-UV. Various substituted benzophenones have been found to stabilize against γ-irradiation, especially when used in conjunction with a substituted piperidine.[51] In PP films, we have found that benzophenone itself accelerates post-γ-degradation, whereas 2-hydroxy-4-octoxybenzophenone is very effective in its prevention when combined with a piperidyl compound (Table 7). It has been suggested that the nitroxyl from a substituted piperidine can scavenge benzophenone products to give a hydroxylamine, reactions (13) and (14).[52] Hydroxylamines are excellent peroxyl radical scavengers as well as hydroperoxide decomposers.

$$PP \xrightarrow{\gamma} PP^* \xrightarrow{PhCOPh} PP + PhCOPh^*$$

$$\xrightarrow{PP} Ph\dot{C}(OH)Ph + PP^{\cdot} \xrightarrow{O_2} \text{oxidation} \quad (13)$$

$$\underset{|}{\overset{OH}{Ph\dot{C}}}-Ph + {>}NO^{\cdot} \longrightarrow PhCOPh + {>}NOH \quad (14)$$

Table 7
Effects of additives on oxidation products from γ-irradiation polypropylene[a] (Ref. 50)

Additive	Concentration (wt%)	$[PPO_2^\cdot]_0$[b] $[(mol\,kg^{-1}) \times 10^3]$	$-d[PPO_2^\cdot]_0/dt$[c] $[(mol\,kg^{-1}\,s^{-1}) \times 10^6]$	Post-irradiation yields[d] $100[PPO_2^\cdot]_t/[PPO_2^\cdot]_0$ (%)	[tert PPOOH] $[(mol\,kg^{-1}) \times 10^3]$	Storage failure (days)[e]
None	—	6·0	48	2	29	12
Mobilizer oil (Witco 300)	3·5	6·7	85	4	25	18
Bis(2,2,6,6-tetramethyl-4-piperidyl)sebacate(I)	0·12	6·5	54	4	16	200
Mobilizer oil + I	3·5 / 0·10	6·3	80	3	15	320
2-Hydroxy-4-octoxybenzophenone (II)	0·1	6·8	45	2	23	10
II + I	0·1 / 0·1	6·0	50	3	13	≥450
Bis(4-methylphenyl)carbinol (III)	0·1	7·1	77	1·2	27	42
III + I	0·1 / 0·1	7·2	80	1·7	9	450

[a] PP film (30 μm) γ-irradiated at −78°C to 2·5 Mrad.
[b] From double integration of the ESR signal. Experimental precision ~±15%.
[c] Initial rate of peroxyl decay at 22°C.
[d] After 120 h at 22°C following γ-irradiation.
[e] Storage at 60°C in air after irradiation. Failure defined as 50% loss of initial impact strength.

In addition, amines are known to deactivate excited benzophenones by a charge transfer process.[53] Alternatively, the benzophenones may undergo radical or ion reactions to form a benzhydrol which is extremely effective in irradiated PP (Table 7). The 2-hydroxybenzophenones are extremely stable when excited either directly or indirectly, but also have some phenolic-like radical-scavenging ability.[54,55]

Dunn et al. have suggested a correlation between the yield of radicals formed from phenolic or amine antioxidants in model peroxide-containing liquids with the radiation protection offered by these additives in polypropylene.[56] However, this O_2-free model system seems to be very different from the O_2-containing polymer.

4.4 Plasticization Effects

As has been pointed out in the previous sections, only a few additives have been found in practice to reduce the number of radicals generated in a polymer by a given dose. In the presence of O_2, at modest dose rates (≤ 1 Mrad h^{-1}) and fairly thin cross-section (≤ 200 μm), all macroalkyl radicals convert quantitatively to peroxyl radicals. An alternative approach to radiation stabilization has come from the proposal to use an additive, not to prevent radical formation, but instead to speed the (hopefully harmless) decay of these radicals.[16,18,43] This concept was originally based on the (erroneous) assumption that post-irradiation degradation depended on long-lived macroalkyl radicals which must be induced to decay.[18] However, rapid self-reaction of peroxyl radicals will minimize hydroperoxide formation by their propagation reactions and so reduce the level of unstable hydroperoxide available to initiate during subsequent storage.[16,43]

This concept of radical 'mobilization' has been clearly shown to speed radical decay, and to improve long-term stability in the case of PP.[18,43] Williams et al. have shown that the decay of macroalkyl radicals, formed from irradiation of PP under vacuum, is accelerated by PE waxes, atactic polypropylene and hydrocarbon oils, in order of increasing effectiveness.[18,43] Our data show that a hydrocarbon clearly accelerates decay of macroperoxyl radicals (Fig. 4). In addition, calcium stearate (the normal acid scavenger present in PP formulations) speeds peroxyl decay, although this might reflect morphological changes induced in the polymer by this salt. The piperidyl compound did not alter the rate of peroxyl decay, but did generate nitroxyl radicals during this decay. Nitroxyl yield was suppressed by the presence of the mobilizer oil.[50]

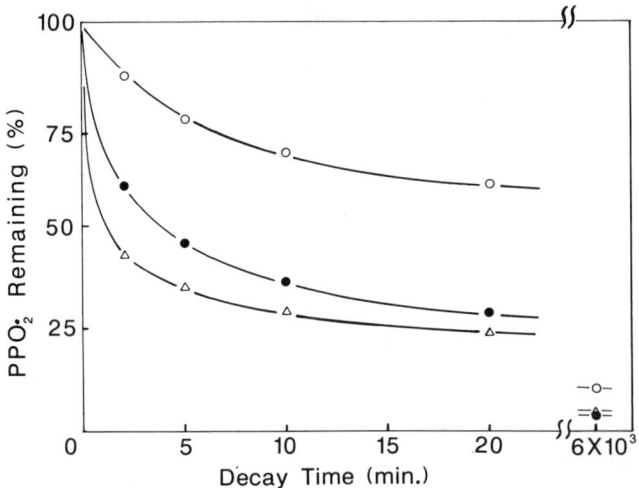

Fig. 4. Decay of macroperoxyl (PPO$_2^\bullet$) radicals in γ-irradiated polypropylene films.[50] Films (30 μm) irradiated in air at −78°C to 2·5 Mrad. Radical decay started by heating to 22°C and storage at this temperature. ○, Additive-free film (all other films contain 0·1 wt% calcium stearate); ●, no other additives, or with 0·1 wt% bis(2,2,6,6-tetramethyl-4-piperidyl)sebacate; △, with mobilizer oil (saturated alkane mixture, Witco 300, 3·5 wt%).

For tensile loss of PP irradiated in air, the elongation to break, both immediately after irradiation and after four months' post-irradiation storage, was preserved by the presence of a hydrocarbon oil (Table 8).[18] Similarly, our data show extended aging protection especially when combined with a piperidyl compound (Table 7).

4.5 New Radiation Stabilizer System
In recent years, several patents have been published which claim radiation protection by groups of compounds originally promoted as nucleating agents for the control of polyolefin morphology.[51,57,58] These compounds usually have the general form of structures **IV** and **V**, where R_1, R_2 denote alkyl or H and R_3 alkyl groups. For example, bis(4-methylphenyl)carbinol (**III**) gave a superior retention of elongation to break in 10-Mrad γ-irradiated PP plaques as compared with a multifunctional hindered phenol [tris(4-hydroxy-3,5-di-tert-butylbenzyl) isocyanurate)].[57] Furthermore, no yellowing was found with the benzhydrol, whereas the plaques containing the phenol became

Table 8
Effect of 'mobilizer' on PP γ-stability[a]

Mobilizer (wt%)	Dose (Mrad) Work to break[b] [(dyne cm^{-1}) × 10^7]			
	0	2	5	10
0	1·55 (1·4)	1·50 (1·0)	1·6 (0·15)	0·2 (0·02)
1·2	1·50 (1·45)	1·6 (1·2)	1·2 (0·30)	0·5 (0·05)
4·0	1·45 (1·42)	1·42 (1·40)	1·3 (0·42)	0·20 (0·02)

[a] Irradiated in air; mobilizer is a hydrocarbon oil. From Ref. 18.
[b] Immediately after irradiation; values in parentheses after 4 months' storage at room temperature.

intensely yellow. These compounds are claimed to be particularly effective when teamed with the substituted piperidines.[51] This is clearly shown by our impact retention data in Table 7.

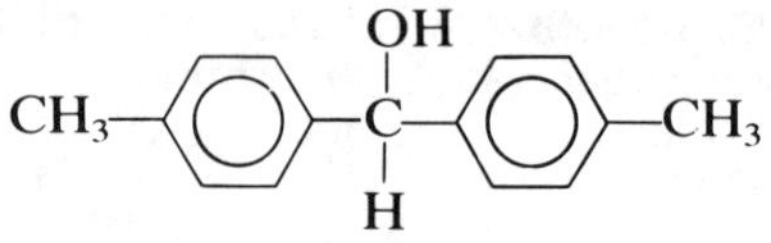

$$CH_3-\underset{}{\bigcirc}-\underset{H}{\overset{OH}{C}}-\underset{}{\bigcirc}-CH_3$$

III

$$\underset{}{\bigcirc}-\underset{R_1}{\overset{OH}{\underset{|}{C}}}-R_3 \qquad \underset{}{\bigcirc}-\underset{R_2}{\overset{R_1}{\underset{|}{C}}}-O-R_3$$

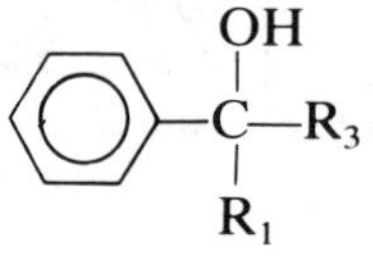

IV 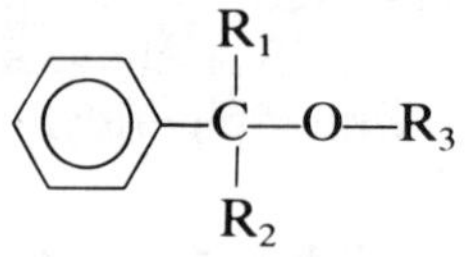   **V**

In order to unravel the mechanisms by which compounds stabilize against high-energy radiation, a detailed analytical study is required. This must include:

(1) Measurement of the radical yield (peroxyl, alkyl, or other) immediately after irradiation, preferably under conditions where no radical decay occurs.
(2) Following the rate of peroxyl decay at room temperature.

(3) Quantification of the labile oxidation products produced by the radiation-generated radical population as the radicals decay.
(4) Analysis of the amount of any one additive, or its active forms, which survives the radiation dose.
(5) Characterization of the changes in physical properties which occur as a direct result of irradiation, and then during the post-irradiation period.

We have recently addressed several of these questions for some types of potential stabilizers for polypropylene film γ-irradiated in air.[50] These data are collected in Table 7. In all cases, the yield of peroxyl radicals ($[PPO_2^.]_0$) generated from γ-irradiation at $-78°C$ is constant and independent of the presence of additives within experimental error. In separate experiments, we have found identical radical yields at $-200°C$ and at $-78°C$. At higher temperatures, rapid radical propagation and termination invalidate any estimates of total radicals generated, but we assume $[PPO_2^.]_0$ will be little different from the $-78°C$ data. Thus, none of these additives is preventing ion–electron reactions or deactivation of excited states as they are formed during the irradiation. Apparently, all stabilize by acting in the post-irradiation period. In particular, for the piperidine-containing films, nitroxyl was not found immediately after irradiation, and began to accumulate only during the $PPO_2^.$ decay phase. Upon warming a sample to room temperature after γ-irradiation at $-78°C$, rapid peroxyl loss occurs (Fig. 4). The initial rates of loss ($-d[PPO_2^.]_0/dt$) and the residual level of 'long-lived' (>120 h) radicals vary with the additive type. The mobilizer oil dramatically accelerates $PPO_2^.$ loss, as reported previously, and does confer some post-irradiation protection (measured in an accelerated storage test at 60°C) (Table 7). However, the stabilization is equally good for other additives which neither accelerate the decay nor reduce the 'long-lived' radical level. This implies that the long-lived radicals are ineffective in promoting oxidation, as we have found previously from $PP^{17}O_2^.$ decay experiments.[19]

From Table 7 the effective 'radiation stabilizers' for polypropylene (actually post-irradiation stabilizers) suppress formation of the unstable tertiary hydroperoxide during the propagation of the bulk of the $PPO_2^.$ radicals. Consequently, all are acting as radical-scavenging antioxidants, which are themselves sufficiently resistant to γ-irradiation.

4.6 Ozone Effects

The high-energy irradiation of O_2 produces atomic oxygen (extremely short-lived) and ozone which can persist for the order of days. Consequently, O_3–polymer reactions may occur subsequently to irradiation, especially if the article is inside a sealed, air-filled package. Some stabilization by the use of an external O_3 scavenger has been reported.[16] In addition, some aromatic amines are excellent antiozonants.

4.7 Radiation Effects on Additives

Most radiation protection mechanisms require the consumption of the initial compound. In addition, direct radiation interaction with an additive, even when present at a low concentration, cannot be ruled out. The data in Table 5 clearly show extensive consumption of all four types of additives although it was not reported at which stage of the deterioration this occurred.[47]

The burgeoning worldwide interest in the γ-ray treatment of foodstuffs to enhance storage life has increased interest in radiation stability of packaging. In particular, the prospect of irradiating packaged foods emphasizes the questions of how rapidly additives are consumed, what their conversion products are and what hazards these products could present if they migrate into the foodstuffs. Allen *et al.* have reported some preliminary studies on additives in poly(vinyl chloride), polyethylene and polypropylene.[59,60] Progressive conversion of dialkyltin diesters through intermediates to $SnCl_4$ was observed. For example, a 14% conversion to $SnCl_4$ after a 2·5-Mrad exposure of PVC, containing dibutyltin bis(iso-octyl maleate) was found. For phenolic antioxidants in PVC, ~50% loss was found after 2·5 Mrad. A similar loss of phenols in irradiated polyethylene was also found but, surprisingly, in polypropylene phenolic loss was much less although strongly concentration-dependent. For the phenolics, it is possible that most of the conversion products are grafted to the host polymer after irradiation, and so locked in the packaging.

5 CONCLUSIONS

The complex sequence of reactions which lead to immediate degradation of properties upon irradiation and post-radiation deterioration are complemented by a range of stabilization mechanisms. Some of these

Table 9
Polyolefin radiation stabilizers and their possible mechanisms[a]

Stabilizer	Mechanism(?)
Polynuclear aromatics, e.g. propylfluoranthene, acenaphthene. Polystyrene, styrene (copolymerized)	Energy transfer Ion-electron scavenge
Aromatic amines	Ion-electron scavenging
Mobilizer/plasticizer	'Increase of free volume' Promotion of peroxyl self-reactions
Phenols, piperidines, aromatic amines	Macroalkyl, macroperoxyl scavengers
Thio-compounds, phosphites	Hydroperoxide decomposition
'Nucleators' i.e. 'nucleators' with piperidines or with substituted benzophenones	Enhanced radical scavenging?

[a] Stabilization during irradiation and in post-irradiation storage.

are itemized in Table 9. It must be emphasized that this table is intended as a general guide because data in the literature are often contradictory. For example, phosphites are reported to be ineffective (Table 5) and effective (Table 4) in protecting polypropylene against γ-irradiation. In practical terms, only some of these routes are feasible at reasonable additive loadings. Furthermore, some of the most recently discovered additives have yet to be fitted within the general degradative scheme.

REFERENCES

1. Charlesby, A., *Atomic Radiation and Polymers*. Pergamon Press, Oxford, 1960.
2. Chapiro, A., *Radiation Chemistry of Polymeric Systems*. Interscience Publishers, New York, 1962.
3. Dole, M., In *Crystalline Olefin Polymers*. Part 1. ed. R. A. V. Raff & K. W. Doak. John Wiley and Sons, New York, 1965, Chapter 16.
4. Dole, M., *Adv. Radiation Chem.*, **4** (1974) 307.
5. Bohm, G. G. A. & Tveekrem, J. O., *Rubber Chem. Technol.*, **55** (1982) 669.
6. Hagiwara, M. & Kagiya, T., in *Degradation and Stabilization of Polymers*, Vol. 1. ed. H. H. G. Jellinek. Elsevier, Amsterdam, 1983, Chapter 8.

7. Greenstock, C. L., *Adv. Radiation Biol.*, **11** (1984) 269 and many other papers in this journal.
8. Yashunskii, V. G. & Kovtum, V. Y., *Russian Chem. Rev.*, **54** (1985) 76.
9. Klassen, N. V. & Teather, G. G., *J. Phys. Chem.*, **89** (1985) 2048.
10. Makhlis, F. A., *Radiation Physics and Chemistry of Polymers*. John Wiley and Sons, New York, 1975.
11. Partridge, R. H., *J. Chem. Phys.*, **52** (1970) 2485.
12. Tsuda, M. & Oikawa, S., *J. Polym. Sci., Polym. Chem. Ed.*, **17** (1979) 3759.
13. Falconer, W. E. & Salovey, R., *J. Chem. Phys.*, **44** (1966) 3151.
14. Bowden, M. J. & O'Donnell, J. H., *Dev. Polym. Deg.*, **6** (1985) 21.
15. Carlsson, D. J. & Wiles, D. M., *J. Macromol. Sci., Rev. Macromol. Chem.*, **C14** (1976) 155.
16. Carlsson, D. J., Dobbin, C. J. B., Jensen, J. P. T. & Wiles, D. M., *Am. Chem. Soc., Symp. Ser.*, **280** (1985) 359.
17. Dunn, T. S., Williams, E. E. & Williams, J. L., *Radiat. Phys. Chem.*, **19** (1982) 287.
18. Williams, J. L., Dunn, T. S. & Stannett, V. T., *Radiat. Phys. Chem.*, **19** (1982) 291.
19. Carlsson, D. J., Dobbin, C. J. B. & Wiles, D. M., *Macromolecules*, **18** (1985) 1791.
20. Wilski, H., *Radiat. Phys. Chem.*, **29** (1987) 1.
21. Gillen, K. T. & Clough, R. L., *J. Polym. Sci., Polym. Chem. Ed.*, **23** (1985) 2683.
22. Yoshii, F., Sasaki, T., Makuvchi, K. & Tamura, N., *J. Appl. Polym. Sci.*, **31** (1986) 1343.
23. Wundrich, K., *Radiat. Phys. Chem.*, **24** (1985) 503.
24. Sasuga, T., Hayakawa, N., Yoshida, K. & Hagiwara, M., *Polymer*, **26** (1985) 1039.
25. Schönbacher, H., *Modern Plastics* (Dec. 1985) 64.
26. Carlsson, D. J., Colin, G., Chmela, S. & Wiles, D. M., *Tex. Res. J.*, **58** (1988) 520.
27. Petruj, J. & Marchal, J., *Radiat. Phys. Chem.*, **16** (1980) 27.
28. Cheng, H. N., Schilling, F. C. & Bovey, F. A., *Macromolecules*, **9** (1976) 363.
29. Shinde, A. & Salovey, R., *J. Polym. Sci., Polym. Phys. Ed.*, **23** (1985) 1681.
30. Carlsson, D. J., Bazan, G., Chmela, S. & Wiles, D. M., *J. Polym. Degrad. Stab.*, **19** (1987) 185.
31. Lyons, B. J. & Lanza, V. L. In *Polymer Stabilization*, ed. W. L. Hawkins. Wiley-Interscience, New York, 1972, Chapter 6.
32. Schnabel, W. In *Aspects of the Degradation and Stabilization of Polymers*, ed. H. H. G. Jellinek. Elsevier, Amsterdam, 1978, Chapter 4.
33. Borovkova, V. A. & Bagdasaryan, K. S., *Khimiya Vysok. Engergii*, **1** (1967) 340.
34. Solar, S., *Radiat. Phys. Chem.*, **23** (1984) 6678.
35. Foure, M. & Rakita, P., *Medical Device and Diagnostic Industry*, **5**(12) (Dec. 1983) 33.

36. Kawanishi, S., Hagiwara, M., Katsumura, Y., Tabata, Y. & Tagawa, S., *Radiat. Phys. Chem.*, **6** (1985) 705.
37. Arakawa, K., Seguchi, T., Hayakawa, N. & Machi, S., *J. Polym. Sci., Polym. Chem. Ed.*, **21** (1983) 1173.
38. Fujimura, T., Arakawa, K., Hayakawa, N. & Kuriyama, I., *J. Appl. Polym. Sci.*, **27** (1982) 2475.
39. Arthur, J. C. & Mares, T., *J. Appl. Polym. Sci.*, **9** (1965) 2581.
40. Menhofer, H. & Heusinger, H., *Int. J. Radiat. Phys. Chem.*, **29** (1987) 243.
41. Fock, J., *J. Polym. Sci., Polym. Lett. Ed.*, **5** (1967) 635.
42. Donnet, J. P. & Voet, A., *Carbon Black*. Dekker, New York, 1976.
43. Dunn, T. S. & Williams, J. L., *J. Indust. Irradiat. Tech.*, **1** (1983) 33.
44. Pospisil, J., *Adv. Polym. Sci.*, **36** (1980) 69.
45. Williams, J. L. & Dunn, T. S., *Radiat. Phys. Chem.*, **15** (1980) 59.
46. Kirguskin, S. G. & Shlyapnikov, Y. A., *Polym. Sci. USSR* **23** (1981) 617.
47. Horng, P. & Klemchuk, P., *Plastics Engineering* (April 1984) 35.
48. Williams, J. L., Williams, E. E. & Dunn, T. S., *Radiat. Phys. Chem.*, **19** (1982) 189.
49. Whelan, D. J., *Chem. Rev.*, **69** (1969) 179.
50. Becker, R. F., Carlsson, D. J., Cooke, J. M. & Chmela, S., *Polym. Degrad. Stab.*, **22** (1988) 313.
51. Donohue, J., European Patent 0 154 071 (1985).
52. Bauer, D. R., Briggs, L. M. & Gerloch, J. L., *J. Polym. Sci., Part B, Polym. Phys.*, **24** (1986) 1651.
53. Cohen, S. G., Parola, A. & Parsons, G. H., *Chem. Rev.*, **73** (1973) 141.
54. Irick, G. & Newland, G. C., *Tetrahedron Lett.* (1970) 4151.
55. Vink, P. & Van Veen, T. J., *Europ. Polym. J.*, **14** (1978) 533.
56. Dunn, T. S., Williams, E. E. & Williams, J. L., *J. Polym. Sci., Polym. Chem. Ed.*, **20** (1982) 1599.
57. Rekers, J. W., US Patents 4 431 497, 4 460 445 (1984).
58. Mitsui Toatsu Chemcials. Japanese Tokkyo Koho 60 99147 (1985).
59. Allen, D. W., Brooks, J. S., Unwin, J. & McGuinness, J. D., *Chem. Ind. (London)* (1985) 524.
60. Allen, D. W., Leathard, D. A. & Smith, C., *Chem. Ind. (London)* (1987) 198.

Chapter 5

Photodegradation and Stabilization of PPO® Resin Blends

JAMES E. PICKETT

General Electric Company, Corporate Research and Development, Schenectady, USA

ABSTRACT

The photodegradation of PPO® resin†/high-impact polystyrene blends is due mainly to the photo-oxidation of the PPO resin. PPO resin undergoes photo-oxidation largely through a self-sensitized electron transfer oxidation of the aromatic ring leading to chain scission and a large number of yellow products. Evidence for this mechanism includes sensitization and quenching experiments, trapping of the superoxide ion intermediate, flash photolysis results, and model compound studies. The blend of PPO resin and HIPS was found to be most sensitive to the 313-nm line present in fluorescent lamps, but the longer-wavelength light bleaches some of the yellow photoproducts generated at 313 nm. The photoyellowing process also exhibits a temperature and light-intensity dependence that indicates a branch point in the photoyellowing process. Hindered-amine light stabilizers (HALS) apparently act on a thermally controlled branch of this process. Under the high light intensities of some testing schemes the importance of this branch is

† PPO® and Noryl® are registered trade marks of the General Electric Company.

diminished and the apparent effectiveness of the HALS is underestimated. UV screeners are effective in reducing the rate of photoyellowing by the amount calculated from a simple light absorption mechanism. Their effectiveness is inherently limited by the fact that the degradation is confined to a thin surface layer.

1 INTRODUCTION

Poly(2,6-dimethyl-1,4-phenylene oxide), PPO® resin, is commonly blended with high-impact polystyrene (HIPS) to make engineering thermoplastic molding compounds such as Noryl® resin. These resins have been available commercially since 1965. The good impact resistance, high heat distortion temperatures, and easy moldability of these blends has led to applications such as computer housings and automobile instrument panels. Commonly, resins for these applications contain 10–15% of a triaryl phosphate flame retardant as well as pigments and minor amounts of thermal stabilizers.[1]

Many of the applications of these resins involve exposure of unpainted surfaces to sunlight or to indoor lighting. While indoor weathering usually does not have a large effect on the impact strength or other physical properties, the surfaces undergo a color change. The photoyellowing is confined to a layer extending only 1–2 μm into the surface and is easily washed off with acetone. The surface remains glossy after indoor exposure, but exposure outdoors results in a loss of gloss due to erosion of the photo-oxidized layer. For unstabilized resins, the discoloration can be noticeable after a few weeks or months of outdoor exposure and after about a year indoors under a combination of fluorescent and window lighting.

The wavelength distribution of sunlight is well known.[2] There is very little light at wavelengths less than 295 nm, even in the summer. Indoor lighting is more complicated (Fig. 1). Sunlight through window glass as well as light from incandescent lamps is cut off below 320 nm. However, most fluorescent lamps have small emissions at 313 and 335 nm and strong emissions at 365 and 405 nm due to leakage of the mercury lines, as well as a continuum from the phosphor emission which begins at about 340 nm. Test methods such as the Atlas HPUV test (Atlas Electrical Devices Company, 4114 N. Ravenswood Ave., Chicago, IL 60613, USA; see also ASTM D4674-87) have been developed to account for the effects of fluorescent lamps on materials

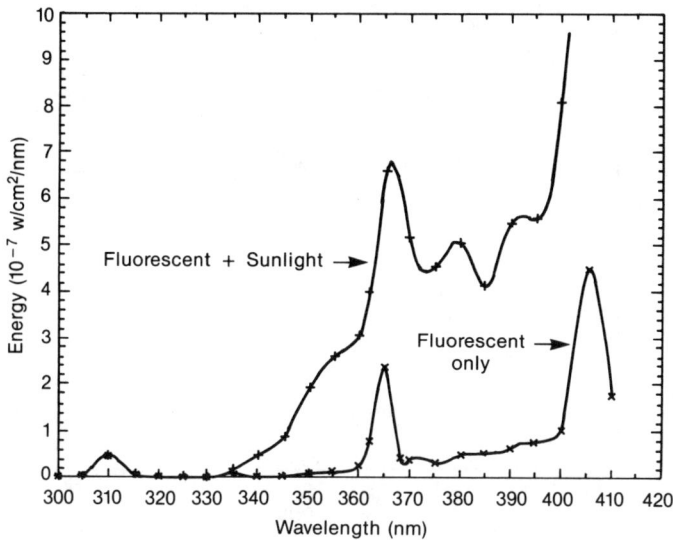

Fig. 1. Light energy as measured by a spectroradiometer in a laboratory office at General Electric Corporate R&D in Schenectady. The measurements were made about 2 m from a south-facing window during the summer. The values are typical of office lighting.

intended for indoor use. These tests use a bank of fluorescent lamps and glass-filtered fluorescent sunlamps to effect a 40–80-fold increase in light intensity over typical business offices.

Both the PPO resin and the HIPS resin components of the blend are known to be susceptible to photodegradation at wavelengths greater than 300 nm. Photodegradation of HIPS at these wavelengths is due mostly to photo-oxidation of the butadiene rubber. However, the photoyellowing of the PPO resin/HIPS blend is due almost entirely to the PPO resin. Figure 2 shows the yellowing as measured by the change in the yellowness index (ΔYI, ASTM D-1925) of blends made with HIPS and with crystal polystyrene (XPS) upon exposure in the HPUV tester. The nearly identical yellowing rates with and without the butadiene rubber component show that the rubber is only a minor contributor to the photoyellowing. Presumably the formation of highly absorbing products from the photo-oxidation of the PPO resin restricts the degradation to a thin surface layer. While all of the rubber in that degraded layer is probably oxidized, the small amount present does not contribute significantly to the total amount of colored products

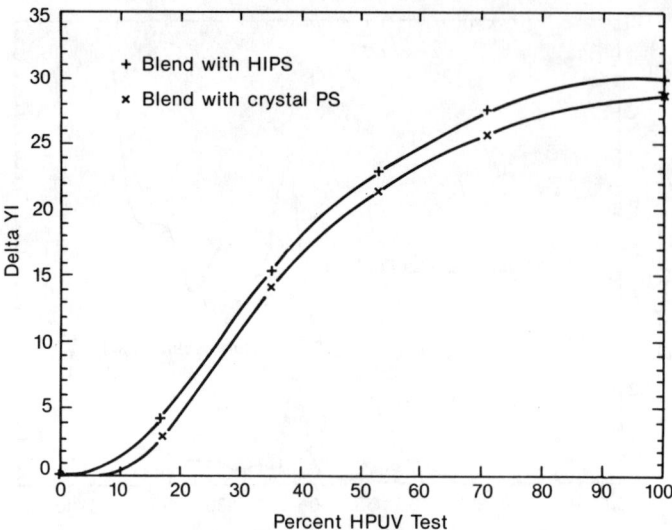

Fig. 2. Photoyellowing on the HPUV tester of a blend of 55 parts PPO® resin, 45 parts polystyrene, 13 parts triaryl phosphate, and 3 parts TiO_2. The blend containing no butadiene rubber photoyellows at the same rate as the HIPS blend, indicating that the rubber plays little role in the photoyellowing process of these blends.

formed. The study of photodegradation of PPO resin/HIPS blends under indoor lighting conditions is therefore the study of PPO resin photo-oxidation.

2 MECHANISM OF PPO RESIN PHOTOOXIDATION

There is considerable published literature on the subject of PPO resin photo-oxidation. Kelleher et al.[3] at Bell Labs exposed chloroform-cast films to unfiltered S-1 and RS Sunlamps (wavelengths >280 nm) and analyzed the degraded polymer. The infrared spectrum showed the formation of broad carbonyl and hydroxyl bands, but no significant change was observed in the NMR spectrum. They detected CO_2, N_2, and H_2 as gaseous products. ESR spectroscopy showed phenoxyl radical formation after irradiation either in air or *in vacuo*.

The authors proposed that both the thermal and photo-oxidations were free-radical oxidations of the methyl groups leading to hydroper-

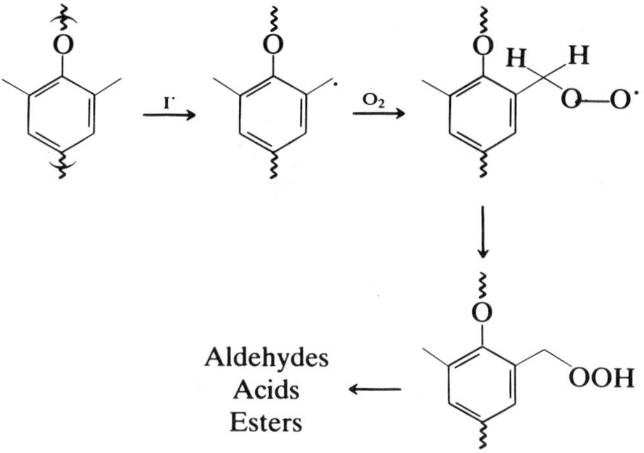

Scheme 1. Hydroperoxide-mediated free-radical oxidation.

oxides and eventually to alcohols, aldehydes, and carboxylic acids (Scheme 1). There is considerable evidence that this is the mechanism for the thermal oxidation, and the apparent similarity of the IR spectra of the thermal and photo-oxidized films suggested to the authors that the reactions proceeded through similar mechanisms. Jerussi[4] at General Electric reported similar results and came to much the same conclusions.

Several other groups have investigated radical formation in PPO resin by ESR spectroscopy. Symons & Yandell[5] investigated radical formation induced by Pyrex®-filtered mercury light ($\lambda > 300$ nm) and by γ-radiation. Unirradiated polymer showed a weak, eight-line spectrum centered at $g = 2 \cdot 005$ due to phenoxyl radicals, as had been observed by Factor (General Electric Company, unpublished) at General Electric. Irradiation with $\lambda > 300$ nm in air or *in vacuo* at room temperature and at 77K resulted only in an increased phenoxyl radical signal. γ-Radiation, however, produced more complex results. Room temperature irradiation in air gave the phenoxyl radical spectrum. At 77K peroxyl radicals ($ArCH_2-OO^{\cdot}$) were observed in the presence of air while benzyl radicals ($ArCH_2^{\cdot}$) were observed *in vacuo*. No benzyl radicals were ever observed under ultraviolet irradiation conditions.

A subsequent study by Tsuji & Seiki[6,7] substantially confirmed these results in the ultraviolet. Irradiation of PPO resin with an unfiltered

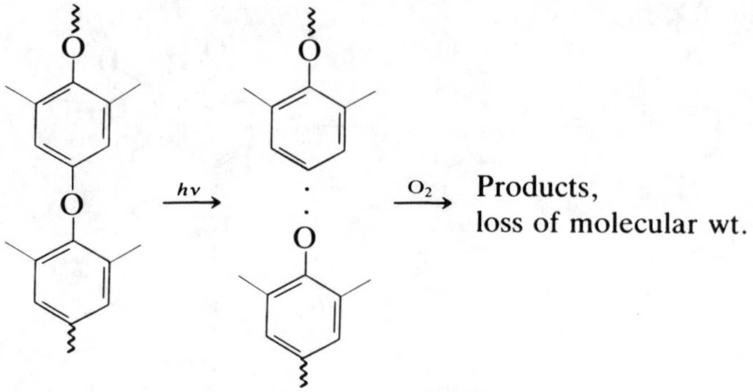

Scheme 2. Direct photolytic cleavage.

xenon arc lamp ($\lambda > 200$ nm) in air or *in vacuo* increased the phenoxyl signal. At 77K in air peroxyl radicals were observed. No signals attributable to benzyl radicals were observed. The authors suggested that direct aryl ether bond cleavage occurred at these wavelengths (Scheme 2). This mechanism received support by the work of Wandelt *et al.*[8] Statistical chain scission was observed after exposure of PPO polymer to 254 nm light. Chandra *et al.*[9-16] and Tovborg-Jensen & Kops[17,18] have also investigated the photo-oxidation at 254 nm, but the short-wavelength chemistry may not have much relevance to the degradation and stabilization of blends at wavelengths greater than 300 nm.

Allen & McKellar[19] in 1979 reported experiments performed in thick PPO resin films. Pressed films (16 mils, 0·4 mm) exposed to a filtered xenon arc ($\lambda > 300$ nm) exhibited a photobleaching effect. Fluorescence studies showed that impurities evidently resulting from processing were being destroyed. Flash photolysis in chloroform solution showed a transient with a half-life of 50 ms that the authors attributed to the phenoxyl radical.

Slama *et al.*[20-22] have extensively investigated the thermal and photo-oxidation of PPO resin and blends and have reviewed the field. Their IR results on films after exposure in a Xenotest-150 apparatus ($\lambda > 300$ nm) are qualitatively similar to Kelleher's. They observed little or no 'induction period' in the formation of carbonyl or hydroxyl groups. Since the hydroperoxide content in fresh PPO resin is very low, they suggested that some other impurity such as a quinone was sensitizing the photo-oxidation but that the mechanism was proceeding as shown in Scheme 1. However, there is no spectroscopic evidence for significant amounts of carbonyl-containing impurities in PPO resin. A band at 1690 cm^{-1} in PPO resin is also present in pure model compounds and is presumably an aromatic overtone.

More recent work by Pickett[23] has called into question the presumption of free-radical oxidation of the methyl groups as a major pathway for the photo-oxidation of PPO resin at wavelengths greater than 300 nm. Since this is a substantial departure from the body of the published literature, the evidence against the free-radical mechanism and for an alternative mechanism will be reviewed in some detail.

2.1 Evidence against the Free-Radical Mechanism

The spectroscopic evidence for methyl group oxidation is the observation that carbonyl and hydroxyl groups appear in the IR. This alone, however, does not indicate that the methyl groups are being consumed upon oxidation. Table 1 shows the loss of methyl groups as determined by proton NMR spectroscopy in PPO resin exposed to a Pyrex-filtered mercury lamp ($\lambda > 300$ nm) as a 2% solution in benzene. Oxygen consumption was measured by a gas buret at 1 atm. When 0·5 equivalent of O_2 per repeat unit has been consumed, a 25% loss of methyl groups would be expected if all of the oxidation occurred there. In fact, although the methyl group signal at 2·2 ppm was observed to broaden, the integrated loss was less than one-tenth of the calculated amount. Indeed, the aromatic signal at 6·4 ppm decreased relative to the methyl group signal. A similar result was observed for a cast film. These results were confirmed by attenuated total reflectance IR spectroscopy on the surface of cast films. The area under a band at 960 cm^{-1}, an aromatic band, decreased 60%, while a band at

Table 1
Loss of methyl groups upon photo-oxidation

Extent of oxidation (equiv. O_2/mer)	Loss (%)	
	Observed by NMR	Predicted for methyl oxidation
0	0	—
0·12 (solution)	0	6
0·52 (solution)	2·3	26
0·48 (film)	3·4	24

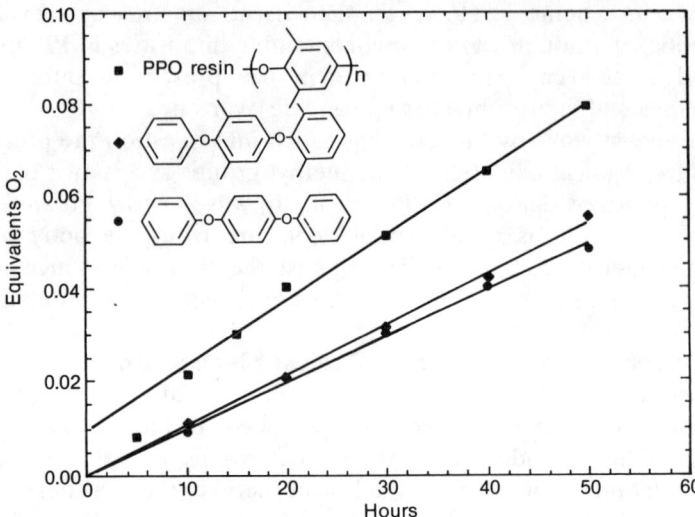

Fig. 3. Photo-oxidation of PPO resin and model compounds in 1 M benzene solution as equivalents of oxygen per disubstituted repeat unit. Pyrex-filtered Hg lamp.

1380 cm^{-1} due to the methyl groups only broadened upon photo-oxidation. Large carbonyl and hydroxyl bands also were formed. This is consistent with the work of Dilks, Clark & Peeling,[24,25] who examined PPO resin films by ESCA after photo-oxidation and found greater oxidation of the surface than could be explained by methyl group oxidation alone.

The fact that neither sensitizing impurities nor methyl groups are necessary for photo-oxidation was shown by comparing the oxidation rates of pure model compounds and commercial polymer (benzene solution, $\lambda > 300$ nm) as shown in Fig. 3. None of the samples showed an 'induction period' to the oxidation. Indeed, the polymer showed an initial more rapid rate that corresponds to about one oxygen molecule per polymer chain. Remarkably, quite similar oxidation rates were observed for the polymer and the model compounds with and without the methyl groups. An examination of the molecular weight of the polymer as a function of oxygen uptake showed 5–6 oxygen molecules were consumed per chain scission. In addition, the rate of disappearance of starting material for the model compounds was only about one-sixth of the rate of oxygen consumption, indicating that the primary oxidation products are much more susceptible to photo-oxidation than the starting material itself. Liquid chromatography showed over 40 oxidation products, all in minor amounts.

The absorption spectra of the polymer and model compounds are shown in Fig. 4. The aryl ether structure has a strong absorption band around 280 nm that extends to approximately 320 nm. Some light in the near-UV, especially the 313-nm mercury line, is capable of being absorbed by the backbone of the polymer itself. The polymer exhibits an additional 'tail' beyond 330 nm, but the similarity of oxidation rates of the pure model compounds and the polymer indicates that the absorption tail is not primarily responsible for the photo-oxidation. Clearly the phenolic end groups and any adventitious impurities do not play much of a role either.

The direct photolytic cleavage mechanism (Scheme 2) that apparently operates at 254 nm does not seem to occur at wavelengths greater than 300 nm. In this mechanism, each primary photochemical event would result in chain cleavage to form a pair of radicals that would be capable of reacting with oxygen or recombining. However, these radicals should be capable of abstracting hydrogen atoms from

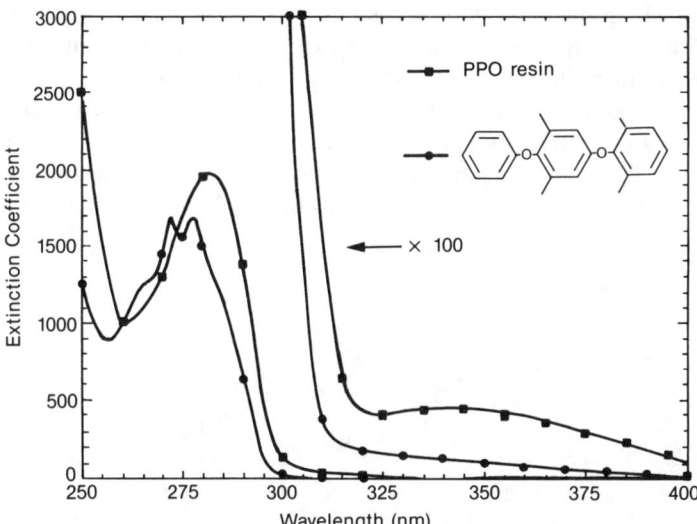

Fig. 4. Absorption spectra of commercial PPO resin and a pure model compound in chloroform solution.

Table 2
Changes in PPO resin molecular weight upon photolysis using a Pyrex-filtered Hg lamp

Irradiation	M_n	M_w
None	25 000	53 000
5 days, N_2, benzene, iPrOH	27 000	63 000
5 days, O_2, benzene	13 000	30 000

isopropyl alcohol to result in irreversible chain cleavage even in the absence of oxygen. Table 2 shows the effect of irradiation at $\lambda > 300$ nm in the presence and absence of oxygen and isopropyl alcohol. The results show that chain cleavage occurs only in the presence of oxygen. The aryl ether cleavage mechanism therefore appears unlikely at these wavelengths.

2.2 Electron Transfer Mechanism

A mechanism consistent with the results is shown in Scheme 3. In this mechanism an excited polymer repeat unit undergoes electron transfer with another unit to generate a radical cation and radical anion pair. The radical anion rapidly transfers an electron to oxygen to make superoxide, O_2^-. Superoxide and the radical cation can then combine to make an unstable primary oxidation product that undergoes further oxidation to give the final degraded products. Possible structures for the primary oxidation product and subsequent reactions are shown in Scheme 7. This mechanism can be considered to be a self-sensitized version of the electron transfer oxidation described by Foote et al.[26-28]

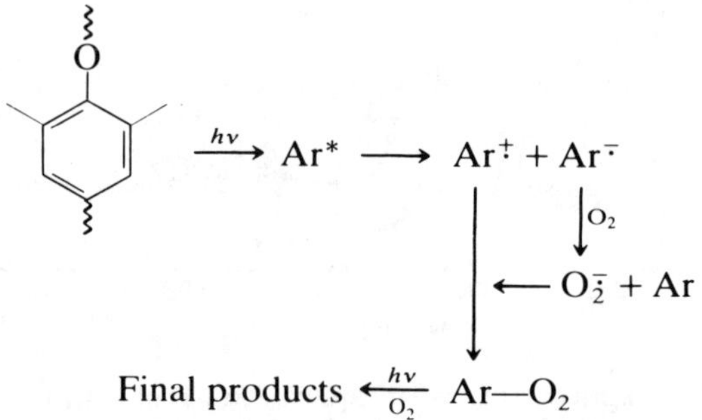

$$\text{Ar} \xrightarrow{h\nu} \text{Ar}^* \longrightarrow \text{Ar}^{\ddagger} + \text{Ar}^{\bar{\,}}$$
$$\downarrow O_2$$
$$\longleftarrow O_2^{\bar{\,}} + \text{Ar}$$
$$\downarrow$$
$$\text{Final products} \xleftarrow{h\nu}{O_2} \text{Ar—O}_2$$

Scheme 3. Electron transfer oxidation.

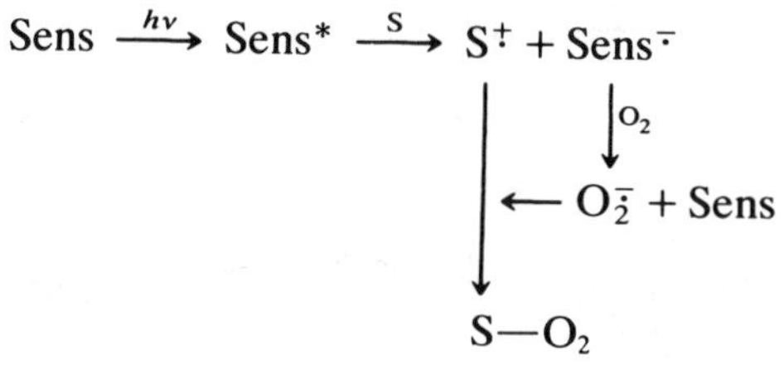

$$\text{Sens} \xrightarrow{h\nu} \text{Sens}^* \xrightarrow{S} S^{\pm} + \text{Sens}^{\mp}$$
$$\downarrow O_2$$
$$\leftarrow O_2^{\bar{}} + \text{Sens}$$
$$\downarrow$$
$$S-O_2$$

Scheme 4. General electron transfer photo-oxidation.

and others[29-31] (Scheme 4). We have obtained evidence supporting this mechanism by performing sensitization and quenching experiments, trapping the superoxide intermediate and observing transients consistent with the radical anion and radical cation in flash photolysis experiments.

Known electron transfer sensitizers promoted the photo-oxidation of PPO resin and model compounds. Thus, 9-cyanoanthracene, 9,10-dicyanoanthracene, and methylene blue sensitized the photo-oxidation at visible wavelengths. Polymer-bound Rose Bengal, which is an efficient singlet oxygen sensitizer but is a poor electron transfer sensitizer, was not effective. This indicates that aryl ethers are capable of undergoing electron transfer oxidation and that singlet oxygen is not involved.

The known electron-transfer quencher 1,2,4-trimethoxybenzene[27] reduced the rate of photo-oxidation of PPO resin as shown in Fig. 5. Diazabicyclo[2.2.2]octane (DABCO) and bis(2,2,6,6-tetramethyl-4-piperidinyl) sebacate (Tinuvin® 770) also quenched the oxidation. Interestingly, only 50% of the oxidation was quenchable in solution, indicating that some of the reaction was proceeding within the solvent cage or by some other mechanism. Similar results were obtained in cast films. Although some of the oxidation could be retarded, even massive amounts of quenchers would not stop it all. These compounds act by transferring an electron to the radical cation and then reacting reversibly with superoxide, resulting ultimately in neutral oxygen and neutral ground-state substrate (Scheme 5).

Direct evidence for the formation of superoxide was obtained by trapping it with 5,5-dimethyl-1-pyrrolidine-N-oxide[32,33] and observing the ESR spectrum of the resulting nitroxyl radical (Scheme 6). Figure 6 shows the ESR spectrum of the adduct formed from authentic superoxide. The adduct degrades thermally or photochemically to give the stable radical with the three-line spectrum shown. A cast film of

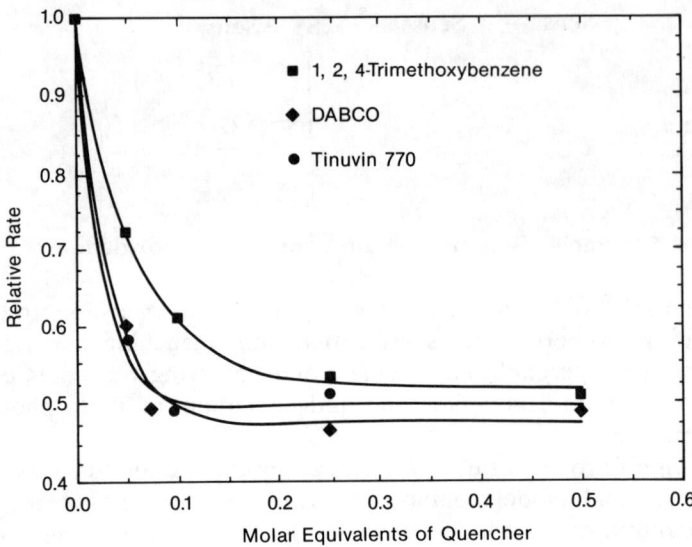

Fig. 5. Effect of electron transfer quenchers on the rate of photo-oxidation of a 2% benzene solution of PPO resin. Pyrex-filtered Hg lamp.

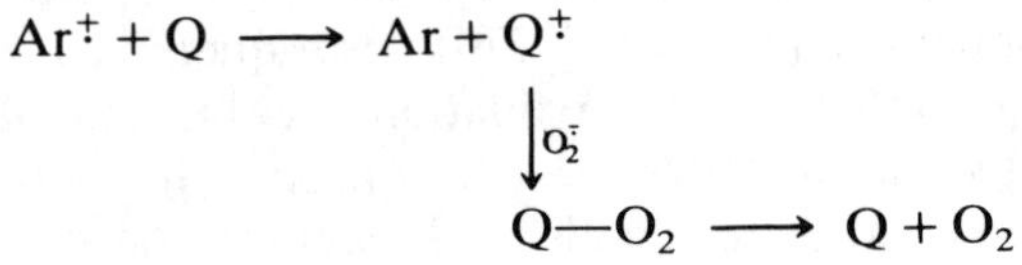

$$Ar^{\ddagger} + Q \longrightarrow Ar + Q^{\ddagger}$$
$$\downarrow O_2^{\bar{\cdot}}$$
$$Q\text{—}O_2 \longrightarrow Q + O_2$$

Scheme 5. Electron transfer quenching mechanism.

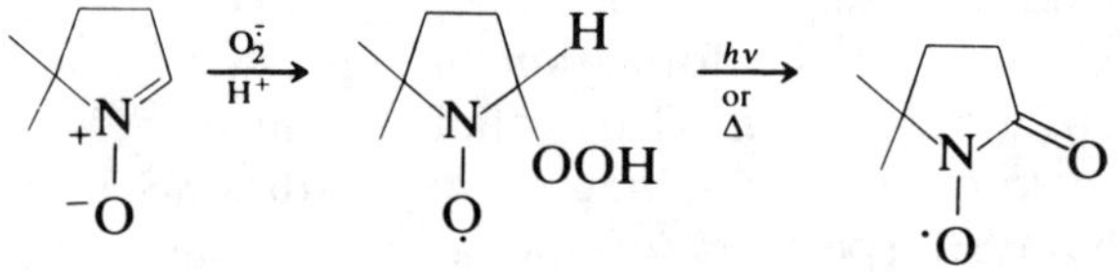

Scheme 6. Reaction of 5,5-dimethyl-1-pyrolidine-N-oxide with superoxide.

the polymer was irradiated in the presence of a methanol solution of the nitrone radical trap. The methanol solution was decanted and was found to have the ESR spectrum shown in Fig. 7. This is the same four-line spectrum as shown in Fig. 6 with some of the three-line dehydrated product formed as well. (Since the adduct is photochemically labile it is not possible to generate it cleanly by a photochemical process.) Upon further irradiation the clean three-line spectrum appeared. No radical signal was generated upon irradiation of the solution in the absence of the PPO resin film, and the film itself showed no ESR signals after washing with fresh methanol. Thus, the radical trap showed the presence of free superoxide in solution but no trappable free radicals on the polymer itself.

Finally, spectroscopic evidence was found for radical cation and anion transient species by xenon arc flash photolysis of a benzene solution of PPO polymer. Table 3 shows the lifetimes of transients at 450 and 510 nm. The different lifetimes of the decays at these two wavelengths indicate the presence of at least two different species. Oxygen effectively reduced the lifetime of the 510-nm transient but did not greatly affect the 450-nm transient lifetime. Conversely, the electron transfer quenchers reduced the lifetimes of the 450-nm transient but actually slightly increased the lifetime of the 510-nm transient in the absence of oxygen. This is consistent with the 450-nm transient being due to a radical cation and the 510-nm transient being due to a radical anion. The anion would react quickly with oxygen while the cation would react with electron-rich quenchers.

We therefore have evidence that the methyl groups are not the site of photo-oxidation of PPO resin at wavelengths greater than 300 nm. It appears that PPO resin can undergo a self-sensitized electron transfer reaction resulting in oxidation of the aromatic ring and not the methyl groups. We have been unable to isolate any of the primary oxidation products, even at low temperature, presumably because these products are more photo-oxidatively unstable than the starting material itself. Since electron transfer oxidations often mimic singlet oxygen reactions, the primary oxidation products are expected to be the result of 1,4- or 1,2-additions across the aromatic ring as shown in Scheme 7.

2.3 Implications of the Mechanism on Stabilization

The electron transfer mechanism has several important implications for those trying to increase the photostability of PPO resin blends.

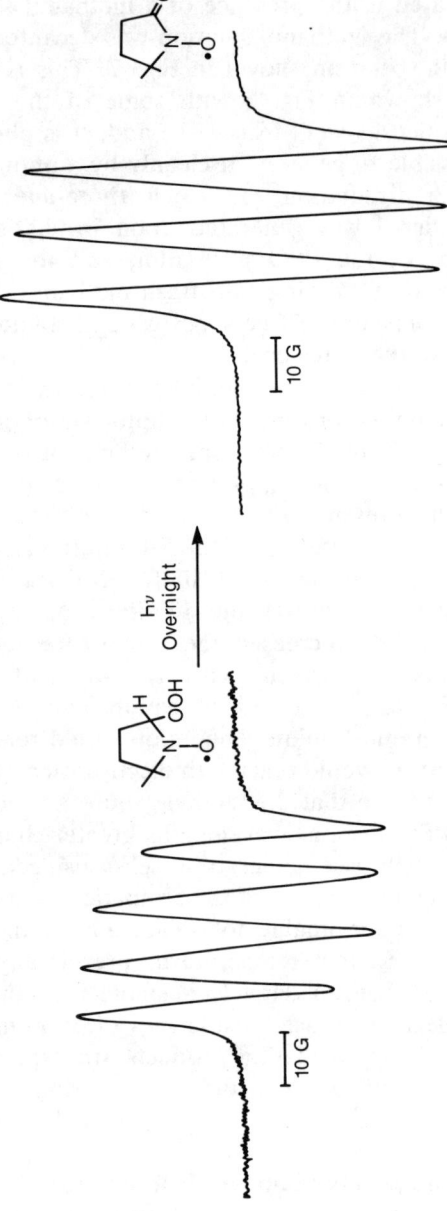

Fig. 6. ESR spectra of authentic nitrone-trapped O_2^- freshly prepared and after photolysis overnight.

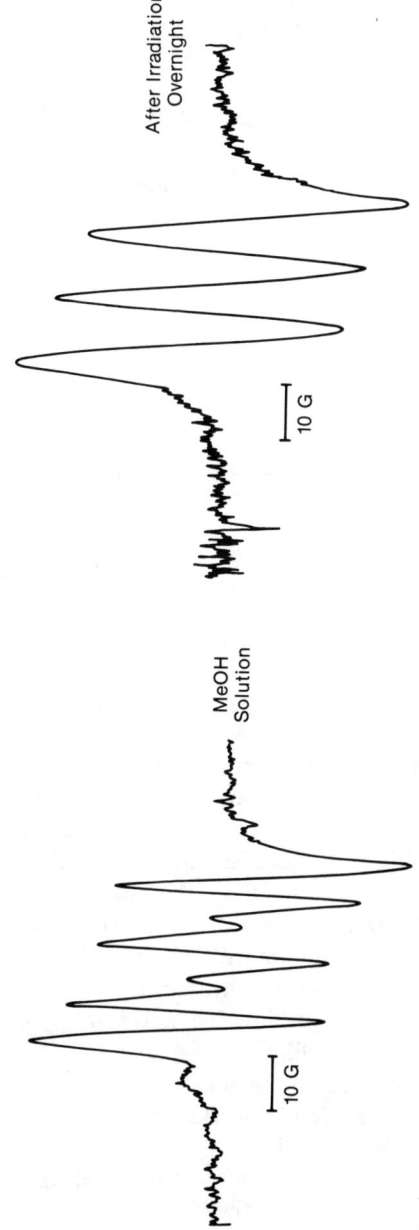

Fig. 7. ESR spectra of polymer-generated, nitrone-trapped O_2^- immediately after preparation and after photolysis overnight.

149

Table 3
Transient lifetimes (µs) after xenon flash of 1·6 M PPO resin solution in benzene

Additive	450 nm	510 nm
None		
Degassed	220	100
Air	180	<50
1,2,4-Trimethoxybenzene (0·6 M)		
Degassed	140	140
Air	120	<50
DABCO (0·8 M)		
Degassed	140	120
Air	140	<50
Tinuvin® 770 (0·6 M)		
Degassed	<50	140
Air	<50	<50

Firstly, the lack of an induction period in the photo-oxidation is not due to an initial concentration of hydroperoxides or to sensitizing impurities. In contrast to the more extensively studied polyolefin polymers, PPO resin oxidation is due to absorption of light by the polymer itself and to chemistry resulting from the excited state of the polymer. The photostability of polyolefins can be greatly enhanced by reducing the amount of catalyst residues and processing-related impurities, thus vastly increasing the induction period of the photo-oxidation. However, efforts to 'clean up' PPO polymer are unlikely to increase the photostability since the rate of oxidation of the commercial polymer is only slightly faster than that of pure model compounds. Indeed, it has been found that 'cleaner' polymer often photoyellows more quickly because there is less colored material to undergo an initial photobleaching reaction.

Secondly, the oxidation does not involve a free-radical chain, so chain-breaking antioxidants are unlikely to be of much help. Even if an antioxidant molecule could intercept an intermediate, it could act only stoichiometrically at best and would be quickly consumed. Similarly, agents that disrupt the oxidation cycle by removing a key intermediate in the way that hindered amines (HALS) stabilize polypropylene are unlikely to be very effective.

Finally, there are no long-lived intermediates to be trapped or

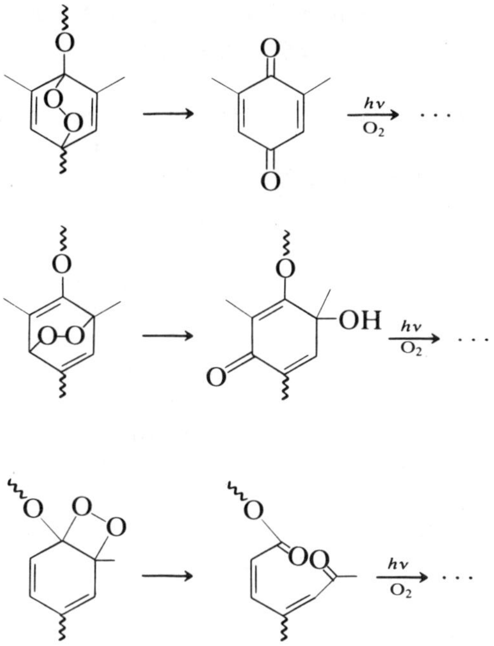

Scheme 7. Possible oxygen–PPO resin primary oxidation products.

quenched. Hindered amines were found to be quenchers in solution, but at best the rate of oxidation could be reduced by only 50%. In the solid phase the effectiveness was even less, and, in practical terms, it is impossible to get enough in a solid resin to have a significant effect on the oxidation rate. Excited-state quenchers such as the nickel complexes are unlikely to work because every repeating unit of the polymer is a potential light-absorbing site, and not enough quencher could possibly be added to be effective.

For cases in which a classical free-radical photo-oxidation mechanism does not operate, at least two other stabilization pathways are open. Firstly, the amount of light reaching the polymer can be reduced through the addition of light absorbers or pigments. Secondly, the effect of the photodegradation can be minimized by diverting the product distribution toward less objectionable, i.e. less yellow, products. Both of these approaches are useful in the stabilization of PPO resin/HIPS blends and are discussed below.

3 ENVIRONMENTAL FACTORS IN THE PHOTOYELLOWING PROCESS

An analysis of the effect of various environmental factors on the rate and nature of photodegradation is a useful exercise for any resin. It points out the important variables in the degradation process, and often leads to useful insight into what can be done to increase photostability. Just as important, however, is the insight it brings into the problems of accelerated testing. Some kind of accelerated testing is necessary for any kind of stabilization program, and it is important that the test method should accurately reflect the performance in the real world. We have examined the effects of several factors important in the degradation and testing of PPO resin blends. (The complete experimental details for the next two sections are being prepared by the author for publication in *Polymer Degradation and Stability*.)

3.1 Light Wavelength

The indoor UV environment, especially an office, is dominated by the mercury-line spectrum leaking from fluorescent lamps (Fig. 1). The

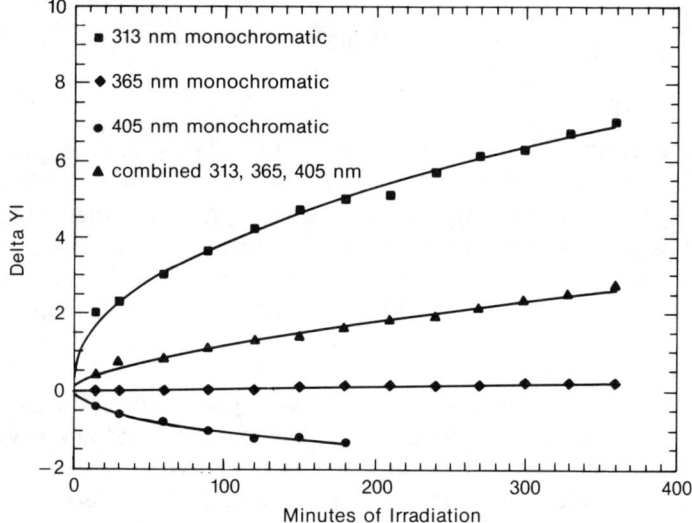

Fig. 8. Effect of isolated mercury lines on the photoyellowing of a 50:50 PPO resin/HIPS blend containing 13% triaryl phosphate and 3% TiO_2. Note how the combined lines result in a slower rate than the 313-nm line alone.

effect of isolated mercury lines from a high-pressure mercury lamp on a 50/50 PPO resin/HIPS blend containing 3% TiO_2 is shown in Fig. 8. The short-wavelength 313-nm line in isolation caused a rapid yellowing reaction while the 365-nm line had little effect and the 405-nm line caused only photobleaching. One might conclude that all of the relevant photochemistry can be studied by exposing samples to only short-wavelength light such as to fluorescent sunlamps in a UV-CON or QUV apparatus. However, the combination of all three mercury lines resulted in a yellowing rate that was slower than for the 313-nm line alone. This is because the 405-nm line is capable of bleaching some of the yellow products that the shorter-wavelength light produces. This is seen most clearly in Fig. 9, in which a sample yellowed to $\Delta YI = 7$ with monochromatic 313-nm light was bleached by monochromatic 405-nm light to result in a ΔYI of only 3. After 17 hours' exposure to 405-nm, the ΔYI was only 1·7. Thus apparently non-actinic visible light plays an important role in the overall

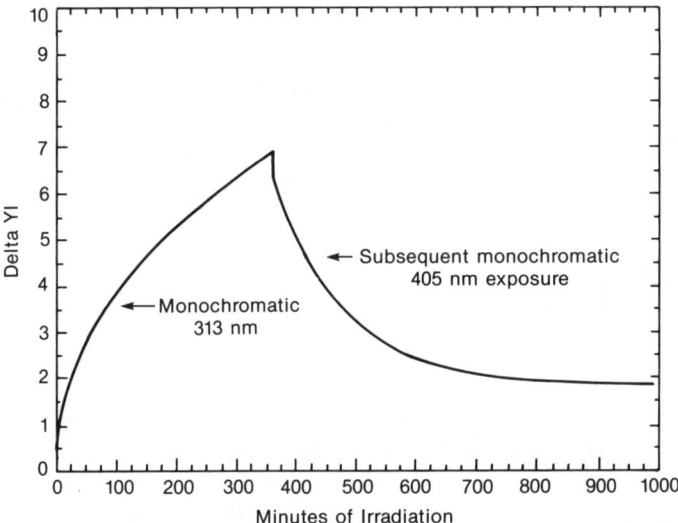

Fig. 9. Exposure of a 50:50 PPO resin/HIPS blend containing 13% triaryl phosphate and 3% TiO_2 to monochromatic 313-nm light and then to monochromatic 405-nm light. The longer-wavelength light bleaches some of the photo products formed at 313-nm.

photoyellowing process, and the complete wavelength distribution of the environment must be reproduced in order to do accurate accelerated aging tests.

3.2 Temperature Effects

Samples of 50/50 PPO resin/HIPS were exposed to a bank of fluorescent lamps at temperatures ranging from 30 to 70°C. The results, shown in an Arrhenius plot in Fig. 10, indicated a small activation energy of about 4 kcal/mol. That is, an increase in temperature of 30°C would double the rate of photoyellowing. This is important because some applications result in the surface being warm, and some accelerated tests involve elevated temperatures. Proper consideration of use and test temperatures must be made in the evaluation of photostability.

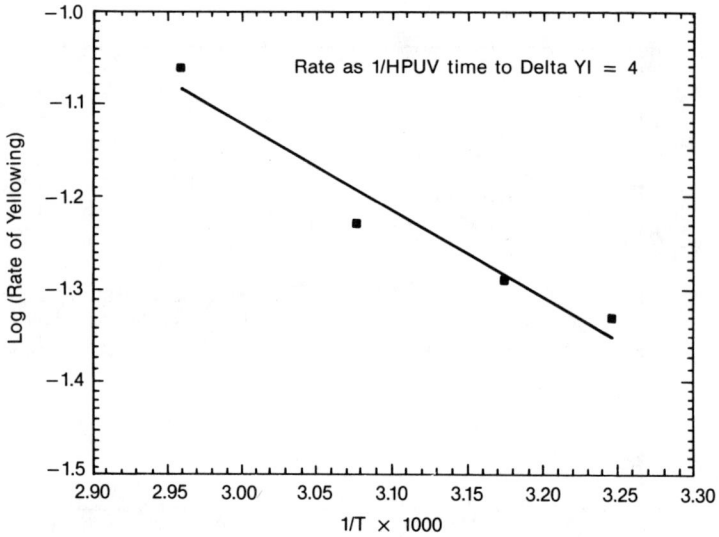

Fig. 10. Arrhenius plot of the photoyellowing rate of a 50:50 PPO resin/HIPS blend containing 13% triaryl phosphate and 3% TiO_2 in an HPUV tester at 35–65°C. Rate as 1/HPUV time to $\Delta YI = 4$ (just visibly yellow). Activation energy from the slope is 4·2 kcal/mol (17·6 kJ/mol).

3.3 Light Intensity Effects

An important article of faith in accelerated testing for photostability is that an increase in the light intensity results in a proportional increase in the rate of photodegradation. The effect of light intensity on the rate of yellowing of a blend of PPO resin and HIPS is shown in Fig. 11. The samples were exposed to a bank of fluorescent lamps through various neutral density filters up to a maximum intensity of 20 times ambient. The data are obviously not linear at intensities only a little greater than ambient. The fact that there is a break in the curve for the blend indicates that there is a change of mechanism as the light intensity is increased. Apparently, there is a branch point in the mechanism as shown in Scheme 8.

In this Scheme, the PPO resin undergoes photo-oxidation to give some intermediate product(s) designated as P_1. At this point, one branch would be purely photo-oxidative or photochemical leading to

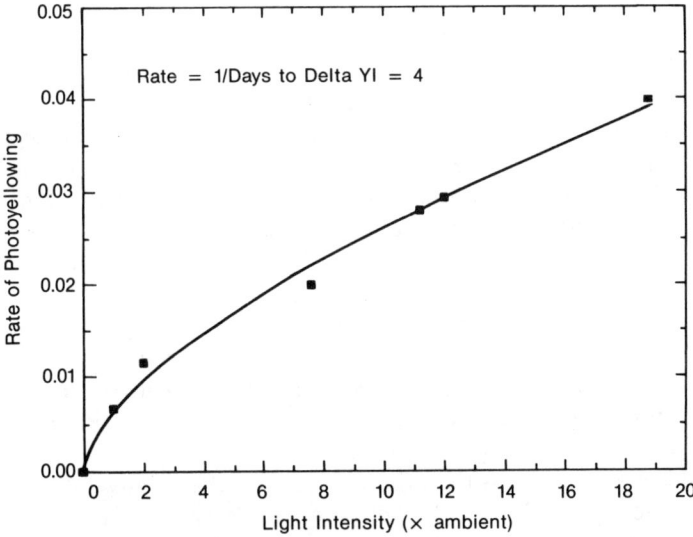

Fig. 11. Photoyellowing of a 50:50 PPO/HIPS blend containing 13% triaryl phosphate and 3% TiO_2 exposed to a bank of fluorescent lamps through neutral density filters. Note the non-linearity of the rate with light intensity. The light intensity is normalized to the office light shown in Fig. 1.

$$\text{PPO resin} + O_2 \xrightarrow{h\nu} P_1 \xrightarrow[O_2]{h\nu} P_2$$
$$\downarrow{\scriptstyle h\nu, O_2}^{\Delta}$$
$$P_3$$

Scheme 8. Branched pathway to ultimate photo products.

one set of products, P_2, while another branch containing at least one thermally controlled step would lead to another set of products, P_3. At low light intensities, the $P_1 \rightarrow P_3$ branch competes with the $P_1 \rightarrow P_2$ branch. As the light intensity is increased, the rates of all the photochemical reactions increase but the rate of the thermal step does not. The $P_1 \rightarrow P_3$ branch therefore has a bottleneck, is no longer competitive, and a greater proportion of the products are P_2. If one set of products is more highly colored than the other, then the rate of photoyellowing will change when the mechanism switches from $P_1 \rightarrow P_3$-dominated to $P_1 \rightarrow P_2$-dominated. The net effect is a change in the slope of the rate vs light intensity curve as the products of one branch are replaced by the products of the other.

There are two important implications of this non-linearity. Firstly, accelerated aging tests at high light intensities will tend to underestimate the rate of photoyellowing at ambient intensity. Expressing the rates as 1/(days of constant exposure to $\Delta YI = 4$) from Fig. 11, the rate at $I/I_0 = 20$ (20 × ambient light intensity) is 0·042. One might expect that dividing this rate by 20 should give a real world rate of 0·0021. The actual rate at this light intensity ($I/I_0 = 1$) is about 0·006. The error arises because the mechanism operating under test conditions is not the same as the mechanism operating in the real world.

A second implication is that additives tested at high light intensities may have different effects from what is observed in the real world. Again, this is due to the change in mechanism at high intensities. If, for example, an additive operates more on one branch of the mechanism in Scheme 8 than the other, then the relative effectiveness of the additive will vary as one branch or the other increases or decreases in importance. This is the case for HALS additives, as will be discussed later.

4 ACCELERATED TESTING OF PPO RESIN BLENDS

If the goal of accelerated testing is to make accurate predictions of real-world performance, then two conditions must be assured. Firstly, the degree of acceleration, i.e. the increase of light intensity over ambient, must be known. Secondly, the mechanism of the degradation process must not be changed. The light intensity experiments described above indicate that the mechanism of photoyellowing in PPO resin blends does change when the light intensity is increased only a few times over ambient levels because thermally controlled reactions cannot keep up with the accelerated photo reactions. This problem presumably can be solved by increasing the temperature as well as the light intensity to bring the mechanisms back into balance.

We have constructed a device in which the light is supplied by fluorescent lamps at an intensity ten times greater than ambient. Ambient is defined as lighting levels measured in offices at General Electric Corporate Research and Development (CRD) in Schenectady, New York. At the same time surface temperatures of the samples were maintained at 60°C by placing them on a thermostated hot plate. These conditions were calculated to produce a ten-fold increase in the rate of photoyellowing, and since the illumination was continuous, about a 20-fold increase over real-world indoor weathering. Thus, nine days in the tester would be equivalent to six months of real-world exposure. We have called this the CRD tester.

Samples were also maintained in a rack about 2 m from a south-facing window in an office illuminated primarily by fluorescent lamps. The results of testing control samples and samples containing HALS and UV screener stabilizers are shown in Fig. 12. The correlation is very good. We have tested over 100 different formulations and have found generally excellent correlation between the tester results and real-world results. Deviations usually occur in highly pigmented samples or in samples which exhibit unusual bleaching behavior.

Other means of accelerated testing for indoor weathering have also been developed. The previously mentioned HPUV test meets the recent ASTM D4674-87 method. Although the ASTM method does not claim any correlation factor with the 'real world', the original test specifications were intended to simulate three years of exposure. Since the lamps decay with time, the total test time varies, and '100%' time was defined as the time in which the samples would receive the same exposure as they would receive in three years in a typical office environment. The test is run at about 35°C. The results of exposure of control and stabilized PPO resin blends (commercial N-190 and UV-180 grades of Noryl® resin) are shown in Fig. 13 with both the CRD and HPUV tests normalized to their equivalent in months of

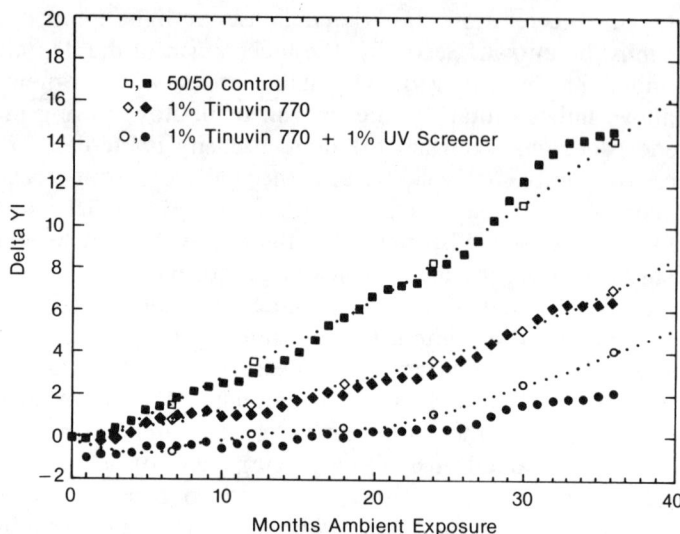

Fig. 12. Change in Yellowness Index of samples exposed to ambient conditions in a laboratory office in Schenectady (closed symbols) and predicted from the CRD test (open symbols, dotted lines) using a 9 test days = 6 months ambient correlation.

real-world exposure times. The results seem to indicate that the HPUV test is at least two-fold harsher than intended. This was confirmed by light-intensity readings made by us. While the light intensity should be 40 times greater than ambient to obtain the exposure intended in the product literature, the light levels were in fact 80–100 times greater than ambient. Thus, in our hands, '100% HPUV time' corresponds to about six years of exposure.

5 PHOTOSTABILIZATION OF PPO RESIN BLENDS

Since the mechanism of PPO resin photo-oxidation does not seem to be a conventional free-radical oxidation, it would not be surprising if standard stabilizers had little effect. Slama[34] has found that a variety of phenolic and sulfur-containing antioxidants have no effect on the photo-oxidation of PPO resin in the Xenotest 150 apparatus. Among the compounds tested were Irganox® 1010, dilauryl thiodipropionate, trilauryl phosphite and 2-mercaptobenzimidazole. UV screeners of

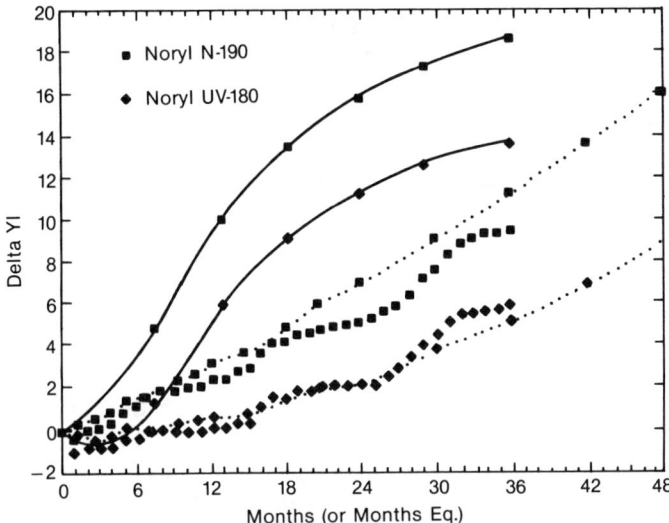

Fig. 13. Normalized photoyellowing data for commercial Noryl® resins N-190 and UV-180 under ambient office exposure, CRD test conditions (dotted lines), and HPUV test conditions (solid lines). The HPUV data were normalized using '100% exposure time' as being equivalent to three years of exposure as described in the HPUV product brochure (Atlas Electric Devices Co., 4114 N. Ravenswood Ave., Chicago, IL 60613, USA).

both the benzotriazole and benzophenone types did reduce the rate of photoyellowing.[35] These results have been confirmed in our laboratories for the photoyellowing of blends.

5.1 Hindered Amines

Although the accepted mechanism of hindered-amine action in polyolefins involves disruption of the oxidation cycle, we have found that these compounds are in fact effective in reducing the rate of photoyellowing in PPO resin blends. The effectiveness is greater than expected from the small amount of quenching that would be possible by low loadings in the solid state. Table 4 shows the increase in test times to 'failure' for 1% of Tinuvin 770 added to a 50/50 PPO resin/HIPS blend. It is notable that the HPUV test underestimated the degree of improvement seen on the other tests and in real world testing. Further experiments have shown that under high-intensity

Table 4
Effect of hindered amine on photoyellowing under various test conditions

Test	Times to $\Delta YI = 4$		
	Control	+1% Tinuvin® 770	Improvement (-fold)
CRD	16 days	32 days	2·0
HPUV	63 h	75 h	1·2
Glass-filtered xenon arc	190 h	330 h	1·7
Ambient exposure	16 months	27 months	1·7

light the extent of rate reduction is dependent on the temperature of the test. Thus, under a bank of fluorescent lamps, 1% added Tinuvin 770 reduced the rate of photoyellowing by a factor of three at 70°C but by only a factor of 1·5 at 30°C. This is apparently because the hindered amine acts on the thermally controlled $P_1 \rightarrow P_3$ branch shown in

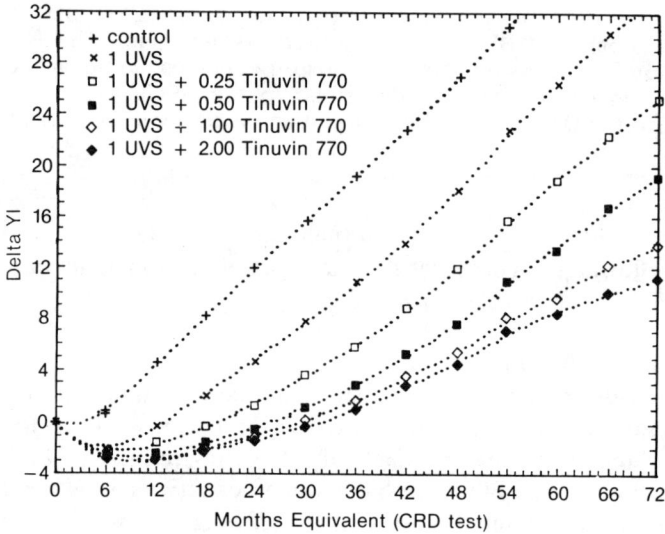

Fig. 14. Effect of increasing Tinuvin® 770 loading on the photoyellowing of 50:50 PPO resin/HIPS blends containing 13% triaryl phosphate and 3% TiO_2. The UV screener was 2-hydroxy-4-octyloxybenzophenone. On the CRD test, 9 days of exposure is equivalent to 6 months of ambient office exposure.

Scheme 8. Under conditions of high light intensity and relatively low temperature this branch is less important and so the additive shows little effect. At higher temperatures or under ambient conditions this branch of the mechanism is more important and the HALS is more effective.

Figure 14 shows the effect of HALS loading on a 50/50 blend containing 1% 2-hydroxy-4-octyloxybenzophenone (HOBP) in the CRD test. The HALS effectiveness does not increase much for loadings greater than 0·5%. Apparently this loading is enough to shut down the $P_1 \rightarrow P_3$ branch and adding more has little further effect. The way in which the amine interferes with this chemistry is not known. It is possible that some step of this branch is acid-catalyzed and the amine acts simply as a photostable base. All of the commercially available HALS stabilizers show at least some effectiveness, although the polymeric HALS seem to be slightly less effective in the CRD test. Simple amines do not seem to work, presumably because they are themselves rapidly destroyed. All of these results have been confirmed in real world testing.

5.2 UV Screeners

The effects of various loadings of HOBP in the presence of 1% of Tinuvin 770 on the HPUV and CRD tests are shown in Figs 15 and 16. Increasing effectiveness is observed for increased loading (even well beyond any practical loadings). All of the 4-alkoxy-2-hydroxybenzophenones and all of the hydroxybenzotrazole-type UV screeners are about equally effective. The benzylidine malonates (e.g. Cyasorb® 1988), the oxamides (e.g. Sanduvor® VSU), and foramidines (e.g. Givsorb® UV-2) do not work as well.

Inspection of Figs 15 and 16 reveals that one aspect of the UV-screener action is the deepening and prolongation of the bleaching period. The bleaching phenomenon can be examined by exposing the samples through a cut-off filter so that they receive only visible light. Figure 17 shows the effect of the UV screener on the bleach in the presence of 1% Tinuvin 770. The amine itself has a little effect, but the UV screener dramatically increases the amount of bleaching. Some of this is probably due to bleaching of color bodies that come from the additive itself during processing. Some is also due to a 'plasticizer effect' we have observed in PPO blends. Any plasticizing additive, including the phosphate flame retardant, increases the amount of bleaching. We have no explanation for this phenomenon at present.

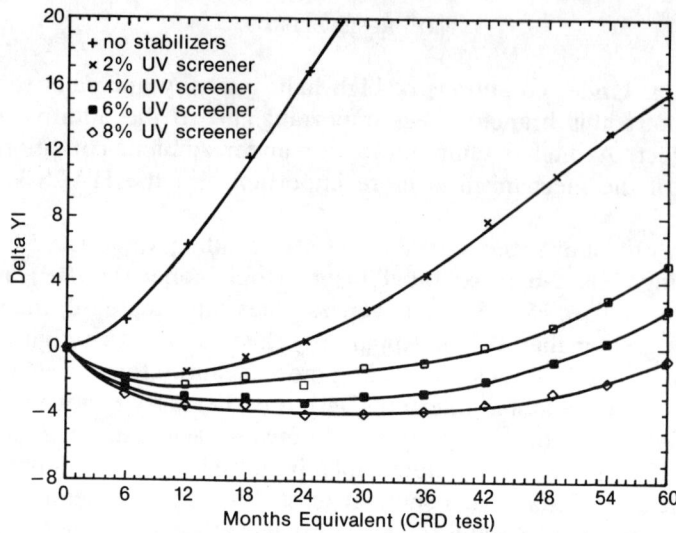

Fig. 15. Effect of increasing UV screener loadings (2-hydroxy-4-octyloxybenzophenone) on photoyellowing in the CRD test. Samples were 55:45 PPO resin/HIPS with 13% triaryl phosphate and 3% TiO_2. The samples containing UV screener also contained 1% Tinuvin 770.

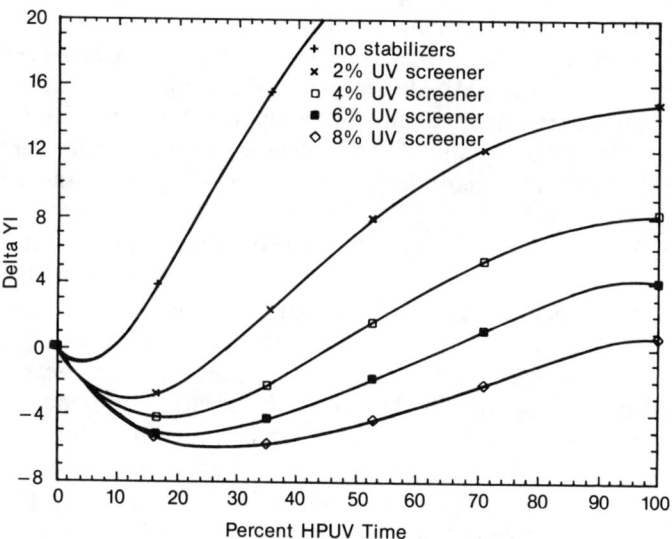

Fig. 16. Effect of increasing UV screener loadings in the HPUV test. The samples were as described in Fig. 15.

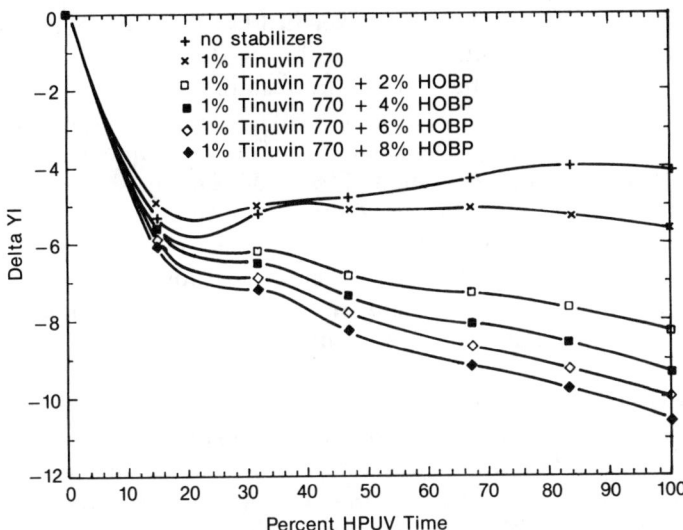

Fig. 17. Effect of stabilizers on the bleaching of 55:45 PPO resin/HIPS blends containing 13% triaryl phosphate and 3% TiO_2. The samples were exposed in an HPUV tester while covered with a 380-nm cut-off filter so that only bleaching and no yellowing could occur. The UV screener (2-hydroxy-4-octyloxybenzophenone) both prolongs and increases the bleaching, thus reducing the net photoyellowing of the resin.

The expected effectiveness of UV screeners can be calculated if the thickness of the degraded layer is known.[36] Equation (1) gives the relative rate of degradation for a stabilized sample as a function of the absorbance (A) and transmittance (T) of the sample where the subscripts s and o denote the presence and absence of UV screener. These values are available from a simple Beer's Law calculation using readily available molecular extinction coefficients of the polymer and additives. The light absorption due to pigments such as TiO_2 can be calculated from eqn (2), where V is the volume fraction of the pigment, l is the degradation thickness and d is the average particle diameter.

$$k_s/k_0 = \frac{(1-T_s)A_0}{(1-T_0)A_s} \qquad (1)$$

$$I_1/I_0 = (1 - 3V/2)^{l/d} \qquad (2)$$

The thickness of the degraded layer in PPO resin/HIPS blends can be determined by washing off a portion of the degraded layer with acetone and measuring the size of the resulting 'step' by means of a Dektac device. This device draws a stylus across the surface and displays the topography it encounters. This experiment on several samples of 50/50 blend containing 3% TiO_2 pigment showed that the degraded layer is approximately 2 μm thick. The plasticizer effect of the UV screener is responsible for an additional rate reduction of 2% per part of additive. Table 5 shows the calculations from eqn (1) for several loadings of HOBP and the rate reductions found upon exposure in the HPUV tester. The calculations were based on absorption at 313 nm for HOBP in a 50/50 PPO resin/HIPS matrix containing 3% TiO_2 and 13% of phosphate flame retardant. The agreement between the calculated and experimental rate reductions is good for a degradation thickness of 2–3 μm. The slightly higher than calculated performance may be due to the unusual effect that the screener has on the bleaching portion of the photoyellowing process. This indicates that the UV screener is acting mostly as a light absorber, and that it is performing as well as it can.

Overall, we have found that the rate of photoyellowing can be reduced by a factor of 3–4 by the addition of 1% HALS and 1% of UV screener in the CRD test and in our limited ambient exposure testing. This kind of formulation does not perform as well in the HPUV test, where the importance of hindered amines is diminished. Formulations optimized for the HPUV test contain more UV screener and less HALS. Xenon arc exposure, which is more popular in

Table 5
Calculated and observed relative rates of yellowing for various UV screener loadings in a 50 PPO resin/50 HIPS/13 phosphate/3 TiO_2 blend

Parts UVS	Calculated for degradation thickness:			Found
	1 μm	2 μm	3 μm	
0	(1)	(1)	(1)	(1)
2	0·89	0·82	0·77	0·66
4	0·78	0·67	0·58	0·56
6	0·68	0·54	0·43	0·45
8	0·59	0·42	0·31	0·37

Europe, tends to emphasize the HALS over the UV screener. Only continued field exposure studies will resolve the question of which tests most accurately reflect the real world performance.

6 CONCLUSIONS

The limited effectiveness of most stabilizers in PPO resin blends can be understood since the mechanism of photo-oxidation apparently is not a free-radical chain oxidation. The self-sensitized electron transfer oxidation mechanism is not amenable to interruption by hydrogen atom donors. Since hydroperoxides are not necessary as initiators, reducing the level of peroxides during processing by sulfur- or phosphorus-containing stabilizers has no effect. While hindered amines can quench up to 50% of the oxidation rate in solution, their effectiveness by this mechanism in the solid state in practical concentrations is limited.

The total photo-oxidation pathway is long, consuming a total of five molecules of oxygen per repeat unit oxidized. There seems to be a branch point after the initial photo-oxidation so that the rate of net photoyellowing is dependent on the temperature and light intensity of the test. Hindered-amine light stabilizers seem to operate only on one of the branches, so their effectiveness is dependent on the test method. This points out the need for as much careful correlation of test results to field exposure results as possible.

UV screeners are effective in reducing the rate of photoyellowing, but their action is limited by the thin degradation layer. The products of PPO resin photo-oxidation are strongly absorbing in the near-UV and build a self-screening layer on the surface, so the highly oxidized layer is only about 2 μm thick. Since UV screeners can absorb only a little of the incident light in such a short pathlength, large amounts are needed to reduce the rate of photoyellowing by any substantial amount.

ACKNOWLEDGMENTS

The author wishes to thank Arnold Factor, Dwain White, Gary Davis, Richard Bopp and Jan Lohmeijer for sharing their knowledge and expertise in PPO chemistry. Steven Rice, Peter Codella, James Carnahan, Sue Weissman and Paul Gundelach provided valuable technical and analytical support.

REFERENCES

1. Shu, P. C. H., US Patent 4555538; *Chem. Abstracts*, **104**: P6608s.
2. Hirt, R. C., Schmitt, R. G., Searle, N. D. & Sullivan, A. P., *J. Opt. Soc. Am.*, **50** (1960) 706.
3. Kelleher, P. G., Jasse, L. B. & Gesner, B. D., *J. Appl. Polym. Sci.*, **11** (1967) 137.
4. Jerussi, R. A., *J. Polym. Sci.*, *A-1* **9** (1971) 2009.
5. Symons, M. C. R. & Yandell, J. K., *J. Chem. Soc.* (A) (1970) 1995.
6. Tsuji, K., *Adv. Polym. Sci.*, **12** (1973) 171.
7. Tsuji, K. & Seiki, T., *Polym. J.*, **4** (1973) 589.
8. Wandelt, B., Jachowicz, J. & Kryszewski, M., *Acta Polym.*, **32** (1981) 637.
9. Chandra, R., Singh, B. P., Singh, S. & Handa, S. P., *Polymer*, **22** (1981) 523.
10. Chandra, R., *Europ. Polym. J.*, **17** (1981) 567.
11. Chandra, R. & Singh, B. P., *Europ. Polym. J.*, **18** (1982) 199.
12. Singh, R. P. & Chandra, R., *Europ. Polym. J.*, **18** (1982) 289.
13. Chandra, R., Singh, S. & Verma, A. K., *Acta Polym.*, **34** (1983) 216.
14. Chandra, R., *Acta Polym.*, **35** (1984) 597.
15. Chandra, R., *Acta Polym.*, **33** (1982) 672.
16. Chandra, R., *Polym. Photochem.*, **3** (1983) 367.
17. Tovborg-Jensen, J. P. & Kops, J., *J. Polym Sci., Polym. Chem. Ed.*, **18** (1980) 2737.
18. Tovborg-Jensen, J. P. & Kops, J., *J. Polym. Sci., Polym. Chem. Ed.*, **19** (1981) 2765.
19. Allen, N. S. & McKellar, J. F., *Makromol. Chem.*, **180** (1979) 2875.
20. Slama, Z., Svejdova, E. & Majer, J., *Makromol. Chem.*, **181** (1980) 2449.
21. Petruj, J. & Slama, Z., *Makromol. Chem.*, **181** (1980) 2461.
22. Slama, Z., *Acta Polym.*, **31** (1980) 746.
23. Pickett, J. E. In *Polymer Stabilization and Degradation*, ACS Symposium Series No. 280, ed. P. Klemchuk American Chemical Society, Washington, DC, 1985, p. 313.
24. Dilks, A. & Clark, D. T., *J. Polym. Sci., Polym. Chem. Ed.*, **19** (1981) 2847.
25. Peeling, J. & Clark, D. T., *J. Appl. Polym. Sci.*, **26** (1981) 3761.
26. Eriksen, J., Foote, C. S. & Parker, T. L., *J. Am. Chem. Soc.*, **99** (1977) 6455.
27. Spada, L. T. & Foote, C. S., *J. Am. Chem. Soc.*, **102** (1980) 391.
28. Manring, L. E., Eriksen, J. & Foote, C. S., *J. Am. Chem. Soc.*, **102** (1980) 4275.
29. Schaap, A. P., Zaklika, K. A., Kashir, B. & Fung, L. W.-M., *J. Am. Chem. Soc.*, **102** (1980) 389.
30. Mattes, S. L. & Farid, S., *J. Chem. Soc., Chem. Comm.* (1980) 457.
31. Santamaria J., *Tetrahedron Lett.*, **22** (1981) 4511.
32. Harbour, J. R., Chow, V. & Bolton, J. R., *Can. J. Chem.*, **52** (1974) 3549.

33. Harbour, J. R. & Bolton, J. R., *Biochem. Biophys. Res. Comm.*, **64** (1975) 803.
34. Slama, Z., *Chemicky Prumysl*, **31/56** (1981) 185.
35. Hageman, H. J. & Huntsjens, F. J., *Lenzinger Berichte*, **44** (1978) 38.
36. Pickett, J. E., *J. Appl. Polym. Sci.*, **33** (1987) 525.

Chapter 6

Photo-oxidation and Stabilization of Polyethylene

FRANÇOIS GUGUMUS
Ciba-Geigy Ltd, CH-4002 Basle, Switzerland

ABSTRACT

The main reactions of ketones and hydroperoxides to be expected on photo-oxidation of polyethylene are reviewed. The performance of various light stabilizers on outdoor exposure of polyethylene is compared. HALS used alone or in combination with benzotriazole-type UV absorbers show by far the best effect.

Possible reaction mechanisms are considered for the interpretation of outdoor exposure data. The mechanisms of quenching of excited ketone carbonyl groups are examined in detail. It is shown that hindered-amine light stabilizers (HALS) do not quench ketone photolysis. However, benzotriazole-type UV absorbers are very efficient quenchers for ketones. The Norrish type I reaction is quenched according to the long-range energy transfer mechanism. The chemical changes on outdoor exposure of 2-mm HDPE plaques were monitored by IR spectroscopy. It is found that build-up of vinyl groups in HALS-stabilized samples starts with the beginning of exposure, concomitantly with loss of tensile impact strength. The reactions responsible for the observed effects are discussed. Finally, the fate of HALS-1 on outdoor exposure of HDPE plaques is examined. Loss of ester groups is shown *to be caused by a side reaction of the nitroxyl radical formed in the stabilization process.*

1 INTRODUCTION

Photo-oxidation and stabilization of polyethylene have been the subject of numerous studies in the 55 years following the first successful polymerization of ethylene on a commercial scale in 1933.[1-5] Nevertheless, many aspects of the photo-oxidation mechanisms are not yet fully understood and the subject of controversy. Moreover, although stabilization of polyethylene against photo-oxidation has made considerable progress in the last decade,[5-9] understanding of the stabilization mechanisms of various UV stabilizers in polyethylene is still limited, and is often more qualitative than quantitative.

In this chapter, present knowledge of photo-oxidation mechanisms of polyethylene is presented, then various aspects of practical UV stabilization of high-density polyethylene (HDPE), low-density polyethylene (LDPE) and linear low-density polyethylene (LLDPE) are examined in detail. Finally, new results concerning stabilization mechanisms are discussed and applied to the interpretation of weathering data.

2 PHOTO-OXIDATION MECHANISMS

Hypothetically, 'pure' polyethylene, by analogy with pure saturated aliphatic hydrocarbons, should not absorb UV sunlight reaching the earth (wavelengths above 295 nm). Therefore, photo-oxidation occurring in 'normal' polyethylene has to be attributed to the presence of chromophoric groups absorbing UV light of wavelengths above 295 nm.

Catalyst residues, hydroperoxides, carbonyl groups and charge transfer complexes were considered as potential chromophores for polyethylene photo-oxidation. The main discussions have centred upon the roles of hydroperoxides and carbonyl compounds in this initiation step. For more details the reader is referred to review articles.[3-5]

For some time, secondary hydroperoxides formed in polyethylene were believed to be the primary initiating species. However, their importance for photo-initiation has recently been challenged. In fact it was found that, on UV exposure, secondary hydroperoxides decompose directly to ketones without inducing any new oxidation chains.[10–14] Even tertiary hydroperoxides, formed in polyethylene in small proproportions, seem to decompose mainly without formation of free radicals.[13,14] It was suggested that initiation may be due, at least in part, to photolysis of alkyl peroxides also formed on oxidation of polyethylene.[14] Another possibility of initiation is deduced from results showing that both the singlet and triplet $n-\pi^*$ states of ketones are quenched by peroxides and hydroperoxides at diffusion-controlled rates and thus lead to sensitized decomposition of hydroperoxides [reaction (1)].[15] The energy transfer is interpreted by assuming the formation of an intermediate exciplex between the excited state of the ketone and the ground state of the peroxide.

$$\begin{array}{c} \text{\Large$>$}C{=}O^* + ROOH \longrightarrow \text{\Large$>$}C{=}O + [ROOH]^* \\ \downarrow \\ RO^{\cdot} + {}^{\cdot}OH \end{array} \quad (1)$$

However, it was shown that it is more common for excited ketones to abstract the hydroperoxidic hydrogen according to reaction (2).[16]

$$\text{\Large$>$}C{=}O^* + ROOH \longrightarrow \text{\Large$>$}C{\diagup}^{OH} + ROO^{\cdot} \quad (2)$$

Although the initiation step is not yet clearly established, photothermal oxidation of polyethylene can be assumed to proceed according to the general scheme involving initiation, propagation and termination. Even the chain-branching step can still be taken into account. However, it will not involve the hydroperoxides usually associated with this step but, rather, dialkylperoxides:

$$POOP \xrightarrow{h\nu} PO^{\cdot} + {}^{\cdot}OP \quad (3)$$

$$PO^{\cdot} + PH \rightarrow POH + P^{\cdot} \quad (4)$$

The photo-oxidation scheme has been discussed previously.[8] However, the reactions accounting for the main photo-oxidation products and polyethylene degradation are reviewed in detail below.

2.1 Photolysis of Carbonyl Compounds

The main reactions considered for photolysis of carbonyl compounds are the well-known Norrish type I and type II reactions. They are shown schematically below as reactions (5) and (6).

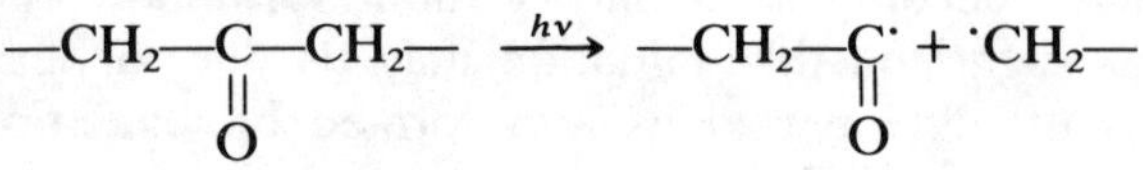

$$-CH_2-\underset{\underset{O}{\|}}{C}-CH_2- \xrightarrow{h\nu} -CH_2-\underset{\underset{O}{\|}}{C}\cdot + \cdot CH_2- \qquad (5)$$

$$-CH_2-\underset{\underset{O}{\|}}{C}-CH_2-CH_2-\underset{\underset{H}{|}}{CH}- \xrightarrow{h\nu} -CH_2-C\underset{O_*}{\overset{CH_2-CH_2}{\diagup}}\underset{H}{\overset{\diagdown}{CH-}}$$

$$\longrightarrow -CH_2-C\underset{O-H}{\overset{CH_2-CH_2}{\diagup}}\cdot CH-$$

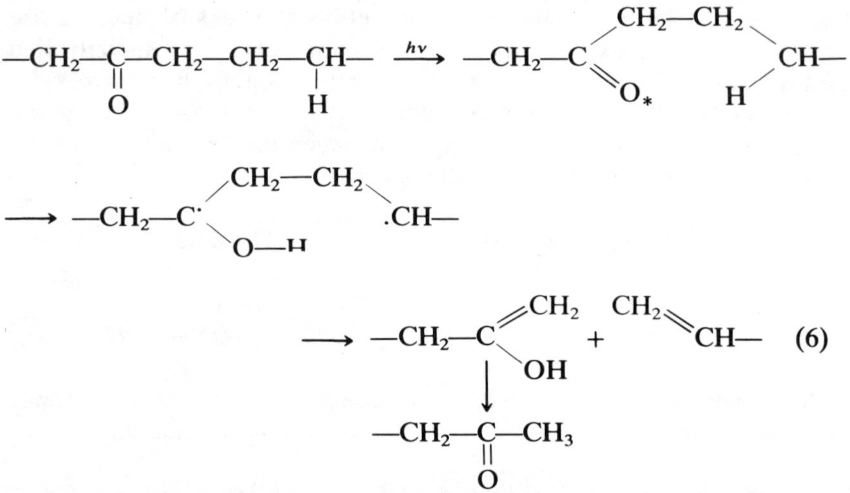

$$\longrightarrow -CH_2-C\underset{OH}{\overset{\diagup\!\!=\!CH_2}{\diagdown}} + \overset{CH_2=\!\diagdown}{}CH- \qquad (6)$$

$$\downarrow$$

$$-CH_2-\underset{\underset{O}{\|}}{C}-CH_3$$

The Norrish type I reaction yields free radicals and can initiate oxidation in this way. Further oxidation of the carbonyl-bearing radical yields carboxylic acids according to the following schematized reaction sequence:

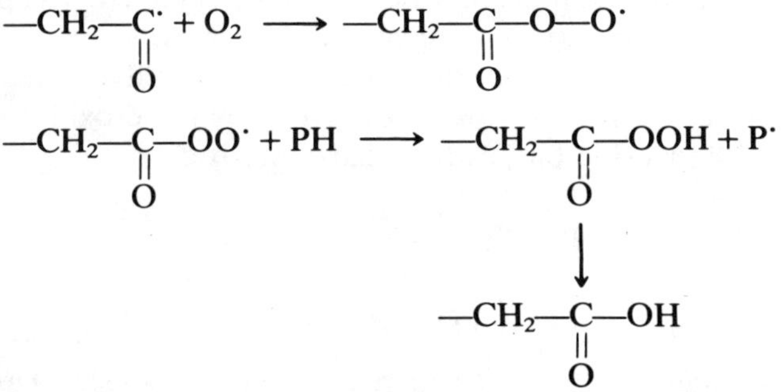

$$-CH_2-\underset{\underset{O}{\|}}{C}\cdot + O_2 \longrightarrow -CH_2-\underset{\underset{O}{\|}}{C}-O-O\cdot$$

$$-CH_2-\underset{\underset{O}{\|}}{C}-OO\cdot + PH \longrightarrow -CH_2-\underset{\underset{O}{\|}}{C}-OOH + P\cdot \qquad (7)$$

$$\downarrow$$

$$-CH_2-\underset{\underset{O}{\|}}{C}-OH$$

Because carboxylic acids are not photosensitive, they accumulate in the system and are one of the main oxidation products formed on lengthy UV exposure of polyethylene. Among the products of the Norrish type II reaction, only the methyl ketone is photosensitive. The vinyl groups are not very reactive and normally accumulate on polyethylene photo-oxidation.

2.2 Photolysis of Hydroperoxides

The classic photochemical decomposition of free hydroperoxides involving homolysis of the hydroperoxy bond was challenged recently.[13,14] Instead, it was suggested that bimolecular reactions of the hydroperoxide groups with neighboring chain segments from the same or another macromolecule could occur. Secondary and primary hydroperoxides are postulated to decompose mainly into ketones and aldehydes, respectively. This is shown for secondary hydroperoxides in reaction (8).

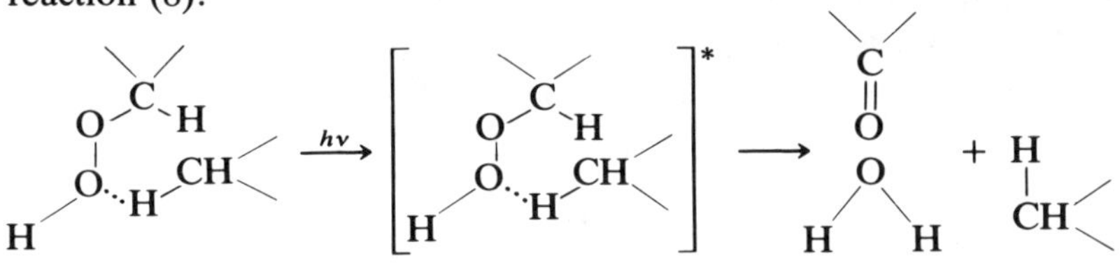

$$\begin{array}{c}\text{[structure with O-H···O and CH groups]} \xrightarrow{h\nu} [\text{excited intermediate}]^* \longrightarrow \text{C=O} + \text{H-CH} \end{array} \quad (8)$$

Tertiary hydroperoxides, also present in polyethylene to a minor extent, are assumed to react with a neighbouring chain segment according to a reaction involving a six-membered transition state to yield another product of polyethylene photo-oxidation: *trans*-vinylene groups [reaction (9)].

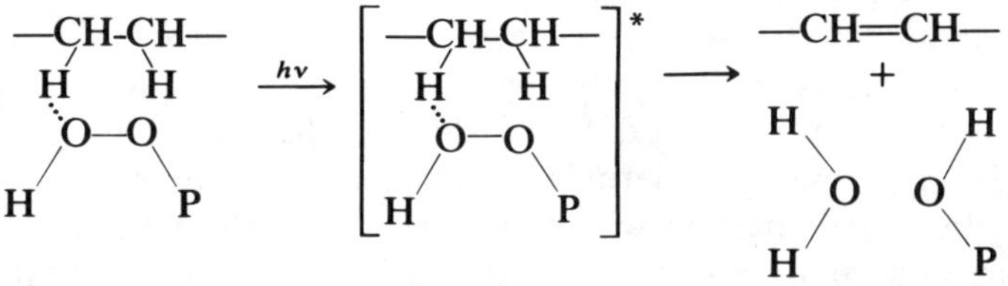

$$\begin{array}{c}\text{—CH-CH—}\\ \text{H} \quad \text{H}\\ \text{O—O}\\ \text{H} \quad \text{P}\end{array} \xrightarrow{h\nu} \left[\begin{array}{c}\text{—CH-CH—}\\ \text{H} \quad \text{H}\\ \text{O—O}\\ \text{H} \quad \text{P}\end{array}\right]^* \longrightarrow \begin{array}{c}\text{—CH=CH—}\\ +\\ \text{H} \quad \text{H}\\ \text{O} \quad \text{O}\\ \text{H} \quad \text{P}\end{array} \quad (9)$$

Reactions (8) and (9) account for the main photolysis products of hydroperoxides in LDPE.[13,14] They do not directly involve chain scissions. Intramolecular hydroperoxide decomposition reactions, also involving six-membered transition states and leading directly to chain scission have also been proposed.[14] The intramolecular decomposition envisaged for secondary hydroperoxides in polyethylene yields a vinyl group and an aldehyde group according to reaction (10).

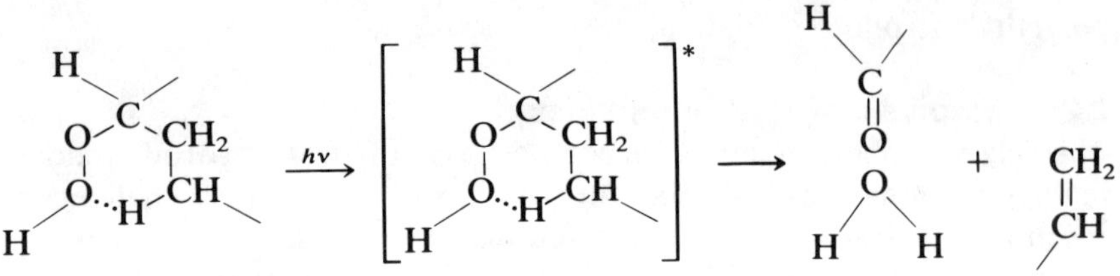

(10)

The corresponding decomposition process for tertiary hydroperoxides in polyethylene yields also a vinyl group and a ketone group according to reaction (11).

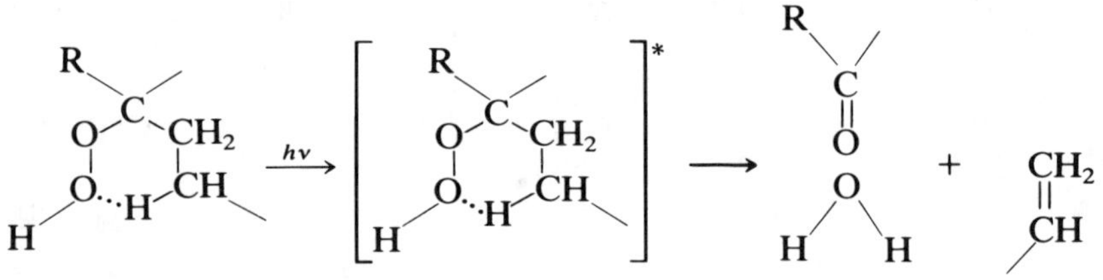

(11)

It is difficult to differentiate between reactions (10) and (11). It has been shown that a small amount of vinyl groups is found on hydroperoxide photolysis above 360 nm where ketone photolysis, according to the Norrish type II reaction, no longer occurs. However, aldehydes that would result from reaction (10) have not yet been identified unambiguously on polyethylene photo-oxidation. Furthermore, experiments with cut-off filters showed that the vinyl groups resulting directly from hydroperoxide photolysis originate mainly from tertiary hydroperoxides.[17] It was concluded that intramolecular decomposition of secondary hydroperoxides according to reaction (10) is of minor importance only in comparison with intramolecular decomposition of tertiary hydroperoxides according to reaction (11). This is thus convincing evidence that the proposed reactions of free hydroperoxides in polyethylene proceed primarily by a non-radical route, although the formation of radicals in a side reaction cannot be ruled out.

3 ULTRAVIOLET STABILIZATION OF POLYETHYLENE

Light stabilization of polyethylene has made considerable progress since the early years of outdoor use of polyethylene when only UV absorbers of the benzophenone and benzotriazole type were available for stabilization of thick sections as well as for thin sections. With the use of nickel stabilizers such as Ni-l,* the light stability of thin sections, especially that of LDPE films, could be increased considerably. The development of hindered-amine light stabilizers (HALS) was a big step forward in polyethylene UV stabilization. Since the first successful attempts to stabilize PE with low-molecular-weight HALS-1,[18] this class of compounds has been increasingly used for PE stabilization. Over the years, low-molecular-weight HALS such as HALS-1 have been complemented by high-molecular-weight HALS such as the polymeric HALS-2 and HALS-3. Today, HALS used alone or in combination with UV absorbers confer optimum stabilization to polyethylene.

3.1 High-Density Polyethylene (HDPE)

The use of HALS in HDPE tapes leads to pronounced improvement in light stability. This is shown in Table 1: HALS-1, HALS-2 and HALS-3 impart much better UV stability to HDPE tapes 50 μm thick than the benzophenone-type UV absorber UVA-1 used at three times the concentration, both in natural and white-pigmented formulations. HALS-3 performs best in this test series. The benzophenone-type UV absorber UVA-1 shows good performance in unpigmented and titanium dioxide-pigmented injection-moulded Ziegler HDPE 2-mm plaques (Table 2). However, HALS-1 and HALS-3 used at a much lower concentration exhibit considerable superiority over UVA-1, both in unpigmented and white-pigmented plaques. HALS-1 and HALS-3 show also pronounced advantage over HALS-2 in unpigmented plaques. This advantage becomes much less pronounced if the HALS are used in combination with UVA-2, and almost vanishes if white-pigmented plaques are considered. An additional interesting point emerging clearly in Table 2 is the fact that with titanium dioxide-pigmented samples, the UV absorber no longer contributes significantly to UV stability of HALS-containing samples.

* See the Appendix for structures of light stabilizers and antioxidants at the end of this chapter.

Table 1
UV stability of HDPE tapes[a]

UV stabilization	E_{50} (kLy)[b]	
	Unpigmented	0·4% TiO_2
Control	100	95
0·3 UVA-1	115	95
0·05% HALS-1	125	180
0·10% HALS-1	210	260
0·05% HALS-2	150	190
0·10% HALS-2	180	225
0·05% HALS-3	220	245
0·10% HALS-3	275	345

[a] Ziegler-type HDPE, +0·1% Ca stearate + 0·05% AO-1; tapes 50 μm thick, draw ratio 1:8·5; pigment, titanium dioxide (coated rutile).
[b] E_{50} = Energy to 50% tensile strength retention on Florida exposure, 45° south, direct (140 kLy year^{-1}), started Nov. 1980.

Table 2
UV stability of HDPE thick sections[a]

UV stabilization	E_{50} (kLy)[b]	
	Unpigmented	0·5% TiO_2
Control	40	60
0·3% UVA-1	140	150
0·05% HALS-1	370	570
0·05% HALS-1 + 0·05% UVA-2	630	650
0·05% HALS-2	70	550
0·05% HALS-2 + 0·05% UVA-2	330	580
0·05% HALS-3	320	600
0·05% HALS-3 + 0·05% UVA-2	470	670

[a] Ziegler-type HDPE + 0·2% Ca stearate + 0·03% AO-1; 2-mm injection-moulded dumbbells; pigment, titanium dioxide (coated rutile).
[b] E_{50} = Energy to 50% retained tensile impact strength on Florida exposure, 45° south, direct (140 kLy year^{-1}), started March 1981.

Table 3
Influence of pigments on UV stability of HDPE thick sections[a]

Light stabilization	E_{50} (kLy)[b]					
	Unpigmented	0·5% TiO_2	0·5% Chromium Yellow	0·5% Cadmium Red	0·5% Phthalo Green	0·5% Phthalo Blue
Control	39	58	59	59	57	61
0·3% UVA-1	140	140	295	280	340	225
0·3% UVA-2	240	210	280	330	295	190
0·1% HALS-1	440	640	790	660	520	720
0·1% HALS-2	135	550	580	620	460	480
0·1% HALS-3	320	680	600	630	660	>800

[a] Ziegler-type HDPE + 0·2% Ca stearate + 0·03% AO-1; 2-mm injection-moulded dumbbells.
[b] E_{50} = Energy to 50% retained tensile impact strength on Florida exposure, 45° south, direct (140 kLy year^{-1}), started March 1981.

Table 3 shows the influence of additional pigments on UV stability of injection-moulded Ziegler-HDPE. It can be seen that UV stability achieved with these pigments in combination with HALS is comparable with that observed with titanium dioxide. However, the performance of the UV absorbers UVA-1 and UVA-2 with these pigments is, as a rule, superior to that observed with unpigmented or titanium dioxide-pigmented samples.

3.2 Low-Density Polyethylene (LDPE)

The effects of type and concentration of the stabilizer on light stability of typical LDPE greenhouse films (200 µm thick) are shown in Table 4. It can be seen that the 'nickel quencher' Ni-1 is significantly superior to the UV absorber UVA-1 but inferior to the polymeric HALS-2. Nevertheless, the latter is not the optimum light stabilizer for LDPE films. The other polymeric HALS included in the test series, HALS-3, shows considerable improvement over HALS-2 across the whole concentration range under consideration.

In practice, nickel stabilizers and polymeric HALS are mostly used in combination with the benzophenone-type UV absorber UVA-1. This is done essentially for economical reasons with Ni-1, and for both economic reasons and enhanced performance with HALS-2. In Table 5 it can be seen that the 1:1 combination HALS-2/UVA-1 used at 0·3% imparts approximately the same UV stability to the films as the

Table 4
Influence of light stabilizer type and concentration on UV stability of LDPE blown films (200 μm)[a]

Light stabilizer concentration	E_{50} (kLy)[b]				
	Control	UVA-1	Ni-1	HALS-2	HALS-3
None	35	—	—	—	—
0·15%	—	90	140	170	440
0·30%	—	115	190	215	500
0·60%	—	160	285	330	700
1·20%	—	—	390	470	850

[a] LDPE homopolymer + 0·03% AO-1, films exposed without backing.
[b] E_{50} = Energy to 50% retained elongation on Florida exposure, 45° south, direct (140 kLy year^{-1}), started May 1980.

1:1 combination Ni-1/UVA-1 used at twice the concentration. The superiority of HALS-3/UVA-1 over the corresponding combination HALS-2/UVA-1 is less pronounced than that observed when HALS-2 and HALS-3 are used alone. This has to be attributed to a marked synergistic effect between HALS-2 and UVA-1, not observed either with HALS-3 or with Ni-1.

Another important feature emerging in Table 5 is the considerable influence of the backing of the films. In fact, greenhouse covers usually fail at the points of contact with the supporting structure. It has been shown[6] that the lifetime of the film on aluminium backing is, on average, only about two-thirds of that of Plexiglass or in the absence of any backing.

It should be kept in mind that the contact with the supporting structure is not the only factor reducing the lifetime of the films. With actual greenhouses, the mechanical strain on mounting and wind stress as well as the chemical effects of pesticides on the film and the stabilizers, may lead to a further pronounced stability decrease. Finally, compared with test results obtained under almost ideal conditions, much higher stabilizer concentrations may be required to achieve the desired film-lifetime with actual greenhouses.

Table 5
Influence of backing on light stability of LDPE films (200 μm)[a]

Light stabilization	E_{50} (kLy)[b]	
	Without backing	Aluminium backing
Control	40	32
0·3% Ni-1 + 0·3% UVA-1	290	180
0·15% HALS-2 + 0·15% UVA-1	260	170
0·3% HALS-2 + 0·3% UVA-1	370	230
0·15% HALS-3 + 0·15% UVA-1	400	300

[a] LDPE homopolymer + 0·03% AO-1.
[b] E_{50} = Energy to 50% retained elongation on Florida exposure, 45° south, direct (140 kLy year^{-1}), started March 1979.

Besides the aforementioned factors, film thickness also has a marked effect on light stability. This is exemplified in Table 6 for HALS-3 and a combination of HALS-3 with UVA-1. It can be seen that the performance of HALS-3 increases considerably with film thickness. Similar behaviour has been observed with the other UV stabilizers.[8]

Film thickness is also a primary factor with regard to the contribution to UV stability of the benzophenone-type UV absorber UVA-1

Table 6
Influence of film thickness on the performance of HALS-3 in blown films[a]

Light stabilization	E_{50} (kLy)[b]		
	50 μm	100 μm	200 μm
Control	27	31	34
0·15% HALS-3	190	260	440
0·15% HALS-3 + 0·15% UVA-1	200	295	535
0·30% HALS-3	280	445	500

[a] LDPE homopolymer + 0·03% AO-1, films exposed without backing.
[b] E_{50} = Energy to 50% retained elongation on Florida exposure, 45° south, direct (140 kLy year^{-1}), started May 1980.

used in combination with other light stabilizers. This is shown in Table 6 for HALS-3. In combination with HALS-3, UVA-1 has practically no effect on the light stability of a film 50 μm thick, but there seems to be a slight contribution of UVA-1 for the 100-μm film. However, for both the 50-μm and 100-μm films the stabilizing efficiency of the 1:1 combination is definitely inferior to that of HALS-3 used at the same concentration. It is only for films 200 μm thick that the contribution of UVA-1 becomes significant. Then the combination seems slightly superior to HALS-3 used at the same concentration but the difference observed may be due to experimental error.

It has already been mentioned that in practice nickel stabilizers and polymeric HALS are often used in combination with the benzophenone-type UV absorber UVA-1 for LDPE film stabilization. Benzotriazole-type UV absorbers such as UVA-3 are not used because their performance in LDPE films, when used alone, is inferior to that of the benzophenone-type UV absorber UVA-1. Recently it was found that, in combination with polymeric HALS, this is no longer true. In fact, the combinations of UVA-3 with HALS-3 outperform the corresponding combinations UVA-1/HALS-3 (Table 7).

Table 7
Effect of combinations of UVA-1 and UVA-3 with HALS-3 on light stability of LDPE blown films (200 μm)[a]

Light stabilization	E_{50} (kLy)[b]	
	Without backing	On aluminium
0·15% HALS-3	400	310
0·30% HALS-3	500	450
0·15% HALS-3 + 0·15% UVA-1	465	365
0·15% HALS-3 + 0·15% UVA-3	540	475
0·30% HALS-3 + 0·30% UVA-1	625	475
0·30% HALS-3 + 0·30% UVA-3	725	625
0·60% HALS-3 + 0·60% UVA-1	675	475
0·60% HALS-3 + 0·60% UVA-3	>800 (75%)	≥750

[a] LDPE homopolymer + 0·03% AO-1.
[b] E_{50} = Energy to 50% retained elongation on Florida exposure, 45° south, direct (140 kLy year^{-1}), started May 1981.

In an EVA copolymer the ranking of the UV stabilizers according to their efficiency is the same as that in LDPE homopolymer, i.e.

HALS-3 > HALS-2 > Ni-1 > UVA-1

However, the superiority of the polymeric HALS over the nickel quencher Ni-1 and the UV absorber UVA-1 is even more pronounced in EVA films. It is as advantageous to use combinations of Ni-1 or HALS-2 with UVA-1 for EVA as for LDPE homopolymer. Again, the combinations with HALS-2 are much superior to the corresponding combinations with Ni-1 (Table 8). The combinations of HALS-3 with UVA-1 show by far the best performance (Table 8).

Mineral fillers (talc, chalk, kaolin, etc.) may be added to LDPE as opacifiers or infrared barriers in greenhouse films. Their effect on light stability has to be checked. The detrimental effect of china clay (kaolin) on UV stability of LDPE films is well documented.[7] This is observed with all light-stabilizer classes. Nevertheless, it is possible to achieve excellent UV stability by choosing an adequate stabilizer system. This is shown quite convincingly with HALS-3 in Table 9.

Preliminary results with unpigmented 2-mm LDPE dumbbells show

Table 8
Light stability of EVA copolymer blown films (200 μm)[a]

Light stabilization	E_{50} (kLy)[b]	
	Without backing	Aluminium backing
Control	50	40
0.15% Ni-1 + 0.15% UVA-1	170	110
0.3% Ni-1 + 0.3% UVA-1	260	210
0.15% HALS-2 + 0.15% UVA-1	260	190
0.3% HALS-2 + 0.3% UVA-1	385	315
0.15% HALS-3 + 0.15% UVA-1	580	580

[a] EVA copolymer (14% VA) + 0.03% AO-1.
[b] E_{50} = Energy to 50% retained elongation on Florida exposure, 45° south, direct (140 kLy year^{-1}), started March 1981.

Table 9
Effect of HALS concentration in the presence of china clay on UV stability of LDPE blown films (200 μm)[a]

Light stabilization	E_{50} (kLy)[b]	
	Without backing	Aluminium backing
Control	24	20
0·15% HALS-3	215	180
0·30% HALS-3	365	270
0·60% HALS-3	440	380
1·20% HALS-3	710	600

[a] LDPE homopolymer + 0·03% AO-1 + 3% china clay.
[b] E_{50} = Energy to 50% retained elongation on Florida exposure, 45° south, direct (140 kLy year^{-1}), started January 1982.

rather high performance for the benzophenone-type UV absorber UVA-1 on outdoor weathering (Table 10). Again, the polymeric HALS-2 and HALS-3 are markedly more effective than the UV absorber. Yet, even after five years' exposure in Florida, the superiority of the HALS cannot be estimated correctly.

Table 10
UV stability of LDPE thick sections

Light stabilization	E_{50} (kLy)[b]
Control	95
0·1% UVA-1	265
0·2% UVA-1	335
0·1% HALS-2	700
0·1% HALS-2 + 0·1% UVA-1	>900
0·1% HALS-3	≥900
0·1% HALS-3 + 0·1% UVA-1	>900

[a] LDPE homopolymer + 0·03% AO-1; 2-mm injection-moulded dumbbells, unpigmented.
[b] E_{50} = Energy to 50% retained elongation on Florida exposure, 45° south, direct (140 kLy year^{-1}), started November 1981.

3.3 Linear Low-Density Polyethylene (LLDPE)

The light stabilizers used for LLDPE are the same as those used for LDPE, i.e. UVA-1, Ni-1 and the polymeric HALS-2 and HALS-3. The performance of combinations of HALS-2 and HALS-3 with UVA-1 in a butene copolymer is shown in Table 11. The data in Table 12 show clearly that, with HALS-based stabilization systems, the light

Table 11
Light stability of LLDPE blown films (200 μm)[a]

Light stabilization	E_{50} (kLy)[b]
Control	30
0·15% HALS-2 + 0·15% UVA-1	240
0·25% HALS-2 + 0·25% UVA-1	320
0·15% HALS-3 + 0·15% UVA-1	435
0·25% HALS-3 + 0·25% UVA-1	510

[a] LLDPE (butene copolymer), films exposed on Plexiglass.
[b] E_{50} = Energy to 50% retained elongation on Florida exposure, 45° south, direct (140 kLy year^{-1}), started May 1981.

stabilities achieved with films manufactured from conventional LDPE, EVA copolymer or LLDPE are comparable. In this connection it should be remembered that, the UV stabilizer system remaining the same, UV stability increases with film thickness. Because LLDPE shows improved mechanical properties over LDPE homopolymer, it is often possible to reduce film thickness by using LLDPE instead of LDPE. Then, the resulting loss in UV stability should be compensated for by an improved stabilization system.[7]

4 MECHANISMS OF ULTRAVIOLET STABILIZATION

The general principles of UV stabilization seem to be well established.[5,8,19–24] In fact, UV absorption, excited-state quenching, peroxide decomposition and free-radical scavenging are said to ac-

Table 12
Comparison of the light stability of LDPE, EVA and LLDPE blown films (200 μm)[a]

Light stabilization	E_{50} (kLy)[b]			
	LDPE-1	EVA	LDPE-2	LLDPE
Control	40	50	25	30
0·15% HALS-2 + 0·15% UVA-1	285	260	260	240
0·15% HALS-3 + 0·15% UVA-1	500	580	400	435

[a] LDPE-1 (homopolymer) + 0·03% AO-1 } without backing.
EVA copolymer (14% VA) + 0·03% AO-1 }
LDPE-2 (homopolymer) } on Plexiglass.
LLDPE (butene copolymer) }
[b] E_{50} = Energy to 50% retained elongation on Florida exposure, 45° south, direct (140 kLy year^{-1}); started March 1981 for LDPE-1 and EVA, May 1981 for LDPE-2 and LLDPE.

count for the protection afforded by the different types of UV stabilizers. It should be kept in mind that some light stabilizers or light-stabilizer classes can be active in more than one way.

In the following sections, some specific aspects of these stabilization mechanisms will be taken into consideration with special emphasis on HALS and 2-hydroxyphenylbenzotriazole UV absorbers.

4.1 Quenching of Carbonyl Compounds

4.1.1 Effect of HALS

A possibility of deactivation of excited ketone carbonyls in polyethylene by HALS was proposed some years ago.[9] This involves hydrogen abstraction from the hindered amine by the excited ketone carbonyl, in analogy with the reactions observed with low-molecular-weight ketones and amines. For secondary hindered amines the reaction can be depicted schematically as shown below [reaction (12)]:[9]

$$\text{>C=O}^* + \text{H—N<} \longrightarrow \left[\text{>C=O} \cdots \text{H—N<} \longleftrightarrow \text{>C—O}^{\ominus}\text{H—N}^{\cdot\oplus}\text{<} \right]$$

$$\diagup\begin{matrix} \text{>C=O} + \text{H—N<} \\ \text{>C—OH} + \text{·N<} \end{matrix} \qquad (12)$$

The recombination of the radicals formed in reaction (12) has also been envisaged to explain the stabilizing effect of HALS.[9]

Quenching of acetone singlet and triplet states by hindered amines in solution was shown recently.[25] Model piperidine derivatives, as well as commercial HALS, such as HALS-1, HALS-2, HALS-3, etc., do, in fact, quench singlet and triplet states of acetone. Hindered amines are even better quenchers for the singlet state than for the triplet state. The conclusion is that the Norrish type I and type II reactions are efficiently quenched by hindered piperidine model compounds and commercial HALS.[25]

The effect of HALS on ketone photolysis has been checked in LDPE films containing a high amount of a model dialkyl ketone (1·2% stearone) and increasing amounts of a commercial HALS especially effective in LDPE (HALS-3). The variation of the ketone carbonyl absorbance as a function of exposure time is plotted in Fig. 1. As can be seen, the rate of photolysis is independent of the HALS concentration and, within experimental error, the same as that observed in the absence of HALS during the early exposure period. After this early period, of course, photo-oxidation of the film that does not contain a UV stabilizer leads again to an increase of the carbonyl group concentration. The plot of the absorbance of the vinyl groups formed on ketone photolysis as a function of exposure time in Fig. 2 leads to similar conclusions: the rate of Norrish type II reaction is independent of the HALS concentration and the same as that observed in the absence of HALS-3 in the early stages of photolysis. The ketone concentration used in the experiments is such that polyethylene films that have reached this level of carbonyl content through photo-oxidation have already failed. Therefore, it can be concluded from the results in Figs 1 and 2 that quenching of ketone photolysis by HALS does not contribute significantly, if at all, to HALS stabilization of polyolefins in the solid state.

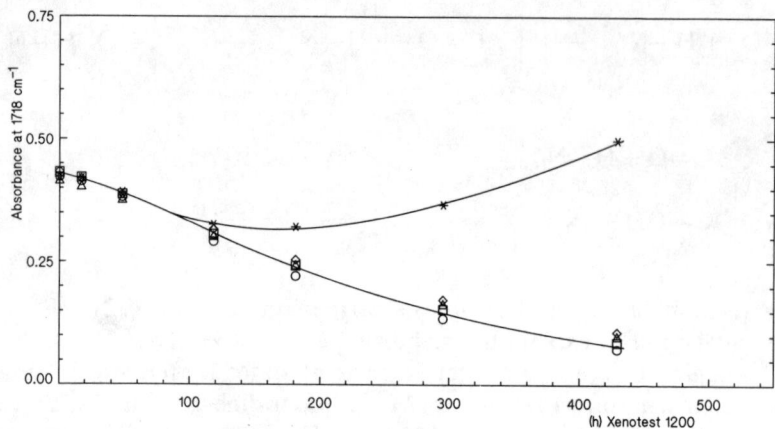

Fig. 1. Effect of HALS on ketone photolysis: 1·2% stearone + 0·1% AO-2 in 0·5-mm LDPE films. ∗, Without HALS-3; ○, 0·15% HALS-3; □, 0·3% HALS-3; △, 0·60% HALS-3; ◇, 1·20% HALS-3.

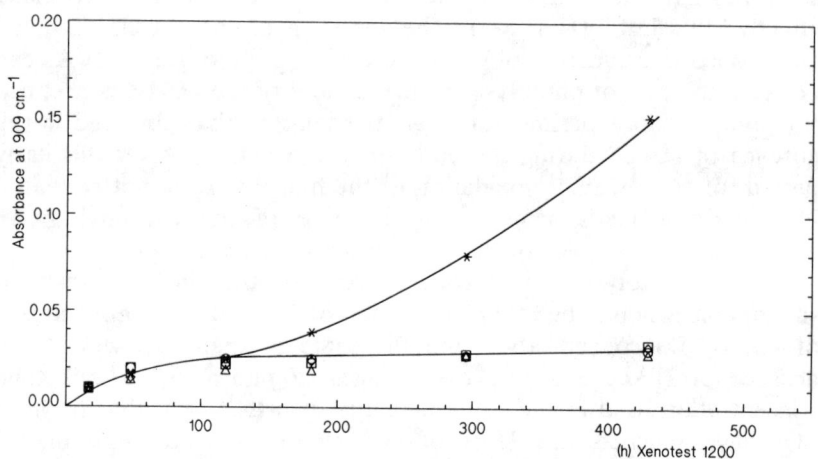

Fig. 2. Effect of HALS on vinyl group formation during ketone photolysis: 1·2% stearone + 0·1% AO-2 in 0·5-mm LDPE films. ∗, Without HALS-3; ○, 0·15% HALS-3; □, 0·3% HALS-3; △, 0·60% HALS-3; ◇, 1·20% HALS-3.

4.1.2 Effect of Benzotriazole-Type Ultraviolet Absorbers

Quenching of carbonyl compounds by typical UV absorbers such as 2-hydroxyphenylbenzotriazoles by long-range energy transfer (Förster mechanism) has been postulated several times.[5,8] However, this has never been shown unequivocally, and in fact has even been excluded on theoretical grounds.[26]

In this connection it is noteworthy that shielding of unstabilized LDPE films by LDPE films stabilized with benzotriazole-type UV absorbers protects polyethylene mainly by UV absorption. This is in contrast with the behaviour of the benzophenone-type UV absorber UVA-1. For this compound it was found that the contribution of UV absorption to the stabilization mechanisms is not important in LDPE and that free-radical scavenging may account for most of the stabilization effects,[6,19,27,28] although other workers have disputed this.[29] There are numerous data obtained with polyolefins containing benzotriazole-type UV absorbers, either alone or in combination with HALS, that can hardly be explained by UV absorption alone (see for example Tables 2 and 7). In fact, screening experiments with LDPE films containing the model dialkyl ketone stearone exposed behind films containing stearone and the benzotriazole-type UV absorber UVA-3 reveal a slightly accelerated decrease of ketone concentration in the shielded film (Fig. 3). Because the energy absorbed by the

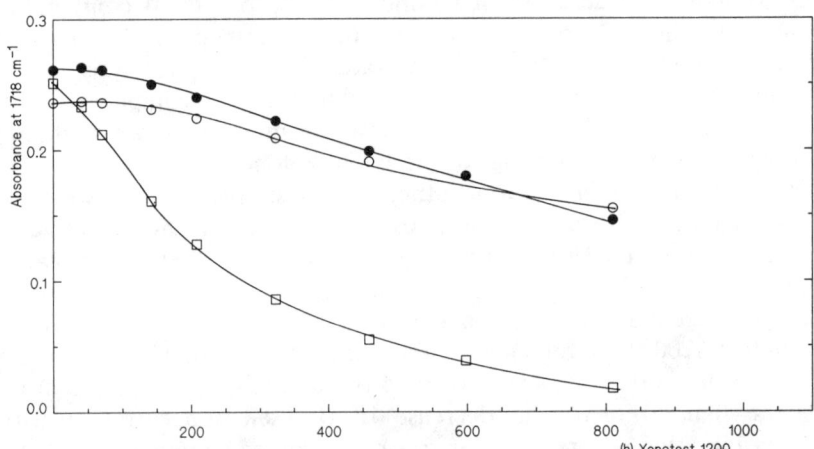

Fig. 3. Comparison of ketone photolysis in stabilized and shielded films: 1·2% stearone + 0·15% HALS-3 in 0·5-mm LDPE films. ○, 0·05% UVA-3; ●, without UVA-3, shielded by preceding film; □, without UVA-3, not shielded.

ketone in the shielded film is much smaller than that absorbed by the ketone in the stabilized film, the conclusion is that ketone photolysis in the UV absorber-containing film is reduced by quenching of ketone excited states by the benzotriazole-type UV absorber UVA-3.

However, these results do not give any clear indication concerning the exact nature of the energy transfer mechanisms in the system under study.

To obtain a better understanding of energy transfer and prove it unambiguously, an experimental set-up used by Förster[30] and previously by Perrin,[31] to prove long-range energy transfer in solution, was adapted to polymer films. Perrin and Förster used cells of thickness inversely proportional to the concentration of fluorescent dye and quencher in solutions obtained by dilution of a mother solution.

The benzotriazole-type UV absorber UVA-3 and the dialkyl ketone were compounded into LDPE in a Brabender plastograph. The ratio ketone/UV absorber was kept constant through the experiment, but the overall concentration was doubled from one blend to the next. The thickness of the films prepared from these blends was accordingly halved on passing from one blend to the following. Then, accepting the validity of the Lambert–Beer law, the UV energy absorbed by the ketone or the UV absorber is the same in all the films. Preliminary results show that decrease of ketone groups proceeds at comparable rates in the films. Because the methyl ketone formed on photolysis of stearone is rather volatile, a lower rate of ketone photolysis in the thinner films (as a consequence of quenching) may be compensated for by a higher rate of loss of the photolysis product whose carbonyl absorption overlaps with that of the initial ketone.

For this reason, monitoring other photolysis products of stearone appears more rewarding. This is especially true for carboxylic acids resulting from the Norrish type I reaction. In the films investigated they are detected as salts with the amine groups of HALS-3. The plot in Fig. 4 shows the salt concentration after 210 h of exposure in Xenotest 1200 as a function of the concentration of UVA-3 in the films. Because the UV light absorbed by the ketone is the same in all of the films studied, the decrease of the salt concentration with increasing UVA-3 concentration can be attributed to quenching of the Norrish type I reaction by the UV absorber. It can be seen in Fig. 4 that a sharp drop of the salt concentration occurs when the concentration of the benzotriazole-type UV absorber is increased from 0·0125%

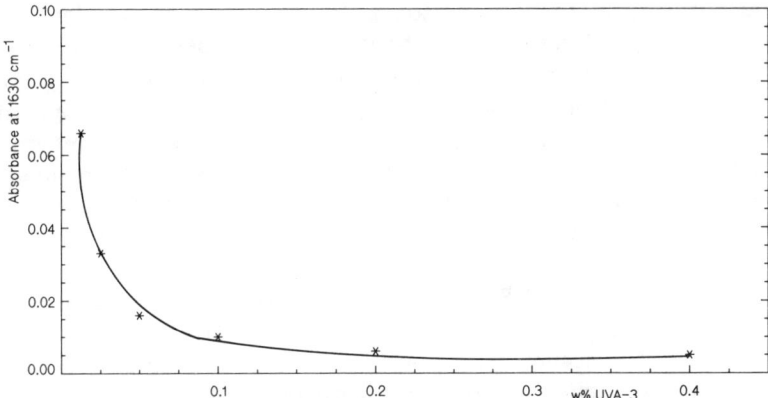

Fig. 4. Concentration of carboxylic acid salts after 210 h of Xenotest 1200 exposure: LDPE homopolymer + 0·1% AO-2 + 0·15% HALS-3 + stearone. Sample thickness and concentrations inversely proportional for isoabsorbance.

to 0·05%. For 0·05% UVA-3, the Norrish type I reaction is quenched to a large extent. Assuming a crystallinity of 50% for LDPE and that ketone, as well as UV absorber, are concentrated in the amorphous phase, it can be deduced from the preceding results that deactivation of the excited states responsible for the Norrish type I reaction occurs over distances of up to 50 Å. This finding leads to the conclusion that the Norrish type I reaction of ketones can be quenched by benzotriazole-type UV absorbers according to the long-range energy transfer mechanism (Förster mechanism). This contradicts the theoretical predictions. Moreover, it can also be concluded that the Norrish type I reaction of a dialkyl ketone is quenched to a large extent by benzotriazole-type UV absorbers used at concentrations as low as 0·05%.

4.1.3 Synergism of HALS/Benzotriazole-Type Ultraviolet Absorbers

It has been shown that HALS-3, a very efficient UV stabilizer for polyethylene, does not quench the photolysis of a model dialkyl ketone of the type formed on polyethylene photo-oxidation. It has also been shown that benzotriazole-type UV absorbers are very efficient quenchers of the Norrish type I reaction. Preliminary results indicate that they also quench the Norrish type II reaction. In addition to their quenching ability they are also, as is well-known, excellent UV absorbers. Thus, benzotriazole-type UV absorbers stabilize very efficiently according to mechanisms totally different from the HALS stabilization mechanisms usually postulated. As a result, the protective power of combinations HALS/benzotriazole-type UV absorbers is more than additive; it is in fact multiplicative. This explains the unique properties of combinations of HALS/benzotriazole-type UV absorbers and also the unexpected synergism observed in polyethylene films.

4.2 Some Aspects of Polyethylene Stabilization with HALS

The efficiency of polyethylene stabilization with HALS regarding the retention of the mechanical properties has been shown already quite convincingly in Section 3 above. In this section, some chemical aspects of HALS stabilization are examined in detail. Afterwards, the fate of HALS-1 on polyethylene photo-oxidation will be discussed.

4.2.1 Chemical Changes in Polyethylene on Photo-oxidation

The chemical changes in 2-mm injection-moulded HDPE plaques on outdoor exposure in Florida have been monitored by IR spectroscopy. The most apparent changes are observed in the spectral regions of the carbonyl group and double-bond absorption. The absorbances at 1718 cm^{-1} (ketone groups), 909 cm^{-1} (vinyl groups) and 967 cm^{-1} (*trans*-vinylene groups) have been plotted as a function of the energy received on Florida exposure.

For the control sample, carbonyl, vinyl and *trans*-vinylene group absorbance develops without an apparent induction period, as shown in Fig. 5. With samples stabilized with HALS such as HALS-1 and HALS-3, carbonyl groups form after a prolonged induction period only. However, formation of vinyl and *trans*-vinylene groups proceeds without induction period. This is illustrated in Fig. 6, where the development of carbonyl, vinyl and *trans*-vinylene groups for 2-mm plaques stabilized by 0·05% HALS-3 is plotted as a function of the energy received on Florida exposure. It can be seen that the rates of formation of vinyl and *trans*-vinylene groups are constant during the induction period of carbonyl group formation and increase after the end of that induction period. Similar results are observed for 0·1% and 0·2% HALS-3 as well as for HALS-1 and HALS-2.

The development of carbonyl, vinyl and *trans*-vinylene groups with exposure time is shown in Figs 7, 8 and 9 respectively for all the concentrations of HALS-3 tested. As expected, Fig. 7 shows that the

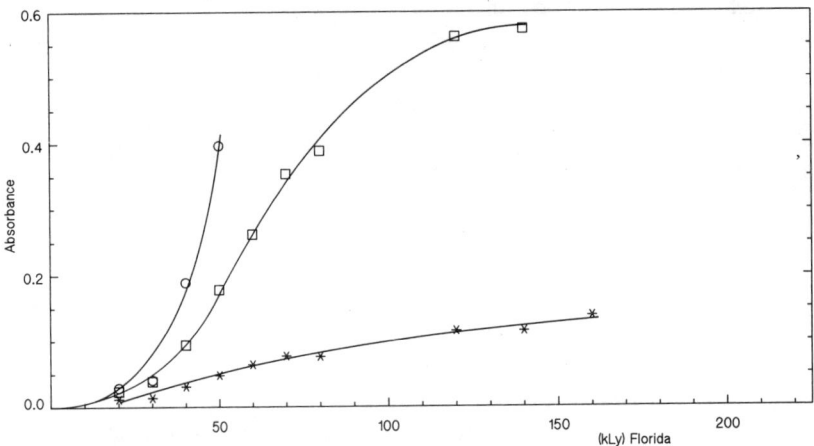

Fig. 5. Chemical groups formed on outdoor exposure of 2-mm HDPE plaques without UV stabilizer. ○, Carbonyl (1718 cm^{-1}); □, vinyl (909 cm^{-1}); ∗, *trans*-vinylene (967 cm^{-1}).

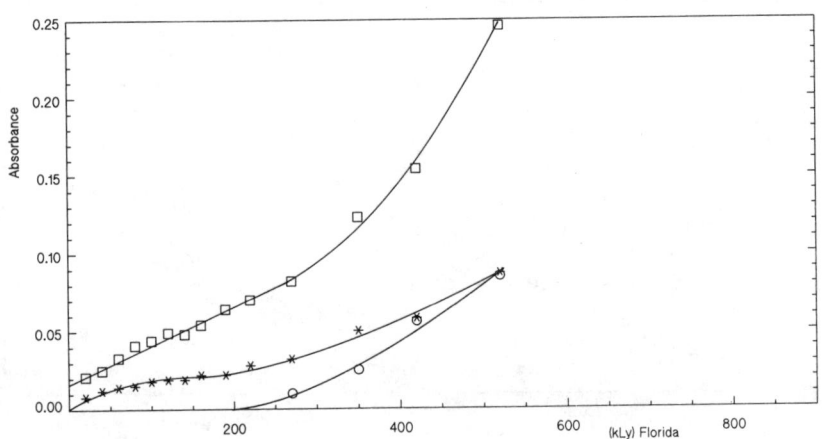

Fig. 6. Chemical groups formed on outdoor exposure of 2-mm HDPE plaques stabilized with 0·05% HALS-3. ○, Carbonyl (1718 cm^{-1}); □, vinyl (909 cm^{-1}); ∗, *trans*-vinylene (967 cm^{-1}).

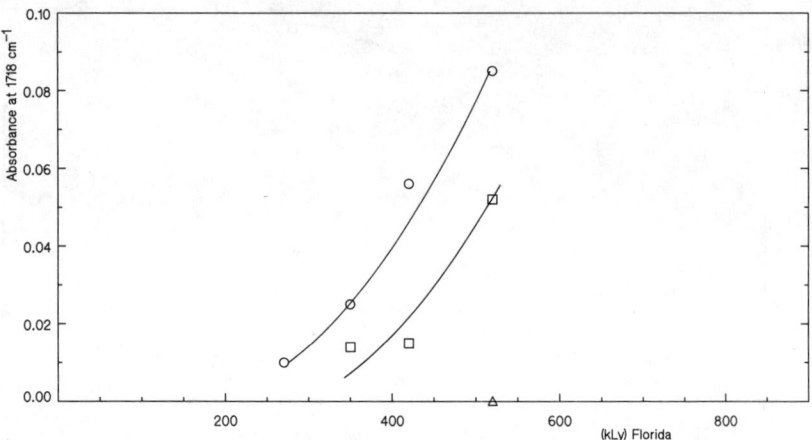

Fig. 7. Carbonyl group formation on outdoor exposure of 2-mm HDPE plaques stabilized with HALS-3. ○, 0·05% HALS-3; □, 0·10% HALS-3; △, 0·20% HALS-3.

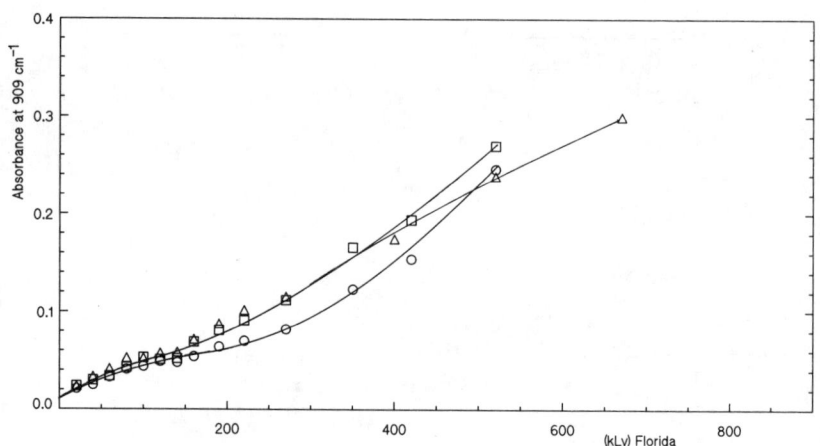

Fig. 8. Vinyl group formation on outdoor exposure of 2-mm HDPE plaques stabilized with HALS-3. ○, 0·05% HALS-3; □, 0·10% HALS-3; △, 0·20% HALS-3.

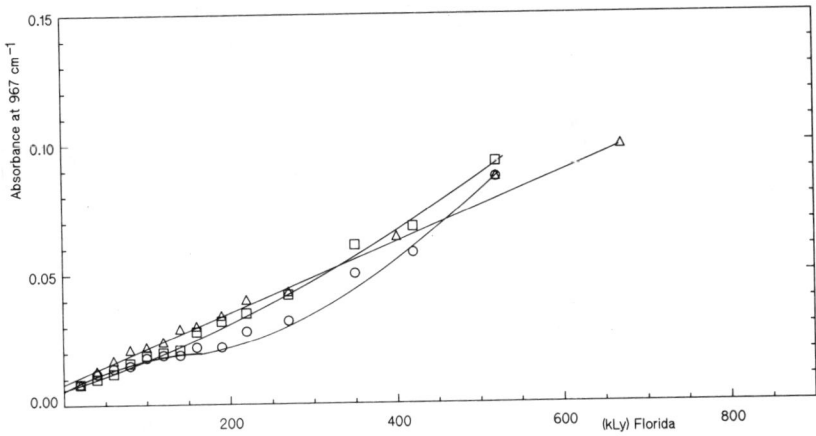

Fig. 9. *trans*-Vinylene group formation on outdoor exposure of 2-mm HDPE plaques stabilized with HALS-3. ○, 0·05% HALS-3; □, 0·10% HALS-3; △, 0·20% HALS-3.

induction period of formation of carbonyl groups increases with HALS-3 concentration. The initial rates of vinyl and *trans*-vinylene group formation do not decrease with increasing HALS-3 concentration; on the contrary, they increase distinctly with HALS-3 concentration (Figs 8 and 9). Similar results are observed with HALS-1. It can be seen in Fig. 10 that the rate of vinyl group formation remains constant up to very high exposure times, even for a concentration of HALS-1 as low as 0·05%. Only for HALS-2 do the initial rates of vinyl group formation not increase with the stabilizer concentration. A decrease is even observed when HALS-2 concentration is raised from 0·05% to 0·1% (Fig. 11). The initial rate of vinyl group formation has been plotted in Fig. 12 as a function of the concentrations of HALS-1, HALS-2 and HALS-3. For comparison purposes the initial rate observed with the control is also reported in Fig. 12. It can be seen that, within experimental error, the initial rate of vinyl group formation is the same for the secondary hindered amines HALS-1 and HALS-3. The tertiary hindered amine HALS-2 shows a different behaviour. The initial rate of *trans*-vinylene group formation shows similar variations.

In the absence of carbonyl groups, vinyl groups are by far the most important groups observed in polyethylene with respect to

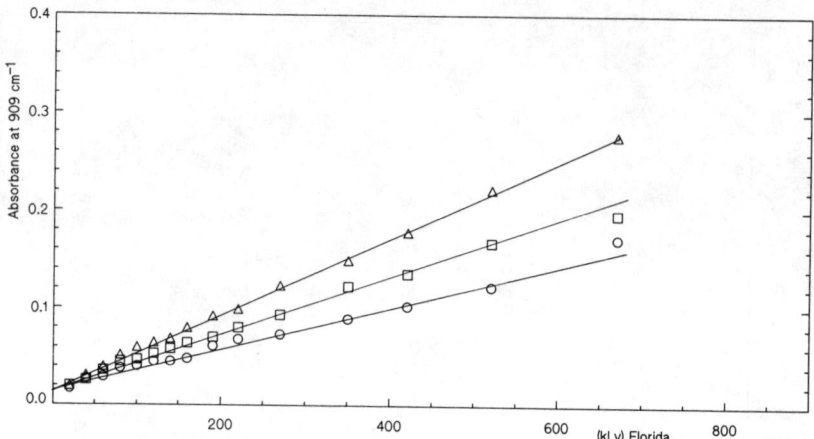

Fig. 10. Vinyl group formation on outdoor exposure of 2-mm HDPE plaques stabilized with HALS-1. ○, 0·05% HALS-1; □, 0·10% HALS-1; △, 0·20% HALS-1.

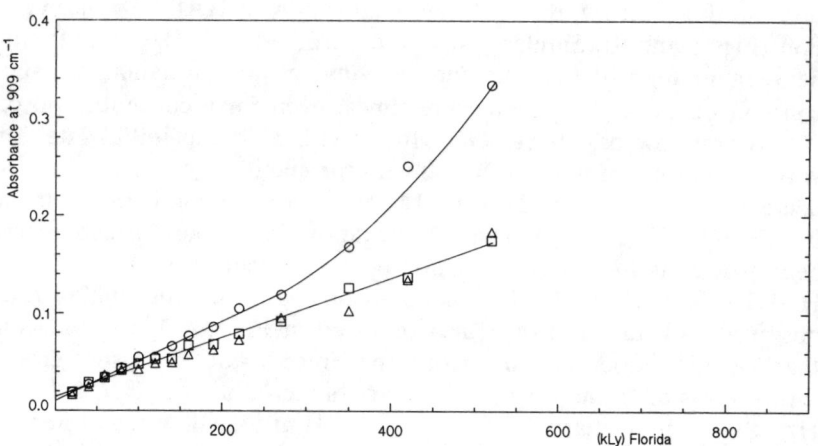

Fig. 11. Vinyl group formation on outdoor exposure of 2-mm HDPE plaques stabilized with HALS-2. ○, 0·05% HALS-2; □, 0·10% HALS-2; △, 0·20% HALS-2.

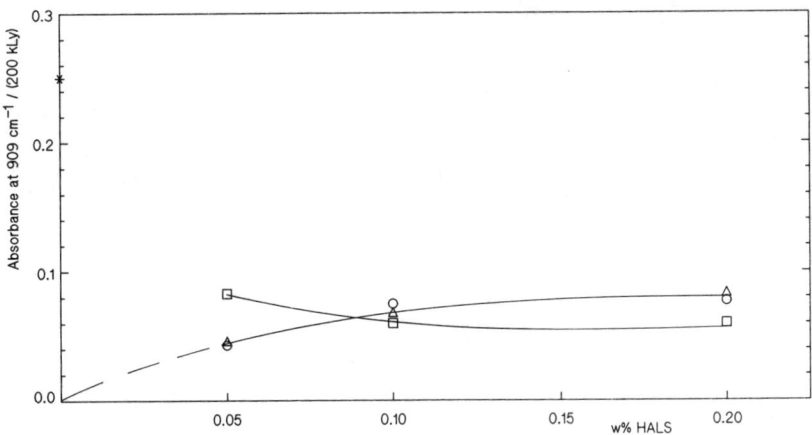

Fig. 12. Initial rate of vinyl group formation as a function of HALS structure and concentration. ○, HALS-1; □, HALS-2; △, HALS-3; *, control.

physical/mechanical properties. As a matter of fact, vinyl groups correspond to chain scissions whereas *trans*-vinylene groups do not. Thus, it is straightforward to compare the loss of mechanical properties with the development of vinyl groups. This comparison is shown in Figs 13, 14 and 15 for 2-mm plaques stabilized with HALS-1, HALS-2 and HALS-3. It can be seen that there is rough parallelism between loss of tensile impact strength and build-up of vinyl groups. However, it can also be seen that 50% retained tensile impact strength cannot be attributed to a constant amount of vinyl groups, i.e. a definite number of chain scissions. This becomes obvious in Fig. 16, where it is shown that the vinyl group concentration corresponding to 50% loss of the initial tensile impact strength is a function of both HALS structure and concentration. For the secondary hindered piperidines HALS-1 and HALS-3, the concentrations (vinyl)$_{50}$ are the same within experimental error. They are also much higher than the corresponding concentrations found with the tertiary HALS-3. The marked increase of (vinyl)$_{50}$ with HALS-1 or HALS-3 concentration is noteworthy. Similar variations are observed if the *trans*-vinylene group concentration corresponding to 50% loss of the initial tensile impact strength is plotted as a function of HALS concentration (Fig. 17). However, the concentration of *trans*-vinylene

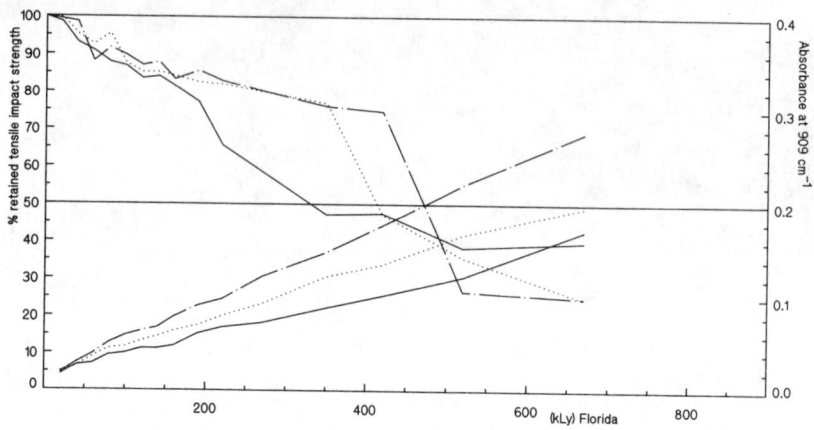

Fig. 13. Comparison of loss of tensile impact strength with formation of vinyl groups for 2-mm HDPE plaques stabilized with HALS-1. —, 0·05% HALS-1; ···, 0·10% HALS-1; —·—, 0·20% HALS-1.

groups with HALS-1 is slightly, but consistently, higher than that with HALS-3.

The chemical changes observed on polyethylene photo-oxidation have to be explained. It was already shown that vinyl groups can result from the Norrish type II reaction [reaction (6)] as well as from the

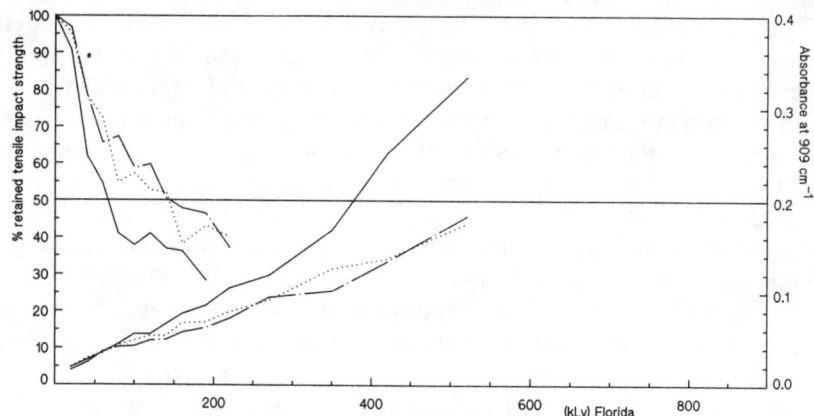

Fig. 14. Comparison of loss of tensile impact strength with formation of vinyl groups for 2-mm HDPE plaques stabilized with HALS-2. —, 0·05% HALS-2; ···, 0·10% HALS-2; —·—, 0·20% HALS-2.

PHOTO-OXIDATION AND STABILIZATION OF POLYETHYLENE 197

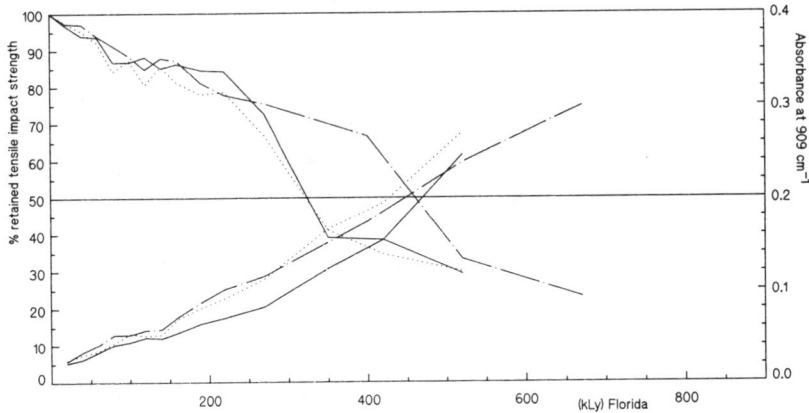

Fig. 15. Comparison of loss of tensile impact strength with formation of vinyl groups for 2-mm HDPE plaques stabilized with HALS-3. —, 0·05% HALS-3; ···, 0·10% HALS-3; –··–, 0·20% HALS-3.

intramolecular hydroperoxide decomposition reactions (10) and (11). Furthermore, it was shown that *trans*-vinylene groups can result from the bimolecular hydroperoxide decomposition reaction (9).

In the presence of HALS other reactions can still be considered, at least in principle. In fact, stabilization of polyolefins by HALS is generally attributed to the sequence of reactions (13)–(15).

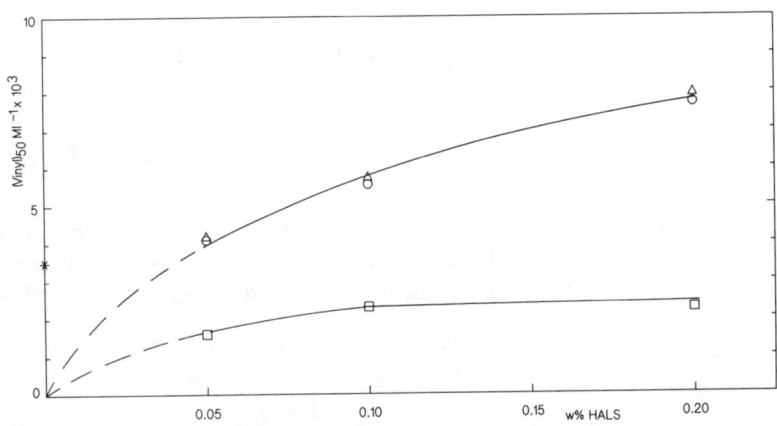

Fig. 16. Vinyl group concentration after loss of 50% tensile impact strength, $(Vinyl)_{50}$ of 2-mm HDPE plaques as a function of HALS structure and concentration. ∗, Control; ○, HALS-1; □, HALS-2; △, HALS-3.

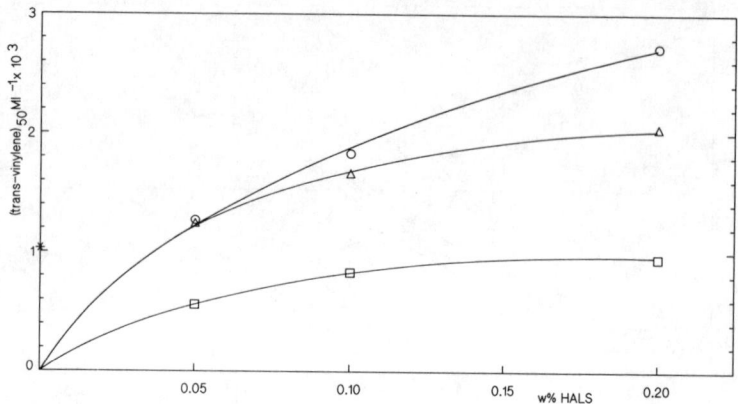

Fig. 17. *trans*-Vinylene group concentration after loss of 50% tensile impact strength, (*trans*-vinylene)$_{50}$ of 2-mm HDPE plaques as a function of HALS structure and concentration. *, Control; ○, HALS-1; □, HALS-2; △, HALS-3.

The reaction of peroxyl radicals with hydroxylamine ethers may also proceed according to reaction (16):[32]

$$\text{>NH} \longrightarrow \text{>N—O}^{\cdot} \tag{13}$$

$$\text{>N—O}^{\cdot} + \text{P}^{\cdot} \longrightarrow \text{>N—OP} \tag{14}$$

$$\text{>N—OP} + \text{PO}_2^{\cdot} \longrightarrow \text{>N—O}^{\cdot} + \text{POOP} \tag{15}$$

$$\text{>N—OP} + \text{PO}_2^{\cdot} \longrightarrow \text{>N—O}^{\cdot} + {-}\overset{|}{\text{C}}{=}\overset{|}{\text{C}}{-} + \text{POOH} \tag{16}$$

This reaction is even faster than the preceding one. With nitroxyl radicals along the polyethylene chain the reaction can explain the formation of *trans*-vinylene groups according to reaction (17) and with

$$\begin{array}{c} -\text{CH}_2-\text{CH}-\text{CH}_2-\text{CH}_2- + \text{RO}_2^{\cdot} \\ | \\ \text{O} \\ | \\ \text{N} \\ \diagup \diagdown \\ \downarrow \\ -\text{CH}_2-\text{CH}{=}\text{CH}-\text{CH}_2- + \text{RO}_2\text{H} + \text{>NO}^{\cdot} \end{array} \tag{17}$$

nitroxyl radicals at the chain ends, the reaction can also explain the formation of vinyl groups according to reaction (18).

$$—CH_2CH_2—O—N{<} + RO_2^{\cdot}$$
$$\downarrow \qquad (18)$$
$$—CH{=}CH_2 + RO_2H + {>}NO^{\cdot}$$

However, it was found that these reactions are rather slow, even above 100°C.[32,33] Moreover, regeneration of nitroxyls from O-(primary alkyl)hydroxylamines such as that involved in the last reaction (18) does not occur even at 130°C.[33] Even O-(secondary alkyl)hydroxylamines such as those involved in reaction (17) were found to be thermally stable at 60°C.[33]

Another reaction has been proposed recently to account for the effect of tetramethylpiperidinyl derivatives on the AIBN-initiated oxidation of cyclohexane in chlorobenzene solution.[34] [Reaction (19).]

Formation of *trans*-vinylene groups by catalytic reactions of nitroxyl radicals has also been envisaged. As an example, it has been suggested that regeneration of inhibitors on hydrocarbon oxidation may in principle proceed by reactions (20) and (21).[36]

$${>}NOC_6H_{11} + C_6H_{11}OO^{\cdot} \longrightarrow {>}NO^{\cdot} + C_6H_{10}O + C_6H_{11}OH \quad (19)$$

$${>}NO^{\cdot} + —CH_2—\overset{\overset{OO^{\cdot}}{|}}{CH}— \longrightarrow {>}NOH + —CH{=}CH— + O_2 \quad (20)$$

$${>}NO^{\cdot} + —\dot{C}H—CH_2— \longrightarrow {>}NOH + —CH{=}CH— \quad (21)$$

However, other reactions should be considered to explain the formation of vinyl and *trans*-vinylene groups. The presence of the O-(secondary alkyl)hydroxylamines points to another possibility of reaction with a peroxyl radical. In fact, the tertiary hydrogen in a position α to an ether group seems particularly suited for attack by a peroxyl radical. A possible reaction sequence is shown in Scheme 1. The tertiary hydroperoxide groups formed can decompose photolytically according to the bimolecular reaction shown [reaction (23)] to yield a *trans*-vinylene group and an alcohol. The latter decomposes into a ketone and a hydroxylamine [reaction (24)]. Photolysis of this

$$-CH_2-CH-CH_2- + RO_2^{\cdot}$$
$$\underset{O-N<}{|}$$
$$\downarrow$$
$$-CH_2-\overset{\cdot}{C}-CH_2- + ROOH$$
$$\underset{O-N<}{|} \quad \Big| O_2, PH \quad\quad\quad\quad (22)$$
$$\underset{OOH}{|}$$
$$\downarrow$$
$$-CH_2-\underset{O-N<}{\overset{|}{C}}-CH_2- + P^{\cdot}$$

$$\underset{HH}{-CH-CH-} \quad\quad -CH=CH-$$
$$\underset{H}{\overset{O-O}{\diagup}}\underset{\diagdown C \diagup}{\overset{O-N<}{\diagup}} \quad H_2O \quad \underset{\diagdown C \diagup}{HOO-N<} \quad (23)$$
$$\xrightarrow{h\nu}$$

$$\underset{\diagup\diagdown}{\overset{OH}{\diagup}C\overset{}{\diagdown}} \longrightarrow {>}C{=}O + HO-N{<} \quad\quad (24)$$
$$O-N{<}$$

Scheme 1. Photo-oxidation of *sec*-alkylhydroxylamines.

ketone according to the Norrish type II reaction will yield a vinyl group and a methyl ketone. Photolysis of the methyl ketone according to the same reaction yields another vinyl group and acetone that is eventually lost by volatilization.

Neglecting in a first approximation the small amount of ketone photolysed according to the Norrish type I reaction, the reaction sequence proposed above would yield approximately two vinyl groups for one *trans*-vinylene group. The experimental ratio of the initial rates of vinyl to *trans*-vinylene group formation is, within experimental error, independent of the stabilizer concentration but depends on the stabilizer structure. This ratio increases from 2·7 for HALS-2 to 3·5 for HALS-1 and 3·7 for HALS-3. The ratios for the secondary hindered amines HALS-1 and HALS-3 are, in fact, very close. From these experimental ratios it can be concluded that the reaction sequence mentioned above does not account for the results in their totality. However, Norrish type II reactions with additional ketones formed on photo-oxidation, e.g. by bimolecular decomposition of secondary hydroperoxides according to reaction (8) or by termination reaction, can explain the formation of more vinyl groups. Intramolecular hydroperoxide decomposition reactions such as (10) and (11) can also yield additional vinyl groups.

The fact that the rate of formation of vinyl and *trans*-vinylene groups increases slightly with HALS concentration can be attributed to the reaction discussed above. In fact, the higher the HALS concentration, the better the protection against oxidation. However, simultaneously and as a direct consequence of a stabilizing step, more hydroxylamine ethers will be formed on the polyethylene chain. These ethers will yield *trans*-vinylene and ketone groups; photolysis of the latter results in chain scission and formation of vinyl groups.

The reactions of Scheme 1 can explain the relatively limited increase in polymer durability with HALS concentration in unpigmented 2-mm HDPE plaques (Table 13). The significantly more pronounced con-

Table 13
Mechanical and chemical changes on outdoor exposure of 2-mm HDPE plaques[a]

Stabilization	E_{50} (kLy)[b]	(Vinyl)$_{50}$ (mol $l^{-1} \times 10^3$)[c]	(trans-Vinylene)$_{50}$ (mol $l^{-1} \times 10^3$)
Control	39	3·5	1·0
0·05% HALS-1	370	4·1	1·3
0·10% HALS-1	420	5·5	1·8
0·20% HALS-1	445	7·7	2·7
0·05% HALS-2	70	1·6	0·6
0·10% HALS-2	135	2·3	0·8
0·20% HALS-2	140	2·3	1·0
0·05% HALS-3	320	4·2	1·2
0·10% HALS-3	320	5·7	1·7
0·20% HALS-3	430	8·0	2·0

[a] Ziegler-type HDPE + 0·2% Ca stearate + 0·03% AO-1, 2-mm injection-moulded plaques.
[b] E_{50} = Energy to 50% retained tensile impact strength on Florida exposure, 45° south, direct (140 kLy year^{-1}), started March 1981.
[c] (vinyl)$_{50}$ and (*trans*-vinylene)$_{50}$ = concentrations corresponding to 50% retained tensile impact strength.

centration effect observed usually with HALS in polypropylene (PP) derives logically from the same argument. As a matter of fact, in PP the nitroxyl radical will be mainly fixed at tertiary carbon atoms where there is no more α-hydrogen available for abstraction.

4.2.2 Fate of HALS-1 During HDPE Photo-oxidation

It has already been mentioned that stabilization of polyolefins with HALS is generally attributed to the sequence of reactions (13)–(15). According to this regenerative cycle, stabilization should go on for ever. This is in contrast with actual practice. The loss of stabilizing properties has been attributed to the formation of various inactive or less active derivatives of the sterically hindered amines.[37] However, one of the most striking features observed with polyolefin films stabilized with HALS such as HALS-1 is the steady decrease of the ester carbonyl groups measured by IR spectroscopy. This is observed with polyethylene as well as with polypropylene from various origins. Thus, it can be assumed to be typical for this stabilizer type.[38]

Two main reactions can be envisaged to explain the disappearance of the ester groups, photolysis and photo-oxidation.[38] Photolysis would involve Norrish type I and II reactions of the ester carbonyl group. Although photolysis is not very likely because absorption of aliphatic esters does not reach into the UV part of sunlight (above 295 nm) it cannot be dismissed completely. In fact, there may be some residual absorption up to those wavelengths resulting in minor photolysis. Moreover, with some accelerated exposure devices, short-wavelength UV radiation is present and may lead to direct photolysis.

To eliminate the uncertainties associated with accelerated exposure devices, polyethylene samples exposed outdoors were studied. The decrease of HALS-1 concentration was monitored by gas chromatography (GC) and infrared spectroscopy (IR) of the ester carbonyl groups. The variation of the concentration of HALS-1 measured by GC, i.e. the concentration of fully intact HALS-1 molecules, is shown in Fig. 18. The corresponding variation at 1738 cm^{-1} is plotted in Fig. 19. It can be seen that the rate of decrease of the ester groups is significantly lower than that of intact HALS-1 molecules. The relative decrease of the ester groups of HALS-1 is plotted in Fig. 20, which shows that the relative decrease is, within experimental error, independent of the initial concentration of HALS-1. If direct photolysis of ester groups were determinant for the observed decrease, it would proceed according to a first-order law. This is not observed. The

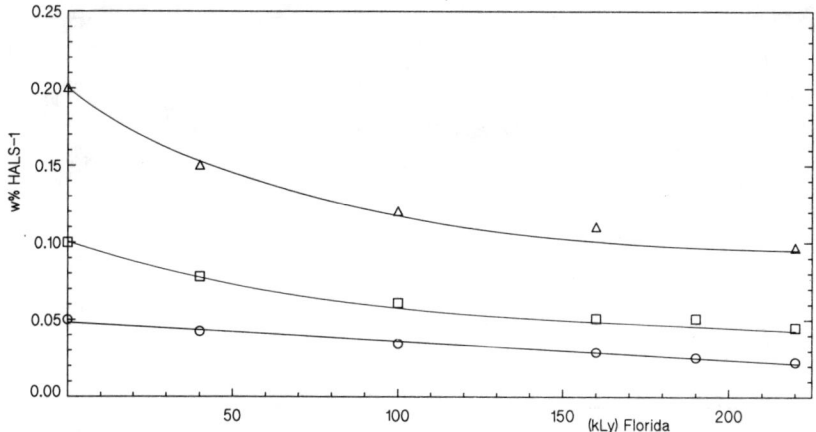

Fig. 18. Decrease of HALS-1 concentration in 2-mm HDPE plaques on outdoor exposure. HALS-1 was measured by gas chromatography. ○, 0·05% HALS-1; □, 0·10% HALS-1; △, 0·20% HALS-1.

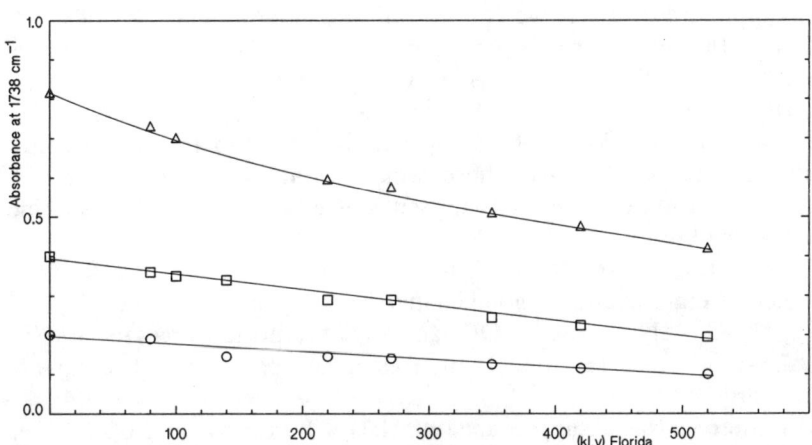

Fig. 19. Decrease of HALS-1 ester groups in 2-mm HDPE plaques on outdoor exposure (IR). ○, 0·05% HALS-1; □, 0·10% HALS-1; △, 0·20% HALS-1.

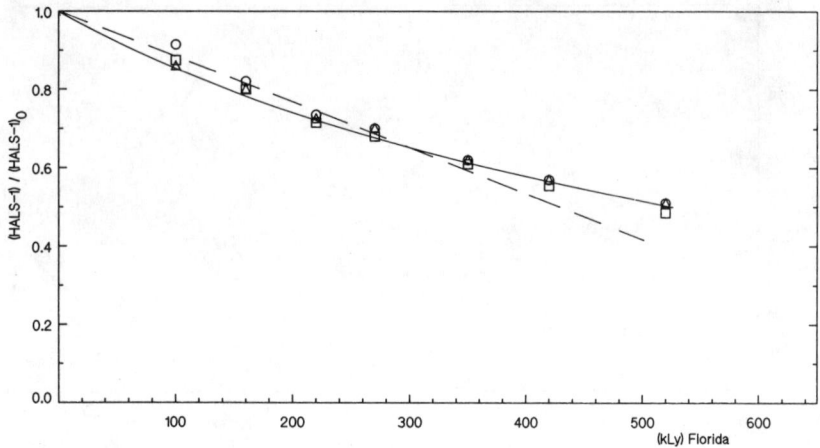

Fig. 20. Decrease of HALS-1 ester groups in 2-mm HDPE plaques on outdoor exposure (IR). ○, 0·05% HALS-1; □, 0·10% HALS-1; △, 0·20% HALS-1.

relative concentration of ester groups corresponding to a relative concentration of intact molecules of 0·5 (after approximately 195 kilolangleys) is 0·78. Statistical loss of ester groups, e.g. by photolysis, would yield a relative concentration of ester groups of 0·71, corresponding to a relative concentration of intact molecules of 0·5.[39] Because the real ester group concentration is higher, the conclusion is that the loss of ester groups is a consequence of the attack and transformation of a HALS-1 molecule and not the reverse. The relative concentration of 0·78 ester groups for 0·5 intact molecules indicates, that on average, there remains practically one ester group per HALS-1 molecule initially present but no longer found by specific analysis by GC.

The analytical results lead to the conclusion that the HALS-1 molecules that have undergone change by transformation of one $>$NH to $>$N—O, $>$NOH or $>$NOR, and thus are no longer detectable as HALS-1, will also lose one of their ester groups more or less rapidly afterwards.

In other words, disappearance of HALS-1 as such, and of its ester groups, is mainly a consequence of its attack by active species; that is, it is a consequence of its stabilizing activity. The conclusion is that oxidation products of HALS-1 initiate reactions leading finally to destruction of the ester groups. A prime candidate in this respect is the nitroxyl radical which is active photochemically. It has been shown that this radical absorbs UV light up to wavelengths of 360 nm.[32,40,41] In its excited state it can abstract hydrogen atoms from hydrocarbons according to reaction (25).

$$>\!\!NO^{\cdot} \xrightarrow{h\nu} \left(>\!\!NO^{\cdot}\right)^{*}$$

$$\left(>\!\!NO^{\cdot}\right)^{*} + PH \longrightarrow >\!\!NOH + P^{\cdot}$$

(25)

The hydrogen atom likely to be abstracted by the photo-excited nitroxyl radical is the tertiary hydrogen in position 4 of HALS-1 and α to the oxygen atom. A possible reaction pathway is presented in Scheme 2.

The formation of acids shown in Scheme 2 is corroborated by fast development of an absorption band at 1630 cm on exposure, indicating the formation of a salt from a carboxylic acid and a secondary hindered-amine group. The *trans*-vinylene groups formed in one of the reactions in Scheme 2 can also explain the excess of *trans*-vinylene groups found with HALS-1 in comparison with HALS-3 (Fig. 17).

The experimental kinetics of HALS-1 decrease observed so far are also in general agreement with a nitroxyl-initiated photo-oxidative destruction of the ester groups. In fact, it was found that in PP the ester groups of HALS-1 are lost at a constant rate, whereas the ester groups of the corresponding bisnitroxyl decrease with first-order kinetics.[38] Similar results were obtained independently in our laboratory. Moreover, depending on the polymer batch and/or the processing conditions, the behaviour of HALS-1 can be similar to that of the nitroxyl, i.e. its concentration can also decrease according to first-order kinetics. All these results are clearly in favour of nitroxyl-initiated loss of ester groups.

It has been shown that the excited nitroxyl radical is quenched by oxygen.[41] This quenching will surely reduce the rate of hydrogen abstraction but not necessarily suppress it. Furthermore, abstraction of the tertiary hydrogen atom in position 4 of HALS-1, e.g. by peroxyl radicals, will lead to the same reactions as those shown in Scheme 2. Hydrogen abstraction by peroxyl radicals may complement excited nitroxyl radical-initiated hydrogen abstraction or even be predominant.

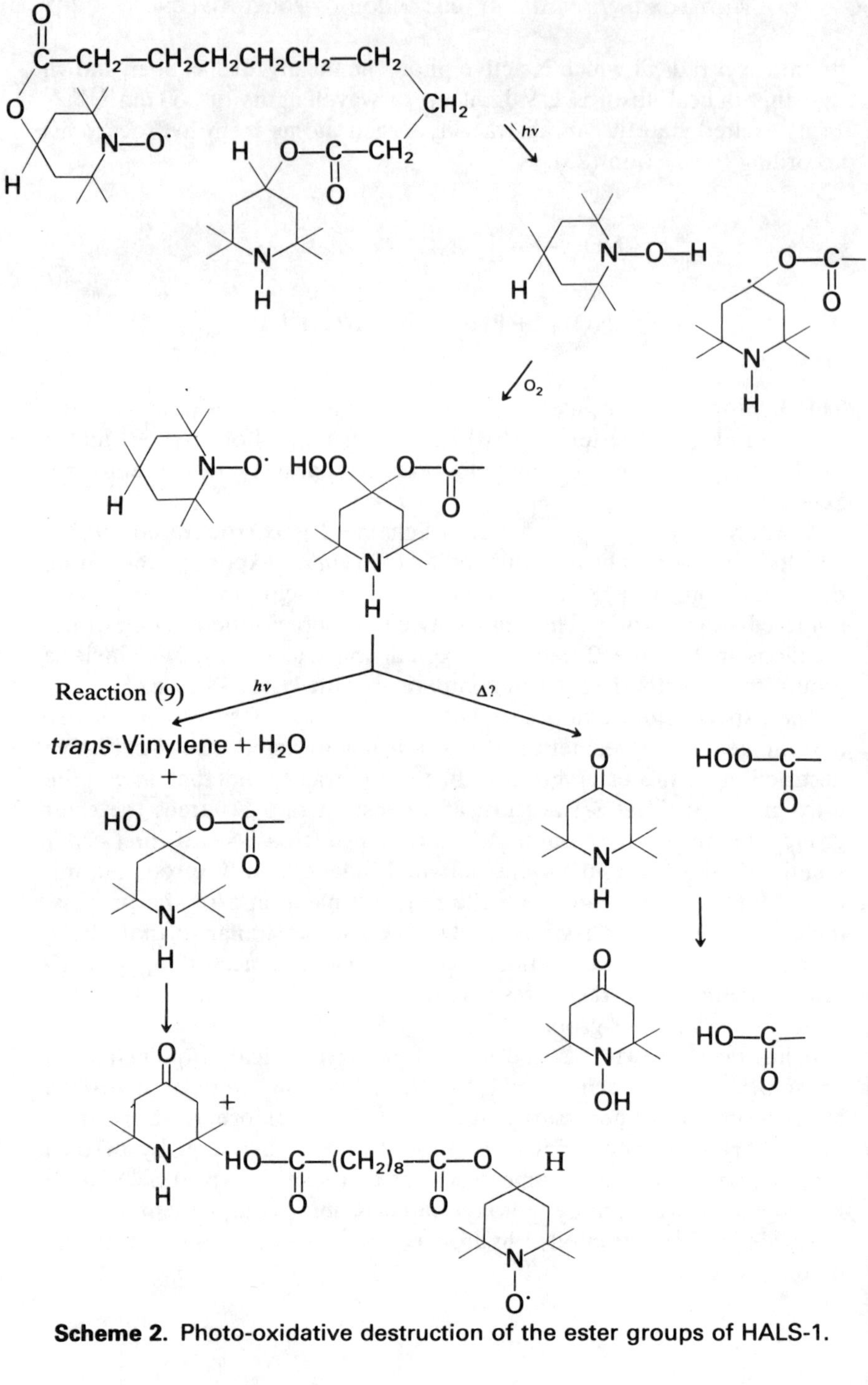

Scheme 2. Photo-oxidative destruction of the ester groups of HALS-1.

5 CONCLUSIONS

The efficiencies of UV absorbers, a 'nickel quencher' and HALS in polyethylene stabilization have been compared. It has been shown that the performance of HALS is outstanding in tapes, films and thick sections. Moreover, it has been established that in unpigmented polyethylene the combination of HALS with benzotriazole-type UV absorbers very often yields optimum stabilization.

In an attempt to interpret these results, possible mechanisms of stabilization have been examined. With respect to quenching of carbonyl compounds, essentially ketones, it has been shown on the one hand that HALS such as HALS-3 are not efficient but, on the other, that benzotriazole-type UV absorbers such as UVA-3 are very efficient quenchers of the Norrish type I reaction according to the long-range energy transfer mechanism. The determination of the chemical changes in 2-mm HDPE plaques on outdoor exposure has led to a better understanding of the loss of mechanical properties. Moreover, it has given some insight into HALS stabilization mechanisms. Finally, it has been shown that the disappearance of the ester groups of HALS-1 on outdoor exposure is initiated by oxidation products of HALS-1. A possible reaction sequence based on the photochemically active nitroxyl radical has been proposed. It has been postulated that such a reaction, and similar ones with HALS not containing ester groups, may be one of the main limitations to HALS activity. Only certain aspects of polyethylene stabilization by HALS are presented. A more detailed account of HALS stabilization mechanisms will be reported in the near future.

ACKNOWLEDGEMENT

The author thanks the management of Ciba–Geigy for permission to publish this paper.

REFERENCES

1. Cicchetti, O., *Adv. Polym. Sci.*, **7** (1970) 70.
2. Winslow, F. H., Matreyek, W. & Trozzollo, A. M., *SPE J.*, **28** (1972) 19.
3. Carlsson, D. J. & Wiles, D. M., *J. Macromol. Sci.—Rev. Macromol. Chem.*, **C14** (1976) 65.

4. Scott, G., In *Ultraviolet Light Induced Reactions in Polymers*, ACS Symp. Series No. 25, ed. S. S. Labana. American Chemical Society, Washington, DC, 1976, p. 340.
5. Gugumus, F., In *Developments in Polymer Stabilization—1*, ed. G. Scott. Applied Science Publishers, London, 1979, p. 260.
6. Gugumus, F., Paper presented at 3rd International Conference on Advances in the Stabilization and Controlled Degradation of Polymers, Luzern, Switzerland, 1981.
7. Gugumus, F., *Kunstst.*, **74** (1984) 620.
8. Gugumus, F., In *Plastics Additives*, ed. R. Gächter and H. Müller. Carl Hanser Verlag, Munich, Vienna, New York, 1987, p. 97.
9. Gugumus, F., Paper No. D 12 presented at the Golden Jubilee Conference, Polyethylenes 1933–1983, London, 8–10 June 1983.
10. Winslow, F. H., *Pure Appl. Chem.*, **49** (1977) 495.
11. Arnaud, R. & Lemaire, J., In *Developments in Polymer Degradation—2*, ed. N. Grassie. Applied Science Publishers, London, 1979, p. 159.
12. Ginhac, J. M., Gardette, J.-L., Arnaud, R. & Lemaire, J., *Makromol. Chem.*, **182** (1981) 1017.
13. Gugumus, F., Paper presented at the 18th Colloquium of Danubian Countries for Natural and Artificial Aging of Plastics, Villach, Austria, 22–26 June 1987.
14. Gugumus, F., Paper presented at the European Symposium on Polymeric Materials, Lyon, France, 14–18 September 1987.
15. Ng, H. C. & Guillet, J. E., *Macromolecules*, **11** (1978) 937.
16. Stewart, L. C., Carlsson, D. J., Wiles, D. M. & Scaiano, J. C., *J. Am. Chem. Soc.*, **105** (1983) 3605.
17. Gugumus, F., Paper presented at 11th Discussion Conference on Macromolecules: Chemical and Physical Phenomena in the Ageing of Polymers, Prague, Czechoslovakia, 11–14 July 1988.
18. Gugumus, F., Paper presented at 4e Conférence Européenne des Plastiques et des Caoutchoucs, Paris, 1974; *Kunstst. Plast.*, **22** (1975) 11; *Caout. Plast.*, **558** (1976) 67.
19. Carlsson, D. J. & Wiles, D. M., *J. Macromol. Sci.—Rev. Macromol. Chem.*, **C14** (1976) 155.
20. Wiles, D. M. & Carlsson, D. J., *Polym. Degrad. Stab.*, **3** (1980–81) 61.
21. Scott, G., *Brit. Polym. J.*, **16** (1984) 271.
22. Scott, G., *Polym. Degrad. Stab.*, **10** (1985) 97.
23. Shlyapintokh, V. Ya., In *Developments in Polymer Photochemistry—2*, ed N. S. Allen. Applied Science Publishers, London, 1981, Chapter 6.
24. Allen, N. S., Chirinos-Padron, A. & Henman, T. J., *Polym. Degrad. Stab.*, **13** (1985) 31.
25. Bortolus, P., Dellonte, S., Faucitano, A. & Gratani, F., *Macromolecules*, **19** (1986) 2916.
26. Klöpfer, W., *Lenzinger Berichte*, **44** (1978) 28.
27. Chakraborty, K. B. & Scott, G., *Europ. Polym. J.*, **13** (1977) 1007.
28. Chakraborty, K. B. & Scott, G., *Europ. Polym. J.*, **15** (1979) 35.
29. Winslow, F. H., *Pure Appl. Chem.*, **49** (1977) 495.
30. Förster, Th., *Z. Naturforsch.*, **4a** (1949) 321.

PHOTO-OXIDATION AND STABILIZATION OF POLYETHYLENE   209

31. Perrin, J., *Compt. Rend.*, **184** (1927) 1097.
32. Schlyapintokh, V. Ya. & Ivanov, V. B., In *Developments in Polymer Stabilization—5*, ed. G. Scott. Applied Science Publishers, London, 1982, p. 41.
33. Bolsman, T. A. B. M., Blok, A. P. & Frijns, J. H. G., *Rec. Trav. Chim. Pays-Bas*, **97** (1978) 313.
34. Klemchuk, P. P. & Gande, M. E., *Polym. Degrad. Stab.*, **22** (1988) 241.
35. Denisov, E. T., In *Developments in Polymer Stabilisation—3*, ed. G. Scott. Applied Science Publishers, London, 1980, Chapter 1.
36. Scott, G. In *Developments in Polymer Stabilisation—7*, ed. G. Scott. Applied Science Publishers, London, 1984, Chapter 2.
37. Carlsson, D. J., Tovborg Jensen, J. P. & Wiles, D. M., *Makromol. Chem., Suppl.* **8** (1984) 79.
38. Hodgeman, D. K. C., In *Developments in Polymer Degradation—4*, ed. N. Grassie. Applied Science Publishers, London, 1982, Chapter 6, p. 189.
39. Gugumus, F., to be published.
40. Keana, J. F. W., Dinerstein, R. J. & Baitis, F., *J. Org. Chem.*, **36** (1971) 209.
41. Bogatyreva, A. I. & Buchachenko, A. L., *Kinet. Katal.*, **12** (1971) 1226.

APPENDIX: CHEMICAL STRUCTURES OF STABILIZERS USED

Abbreviation	Structure	Trade name
HALS-1	H—N⟨piperidine⟩—O—C(=O)—(CH$_2$)$_8$—C(=O)—O—⟨piperidine⟩—N—H	Tinuvin 770
HALS-2	[—O—⟨piperidine⟩—N—CH$_2$—CH$_2$—O—C(=O)—CH$_2$—CH$_2$—C(=O)—]$_n$	Tinuvin 622
HALS-3	[⟨triazine with two N-H piperidinyl groups⟩—N—(CH$_2$)$_6$—N—, with HN—C(CH$_3$)—CH$_2$—C(CH$_3$)—CH$_3$ substituent]$_n$	Chimassorb 944

(continued)

APPENDIX—*contd.*

Abbreviation	Structure	Trade name
UVA-1	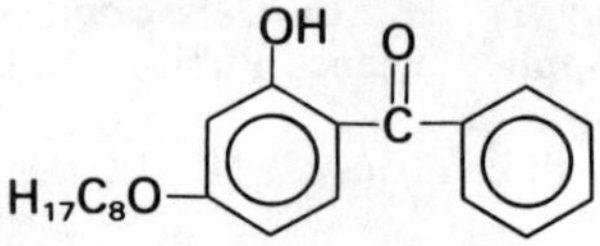 2-hydroxy-4-octyloxybenzophenone ($H_{17}C_8O$-substituted)	Chimassorb 81
UVA-2	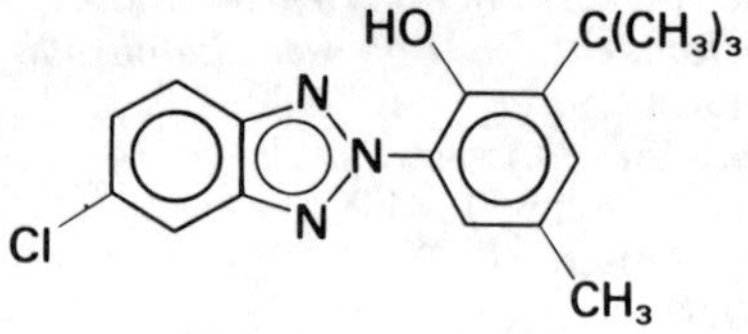 5-chloro-2-(2H-benzotriazol-2-yl)-4-methyl-6-tert-butylphenol	Tinuvin 326
UVA-3	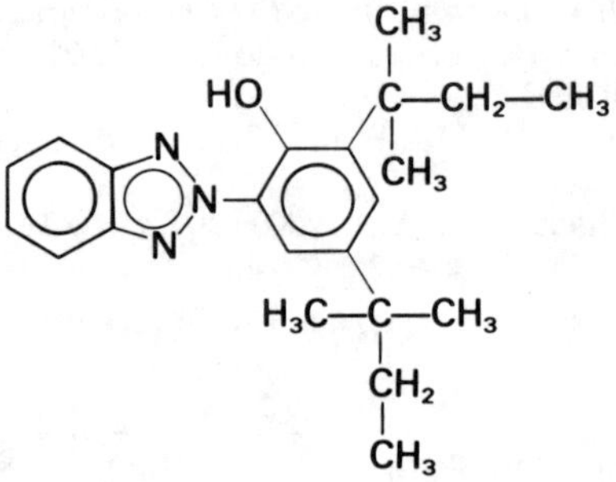 2-(2H-benzotriazol-2-yl)-4,6-bis(tert-pentyl)phenol	Tinuvin 328
Ni-1	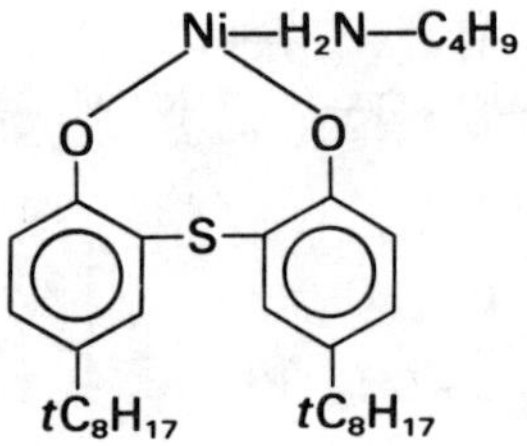 Nickel bis[2,2′-thiobis(4-tert-octylphenolate)] with n-butylamine	Chimassorb N 705
AO-1	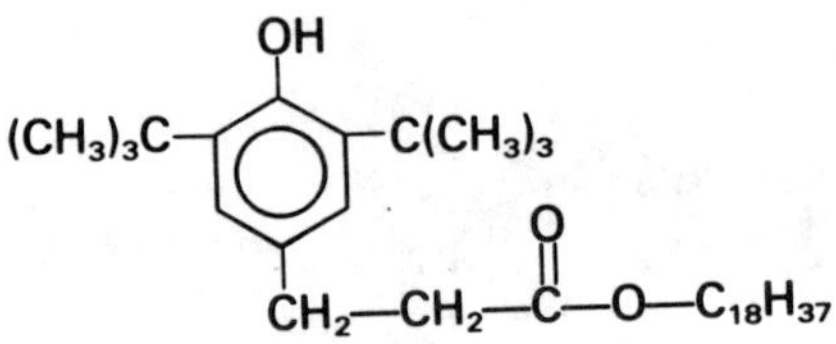 Octadecyl 3-(3,5-di-tert-butyl-4-hydroxyphenyl)propionate	Irganox 1076
AO-2	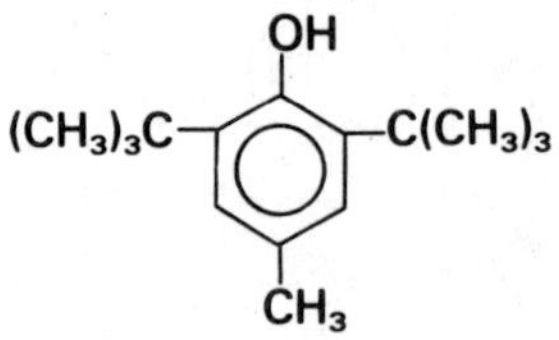 2,6-di-tert-butyl-4-methylphenol	BHT

Chapter 7

Analysis of Antioxidants and Light Stabilisers in Polymers by Modern Liquid Chromatography

DAN MUNTEANU
Chemical Research Institute, Research Centre for Plastics,
Timişoara, Romania

ABSTRACT

The identification and quantification of antioxidants, light stabilisers and other polymer additives is necessary for quality-assurance purposes but it is also useful for other reasons such as screening of unknown or competitors' products and in polymer degradation studies. Polyolefins, the largest-volume family of commercially important high-tonnage thermoplastic polymers, dominate the world market in terms of antioxidant and light-stabiliser usage. Since there is a wide variety and complexity of polyolefin stabilisers, many of them differing only slightly in structure, coupled with the fact that the additive package usually contains more than one component, a technique capable of separating these compounds for qualitative or quantitative analysis is essential. Most polymer additives, and especially today's stabilisers, are compounds of insufficient volatility to be analysed by gas chromatography or they are thermally unstable and decompose under the conditions of gas chromatographic separation. Liquid chromatography (LC) eliminates this disadvantage and provides additional advantages with respect to gas chromatography for the separation, detection and identification of polymer additives. Although the classical LC *techniques, i.e. column, paper and thin-layer chromatography, have been used for the analysis of antioxidants, light stabilisers and other polymer additives, they are today being replaced by modern liquid chromatography, i.e. high-performance liquid chromatography (HPLC) and size-exclusion chromatography (SEC). The preliminary step of additive extraction from polymers drastically influences the success of the chromatographic analysis. Normal versus reversed-phase and isocratic versus gradient-elution HPLC separation of synthetic mixtures of additives and of solvent extracts from polymers are discussed. Some SEC analysis of additives in polymers may be performed without the time-consuming extraction procedure. Recent developments such as high-resolution capillary gas chromatography with cold on-column injection, high-speed LC, microcolumn LC and supercritical fluid chromatography are presented in terms of their application to the analysis of polymer additives. In view of the fact that monitoring of polymer additives has become increasingly important because of stricter laws and regulations, the need for fast and reliable analytical methods has increased.*

1 INTRODUCTION

Chemical additives are frequently used to improve the useful properties and to extend the service life of polymers. Various additives are employed for these purposes, antioxidants and light stabilisers being the most important in protecting and developing desirable physical ahd chemical properties in polymers.[1-5] Polyolefins, the largest-volume family of commercially important high-tonnage thermoplastic polymers, dominate the world usage of antioxidants and light stabilisers. Thus, the polyolefin industry consumes about one-half of the antioxidants and three-quarters of the light stabilisers produced for thermoplastic polymers.[6,7]

Polyolefins are the most widely used thermoplastic polymers and are employed in an extremely wide range of applications. Low-density, high-density, linear low-density polyethylene (LDPE, HDPE, LLDPE) and polypropylene (PP) are the dominant polyolefins but many others are commercially available, e.g. ethylene–propylene copolymers (EPM and EPDM rubbers) and copolymers, especially of ethylene, with higher olefins. Practically all these commercial polyolefins contain at least one antioxidant to protect them against thermal and oxidative stressing during processing and for many end-uses. Because of their relatively poor stability to sunlight, polyolefins, and especially propylene polymers, have to be light-stabilised for outdoor applications.

Table 1 presents the chemical and trade names of the most important antioxidants and light stabilisers whose analysis is discussed in the present contribution. The chemical formulae of these compounds are presented in the Appendix. The vast majority of the antioxidants are various types of compounds containing hindered phenolic groups, i.e. mono-, bis-, tris- and tetrakis-phenols and thiophenols. The most important light stabilisers are benzophenones, benzotriazoles and the low- and high-molecular-weight hindered amines.

Besides antioxidants and light stabilisers, other additives are sometimes incorporated into the polymer matrix depending on the polyolefin type and end-use. Thus, slip agents (usually fatty acid amides) and antiblock agents (usually silica) are incorporated, especially in the thin-film grades of ethylene homo- and co-polymers. Antacid additives (usually calcium stearate) act as scavengers for traces of hydrochloric acid thermally released from the polymer matrix during the processing of polyolefins obtained with Ziegler–Natta-type catalytic systems based on chlorine-containing compounds such as $TiCl_4$ and $AlCl(C_2H_5)_2$. Other additives such as antistatic agents (usually polyol/fatty acid esters), lubricants and mould release agents are sometimes encountered. Other thermoplastic polymers such as poly(vinyl chloride) (PVC) and polyurethanes (PU) also contain varying amounts of additives.

Analysis of antioxidants, light stabilisers and other chemicals in plastics is of increasing importance. Although these additives are incorporated in the polymer matrix at a very low level, typically less than 0·5%, their type, amount and purity drastically influence physical and processing parameters, service life and often the economics of the product. Identification and quantification of these compounds are necessary mainly for quality-assurance tests but also for other purposes, such as screening of unknown or competitors' products and polymer degradation studies. Such analysis is often particularly difficult for three main reasons: (a) polymers usually contain mixtures of additives; (b) these additives occur in wide variety and complexity, and (c) the additives are locked into the polymer matrix and cannot be extracted easily. From all these points of view,

Table 1
Antioxidants and light stabilisers

Trade name	Supplier	Chemical name	FMW	Structure (see Appendix)
BHT[a]	—[a]	2,6-Di-*tert*-butyl-4-methylphenol	220	Ia
Chimassorb 81	Ciba–Geigy	2-Hydroxy-4-*n*-octyloxybenzophenone	326	XVd
Chimassorb N-705	Giba–Geigy	[2,2'-Thiobis(4-*tert*-octylphenol)ato]-*n*-butylamine nickel	572	XXIV
Chimassorb 944	Ciba–Geigy	Polymeric HALS($\bar{M}_n > 2500$)	—	IXf
Cyasorb UV-9	Cyanamid	2-Hydroxy-4-methoxybenzophenone	228	XVc
Cyasorb UV-24	Cyanamid	2,2'-Dihydroxy-4'-methoxybenzophenone	244	XVb
Cyasorb UV-531	Cyanamid	As for Chimassorb 81	326	XVd
Cyasorb UV-1084	Cyanamid	As for Chimassorb N-705	572	XXIV
Cyasorb UV-5441	Cyanamid	2-(2'-Hydroxy-5'-octylphenyl)benzotriazole	443	XVIf
DLTDP[a]	—[a]	Dilauryl thiodipropionate	515	XIa
DMTDP[a]	—[a]	Dimyristyl thiodipropionate	571	XIb
DSTDP[a]	—[a]	Distearyl thiodipropionate	683	XIc
Ethanox 330	Ethyl Corp.	1,3,5-Trimethyl-2,4,6-tris(3',5'-di-*tert*-butyl-4'-hydroxy-benzyl)benzene	726	VIIIc
Ethanox 702	Ethyl Corp.	4,4'-Methylenebis(2,6-di-*tert*-butyl)phenol	424	IIa
Ethanox 754	Ethyl Corp.	4-Hydroxymethyl-2,6-di-*tert*-butylphenol	236	Ib
Goodrite 3114	B. F. Goodrich	1,3,5-Tris(3',5'-di-*tert*-butyl-4'-hydroxybenzyl) isocyanurate	775	VIIIa
Goodrite 3125	B. F. Goodrich	1,3,5-Tris[ethylene(3',5'-di-*tert*-butyl-4'-hydroxyphenyl)propionate] isocyanurate	1041	VIIIb
Hostavin VPN-20	Hoechst	HALS	376	XXa
Inhibitor RMB	Eastman	Resorcinol monobenzoate	169	XXIII
Irgafos 168	Ciba–Geigy	Tris(2,4-di-*tert*-butylphenyl) phosphite	647	XIII
Irgafos TPP	Ciba–Geigy	Triphenyl phosphite	310	XIIa
Irgafos TNPP	Ciba–Geigy	Tris(nonylphenyl) phosphite	689	XIIb
Irganox 245	Ciba–Geigy	Trisethyleneglycol bis[3-(3-*tert*-butyl-4-hydorxy-5-methyl-phenyl)propionate]	587	III
Irganox 259	Ciba–Geigy	1,6-Hexanediol bis[3-(3,5-di-*tert*-butyl-4-hydroxyphenyl)propionate]	639	IIb

Name	Supplier	Chemical name		
Irganox 415	Ciba–Geigy	4,4'-Thiobis(3-methyl-6-*tert*-butylphenol)	322	VIb
Irganox 565	Ciba–Geigy	2,4-Bis(*N*-octylthio)-6-(4-hydroxy-3,5-di-*tert*-butylanilino)-1,3,5-triazine	589	Ie
Irganox 1010	Ciba–Geigy	Pentaerythrityl tetrakis[3-(3,5-di-*tert*-butyl-4-hydroxyphenyl) priopionate]	1178	VIIIe
Irganox 1019	Giba–Geigy	*N*,*N*-Trimethylenebis(3,5-di-*tert*-butyl-4-hydroxyhydrocinnamamide)	594	IId
Irganox 1024 MD	Ciba–Geigy	*N*,*N*'-Bis[3-(3',5'-di-*tert*-butyl-4'-hydroxyphenyl-propan-1-one] hydrazine	553	IIc
Irganox 1035	Ciba–Geigy	2,2'-Thiodiethylenebis[3-(3,5-di-*tert*-butyl-4-hydroxyphenyl)propionate]	643	IIf
Irganox 1076	Ciba–Geigy	Octadecyl 3-(3,5-di-*tert*-butyl-4-hydroxyphenyl)propionate	53i	Id
Irganox 1081	Ciba–Geigy	2,2'-Thiobis(4-methyl-6-*tert*-butylphenol)	359	VII
Irganox 1098	Ciba–Geigy	*N*,*N*'-Hexamethylenebis(3,5-tert-butyl-4-hydroxyhydrocinnamamide)	637	IIe
Irganox 1330	Ciba–Geigy	As for Ethanox 330	726	VIIIc
Irganox 3114	Ciba–Geigy	As for Goodrite 3114	775	VIIIa
Irganox 3125	Ciba–Geigy	As for Goodrite 3125	1041	VIIIb
Irganox PS-800	Ciba–Geigy	As for DLTDP	515	XIa
Irganox PS-801	Ciba–Geigy	As for DMTDP	571	XIb
Irganox PS-802	Ciba–Geigy	As for DSTDP	683	XIc
Nonox WSP	ICI	2,2'-Methylenebis[4-methyl-6-(1-methylcyclohexyl)]phenol	424	IVc
Salol	Dow Chemical	Phenyl salicylate	214	XXIIb
Sandovur EPU	Sandoz	Ethoxy-2-*tert*-butyl-5-ethyl-2'-bisanilideoxalate	368	XVII
Santonox	Monsanto	4,4'-Thiobis(3-methyl-6-*tert*-butylphenol)	322	VIb
Santonox R	Monsanto	4,4'-Thiobis(3-methyl-5-*tert*-butylphenol)	322	VIa
Santowhite Powder	Monsanto	4,4'-Butylenebis(3-methyl-6-*tert*-butylphenol)	382	V
Sunkem MS		Methyl salycilate	152	XXIIa
Tinuvin P	Ciba–Geigy	2-(2'-Hydroxy-5'-methylphenyl)-2*H*-benzotriazole	225	XVIa
Tinuvin 120	Ciba–Geigy	4,6-Di-*tert*-butylphenyl(3,5-di-*tert*-butyl-4-hydroxybenzoate)	438	XVIII
Tinuvin 144	Ciba–Geigy	2-*tert*-Butyl-2-(4-hydroxy-3,5-di-*tert*-butylbenzyl)[bis(methyl-2,2,6,6-tetramethyl-4-piperidinyl)]dipropionate	685	XIXa

(continued)

Table 1—contd.

Tradename	Supplier	Chemical name	FMW	Structure (see Appendix)
Tinuvin 292	Ciba–Geigy	Bis(1-methyl-2,2,6,6-tetramethylpiperidinyl) sebacate	506	XIXc
Tinuvin 315	Ciba–Geigy	As for Sandovur EPU	368	XVII
Tinuvin 320	Ciba–Geigy	2-(2'-Hydroxy-3',5'-di-*tert*-butylphenyl)-2*H*-benzotriazole	323	XIVb
Tinuvin 326	Ciba–Geigy	2(3'-*tert*-butyl-2'-hydroxy-5'-methylphenyl)-2*H*-5-chlorobenzotriazole	316	XVIe
Tinuvin 327	Ciba–Geigy	2(2'-Hydroxy-3',5'-di-*tert*-butyl-phenyl)-2*H*-5-chlorobenzotriazole	358	XVId
Tinuvin 328	Ciba–Geigy	2-2'-Hydroxy-3',5'-di-*tert*-amylphenyl)-2*H*-benzotriazole	351	XVIc
Tinuvin 440	Ciba–Geigy	8-Acetyl-3-dodecyl-7,7,9,9-tetramethyl-1,3,8-triazaspiro[4,5]decane-2,4-dione	436	XX
Tinuvin 622	Ciba–Geigy	Polymeric HALS ($\bar{M}_n > 3000$)	—	XIXe
Tinuvin 765	Ciba–Geigy	HALS	—	
Tinuvin 770	Ciba–Geigy	Bis(2,2,6,6-tetramethyl-4-piperidinyl) sebacate	478	XIXd
Tinuvin 900	Ciba–Geigy	2-(2'-Hydroxy-3',5'-diphenylisopropyl)2-*H*-benzotriazole	477	XVIg
Topanol CA	ICI	1,1,3-Tris(2-methyl-4'-hydroxy-5'-*tert*-butylphenyl)butane	502	IX
Topanol OC	ICI	2,4,6-Tris-*tert*-butylphenol	262	Ic
Uvinul 400	BASF Wyandotte	2,4-Dihydroxybenzophenone	214	XVa
Uvinul M-410	BASF Wyandotte	As for Chimassorb 81	326	XVe
Uvinul N35	BASF Wyandotte	Ethyl2-cyano-3,3-diphenylacrylate	277	XXIa
Uvinul N539	BASF Wyandotte	2-Ethylhexyl2-cyano-3,3-diphenylacrylate	361	XXIb
Vulkanox BKF	Bayer AG	2,2'-Methylenebis(4-methyl-6-*tert*-butylphenol)	340	IVa
Vulkanox NKF	Bayer AG	2,2'-Isobutylidenebis(4,6-dimethylphenol)	298	IVb
Weston 618	Borg Warner	Distearyl pentaerythritol diphosphite	668	XIV

[a] Many commercial names and suppliers.

polyolefins are the most difficult thermoplastic polymers to be analysed. Moreover, taking into account the fact that polyolefins dominate the world thermoplastic market in terms of antioxidant and light-stabiliser usage, it may be concluded that analysis of these stabilisers in polyolefins is representative of stabiliser analysis in thermoplastic polymers.

In some limited cases, rapid additive analyses can be carried out, without extensive pretreatment steps, e.g. extraction, but generally these methods suffer disadvantages due to non-specificity of the tests used. The separation and determination of each additive in an additive package permits the most complete results to be obtained. This generally requires extraction of the additives followed by a separation technique, chromatography being the most important one.

This chapter deals mainly with the analysis of antioxidants and light stabilisers in polyolefins by modern liquid chromatography, i.e. high-performance liquid chromatography and size-exclusion chromatography. Related topics such as analysis of stabilisers in other polymers and materials and analysis of other polymer additives will be presented as appropriate.

2 WHY LIQUID CHROMATOGRAPHIC ANALYSIS

2.1 The Analytical Goal

A primary purpose of the analysis of antioxidants, light stabilisers and other chemicals in polymers is the identification and quantification of these compounds for quality-assurance purposes. To ensure that a single stabiliser or an additive mixture has been added properly to polyolefin batches following synthesis, rapid analytical methods are needed for both quality control and certification analysis. Like other additives, antioxidants and light stabilisers are usually fine powders that are incorporated into the polymer matrix by such techniques as dry blending of the stabiliser powder with polyolefin powder or pellets followed by melt mixing (compounding); melt mixing of the polyolefin with stabiliser masterbatches (concentrates containing a higher level of stabilisers dispersed in the parent polyolefin); dispersion of stabilisers in liquid organic carriers and injected into polyolefins during processing.

Regardless of the stabilisation technique used, the stabilisers have to be uniformly distributed in the polymer matrix. However, the concentration of stabilisers in the polymer may vary from batch to batch and even within the same batch. When polyolefins contain more than one stabiliser, and this is the usual case, analyses are necessary to ensure that the required additives are presented and in the correct ratio. The analysis may be performed on the mixture of stabilisers which will be added to a batch of resin but, more frequently, on the stabilised polyolefin batch. Even if the additive package contains the required additives in the correct ratio, this ratio may change for several reasons during the various stages of polyolefin manufacture and use. Thus, the vast majority of stabilisers are low-molecular-weight compounds and some of them are volatile or even very volatile, e.g. the widely used processing antioxidant BHT. During the melt processing of the polyolefin/stabiliser mixture and also during the subsequent processing of the stabilised polyolefin into finished articles, these stabilisers may be lost from the polymer matrix by volatilisation due to the high processing temperature, which may in some cases be as high as 350°C. Therefore, analytical methods are necessary to check the stabiliser level in the stabilised polyolefin batch or in the finished article. Quality-control and certification analysis allows rejection of the production batches which do not contain the required level of additives, thus avoiding possible future problems during the final processing and the end-use of the polymer. In such analyses the analytical goal consists of the quantification of known stabilisers in the polymer.

A more difficult goal from the analytical point of view is the analysis of stabilisers in unknown or competitors' products. In such analyses, the first and most difficult step is the identification of unknown antioxidants, light stabilisers and other additives in the polymer sample and following identification the quantification of the additives. Such studies need analytical methods able to emphasise the chemical transformation of antioxidants and light stabilisers during the processing and service-life of the polymer, thus providing a better understanding of stabiliser behaviour under oxidative and photo-oxidative conditions. Many antioxidants and stabilisers show very little difference in chemical structure, e.g. the ester-type antioxidants DLTDP, DMTDP, DSTDP and the benzotriazole-type light stabilisers Tinuvin P, 320 and 326–328 (Table 1).

2.2 Spectroscopic Versus Chromatographic Analyses

Many analytical methods have been developed for the analysis of stabilisers in polymers. Crompton's book[8] seems to be the only one dedicated exclusively to the analysis of additives in plastics, and reviews the literature up to 1977. Other books[9-13] concerned with the analysis of plastics or rubbers include some material applicable to analysis of additives. There are also review papers[14-16] concerned with the use of analytical techniques for the determination of additives in polymers.

The analysis of additives in polymers may be performed either by 'direct' techniques, i.e. directly on polymer samples, or by 'indirect' techniques, i.e. by analysing the additives separated from polymer samples by preliminary extraction. The most important work in the analysis of additives in polymers has been performed using spectroscopic and chromatographic methods.

2.2.1 Spectroscopic Analysis

Due to the difficulties which occur in the extraction of additives from polymers, techniques which avoid a prior separation of an additive extract have been investigated. However, of all these direct techniques only the spectroscopic ones have been used with any degree of success. Infrared (IR) spectroscopy is more specific but less sensitive than ultraviolet (UV) spectroscopy. Consequently thicker polymer films (0·2–0·3 mm) have to be used to overcome the disadvantage of its lower sensitivity and IR spectra unsuitable for analytical purposes may result, especially for low levels of stabilisers. The principal advantages of UV over IR direct spectroscopy are greater sensitivity arising from higher extinction coefficients of most antioxidants and light stabilisers, and the lack of interfering absorption from the polyolefin matrix. Thus, UV spectroscopy of thin polymer films is able to determine 0·002–1·0% stabilisers in polyolefins. However, the direct UV analysis of additives in polyolefins is limited by excessive beam dispersion due to light scattering from the crystalline regions of the polymer. At lower levels of stabiliser concentration and correspondingly greater sample thicknesses, the high level of scattering may change the absorption dependence on the wavelength unpredictably.

Direct spectroscopic techniques have limited usefulness and generally allow only the quantification of known stabilisers in the polymer batch but not the identification of unknown stabilisers. Consequently these methods are used mainly for quality-control and certification analysis where rapid and cheap methods are necessary. Direct spectroscopy of polymer films may be very useful for the study of solvent-extraction procedures or stabiliser-ageing processes during simulated processing or end-use conditions. In such studies known amounts of stabilisers are incorporated into the unstabilised polymer and the level of stabiliser remaining in the sample after treatment can be measured directly by the decrease in absorbance of the polymer film at a suitable wavelength.

The main disadvantage of the direct spectroscopic methods is interference between the variety of groups present and hence lack of specificity. Many antioxidants and light stabilisers exhibit very similar or identical spectra because spectroscopy is not able to emphasise small differences in the chemical structure of a compound. The spectrum of an additive package in a polymer is generally very complex and represents the superposition of the spectra of each component. Sometimes, the spectrum emphasises the absorption of a component present in a greater concentration but which is not of interest. Thus, the polymer should exhibit a relatively flat absorption curve in the wavelength range used for the quantitative determination of stabilisers.

The 'indirect' spectroscopic methods, especially UV absorption, are generally preferred even for the quantification of known additives. Stabilisers are extracted from polymers and their concentration is determined in the solvent extract by measuring the characteristic absorption peak and by reference to a prepared calibration curve. The determination of a mixture of stabilisers is also generally not possible.

The spectroscopic methods are generally not able to determine simultaneously more than a single additive in a polymer sample. In such cases they do not permit the identification of the components of the additive package. The quantitative determination of known additives by spectroscopy, both by direct examination of polymer films and in the solvent extract, was extensively reviewed in Crompton's book.[8]

2.2.2 Chromatographic Analysis

For mixtures of antioxidants, light stabilisers and other additives in polymers, the separation and individual determination of each component of the additive package is necessary. This usually requires extraction of additives followed by separation of the extracted additives. The continuous development of chromatography has provided analysts with a very sensitive technique for both separation and identification of many minor or even trace components in very complex mixtures, and it is no wonder that various chromatographic

Table 2
Chromatographic techniques

(A) Gas Chromatography (GC)
1. Gas–solid (adsorption) chromatography (GSC)
2. Gas–liquid (partition) chromatography (GLC)

(B) Liquid Chromatogrpahy (LC)
Traditional Liquid Chromatography
1. Low-pressure column liquid chromatography
2. Planar liquid chromatography
    (a) Paper chromatography
    (b) Thin-layer chromatography (TLC)

Modern Liquid Chromatography
3. High-pressure, high-performance liquid chromatography (HPLC)
    (a) Liquid–solid (adsorption) chromatogjraphy (LSC)
    (b) Liquid–liquid (partition) chromatography (LLC), today bonded-phase chromatography (BPC)
    (c) Ion-exchange chromatography (IEC)
    (d) Ion-pair chromatography (IPC)
4. Size-exclusion chromatography (SEC)
    (a) Low-pressure chromatography (SEC)
    (b) High-pressure, high-performance chromatography (SEC)

(C) Supercritical Fluid Chromatography (SFC)

---

techniques have been used extensively for quali-quantitative analyses of antioxidants, light stabilisers and other additives in plastics, elastomers and other materials. An attempt to classify the chromatographic techniques is presented in Table 2.

A chromatographic analysis consists of two basic steps: separation of the mixture components, and detection of the separated components. Chromatographic separations are based on the distribution of the components between a stationary phase and a mobile phase, which is gas for gas chromatography, liquid for liquid chromatography and compressed gases in the critical temperature range for supercritical fluid chromatography. Detection is based on various physico-chemical properties of the separated compounds, e.g. ultraviolet–visible and infrared absorption, fluorescence, refractive index, thermal conductivity and flame ionisation. Identification and quantification of separated compounds are the final steps of the chromatographic analyses.

2.3 Gas Chromatography Versus Liquid Chromatography

Almost all chromatographic techniques have been used for the analysis of antioxidants, light stabilisers and some other polymer additives.

Most of them are reviewed in Crompton's book[8] and in review papers.[14-16] However, a brief survey of the earlier chromatographic analyses is necessary for a better understanding of the performance of today's methods.

2.3.1 Gas Chromatography

The ability of GC to separate and quantitatively determine very small quantities of compounds in complex mixtures has been widely used in the analysis of stabilisers. Most separations are performed by gas–liquid chromatography according to a partition mechanism. Of all the chromatographic techniques GC provides the highest resolution. However, many antioxidants, light stabilisers and other polymer additives are insufficiently volatile and cannot pass through the column or they are thermally unstable and decompose under the conditions of GC separation. Indeed, it was estimated that only about 20% of known organic compounds can be satisfactorily separated by GC without prior chemical modification of the sample in order to obtain analogous compounds of higher volatility (derivatisation).

Common GC allows the analysis of stabilisers with molecular weights up to about 500. Relatively high-molecular-weight (about 800) stabilisers of very low volatility may be separated at high temperatures or by temperature programming on low stationary-phase loadings with high-flow-rate carrier gas. Despite such developments (see Section 6.1), GC is basically not a suitable general technique for the analysis of stabilisers and other additives.

2.3.2 Liquid Chromatography

Liquid chromatography eliminates the disadvantages of GC when dealing with compounds of low volatility and thermal stability. Another advantage is that the decrease in temperature of the mobile phase generally enhances the chromatographic separation because the intermolecular interactions become more effective. Therefore, most LC techniques perform the separation at room temperature, although in some cases the separation is carried out at higher temperatures. These do not usually exceed 60–80°C, so that the stabilisers are not subjected to thermal degradation. In LC threefold interactions occur; the mobile phase not only moves the sample molecules but also interacts with them and the stationary phase. In GC the mobile phase has almost no interaction with the stationary phase and the sample. Thus, an additional variable for controlling and improving the separation becomes available in LC and a greater variety of fundamentally different stationary phases and detectors are available. Sample recovery is easier than in GC. High-molecular-weight compounds and even oligomers and high polymers may be analysed with some LC techniques.

The practice of LC has undergone continuing development since 1906 when the founder of chromatography, Professor M. Tswett, practised the oldest LC technique—LSC on columns packed with an adsorbent. The advent of the two planar LC techniques, i.e. paper chromatography in the 1940s and thin-layer chromatography in the 1950s, greatly simplified the practice of analytical LC. The power of LC is the wide variety of techniques available to the chromatographer (Table 2). However, traditional column chromatography, paper chromatography and TLC belong to classical LC and improvements in equipment, materials, technique and the application of theory have resulted in the development of modern LC, (i.e. high-performance liquid chromatography and size-exclusion chromatography).

2.4 Classical Liquid Chromatography

The various techniques of classical LC have been used for earlier analyses of antioxidants, light stabilisers and other additives in polymers.[8,14,16] However, today these techniques are practically entirely replaced by modern LC.

2.4.1 Column Chromatography

Traditional column chromatography may perform separations by adsorption, partition or ion exchange. Thus, the LSC the columns are packed with a stationary phase containing polar active sites, usually —OH groups in silica gel or alumina. Separation is performed by successive elution of the sample on the column with a range of solvents of increasing polarity. The fractions in which the separated components are concentrated can be monitored by IR, UV absorption, refractive index and the measurement of thermal effects. A large amount of sample is required, between 50 and 500 mg, making easier the subsequent identification of the separated compounds (e.g. by IR spectroscopy). This leads to an especially critical limitation in the analysis of polymer additives due to their low concentration in the polymer since large extract volumes are required to obtain the sample loading needed for the separation.

Successful separations and subsequent identification of various antioxidants, light stabilisers, elastomer antidegradants and accelerators have been obtained by traditional column chromatography and a large number of papers have been reviewed.[8,14,16] Despite some advances which have helped to reduce the volume of elution solvent, often between 5 and 100 litres were required for the complete separation of the sample components. The quantification of the additives is not satisfactory and the technique is very time-consuming. Today the traditional column LC is entirely replaced by modern LC techniques.

2.4.2 Paper Chromatography

This planar LC technique presents some advantages over traditional column LC, i.e. smaller samples can be analysed and the technique is simpler to use. The better reproducibility of retention factor values allows some basic identification of the separated compounds but it is of no practical value for the direct identification of the great number of antioxidants and light stabilisers now available on the market. Although multiple solvent systems have been used to achieve the necessary specificity, this makes paper chromatography a very time-consuming technique for routine analysis. In spite of the fact that many early stabiliser analyses were performed by paper chromatography this LC technique is today practically not used for this purpose.[8,14,16] However, Sreenivasan[17] recently reported the use of ATR–IR spectroscopy to detect and identify the components separated by paper chromatography and showed that the system is suitable for routine analysis of polymer additives.

2.4.3 Thin-Layer Chromatography

TLC, the other planar LC technique to become very popular, replacing paper chromatography in many applications including the analysis of antioxidants, light stabilisers and other polymer additives, offered some advantages. Thus, TLC is a much more rapid technique with good separation efficiency, high separation speed and a greater potential for improvement of detectors, carrier materials, mobile phases, spray reagents and multistage processes. Therefore, this inexpensive and simple method is able to perform the separation of complex systems and is used even today[18,19] for many analyses of antioxidants, light stabilisers and other chemical additives.[8,14,16] TLC was also used to make traditional column LC less time-consuming. Suitable adsorbents and development solvents for achieving a satisfactory separation were determined by rapid preliminary TLC analysis and then translated to column LC, usually without difficulty. TLC may be also used to verify the purity of compounds separated by column LC.

Although stabilisers extracted from polymers have been frequently separated and identified by TLC, quantitative analysis involves many difficulties. Quantitative TLC suffers from inaccuracy and irreproducibility of sample application and spot quantification. The spots have to be re-extracted from the sorbent with a suitable solvent and determined in the solution. The so-called spectrodensitometric technique, i.e. the elution of substances absorbing UV or visible radiation measured directly in the thin-layer, was applied to quantitative analyses. Thus, the content of the light stabiliser Sandovur EPU in PP was determined by TLC in a $CHCl_3$ extract with a direct (*in situ*) evaluation of the reflectance of the spots.[20] Other progress in planar chromatography resulted in improved versions of TLC such as high-performance and over-pressure thin-layer chromatography (HP-TLC and OP-TLC).[21-23] Thus, HP-TLC was used to determine the antioxidant activities of chemicals such as BHT, BHA, α-tocopherol and some alkyl gallates.[24] However, despite all improvements, the TLC techniques cannot approach the analytical power of modern LC. For the analysis of polymer additives TLC seems to be used today mainly as a general screening method of antioxidants and light stabilisers in polymer samples. Thus, TLC has proved to be a complementary technique that can be used effectively in conjunction with modern LC (see Section 4.3).

2.5 Modern Liquid Chromatography

Today, the classical LC techniques have been almost entirely replaced by modern LC,[25,26] high-performance liquid chromatography (Section 2.5.1) and size-exclusion chromatography (Section 2.5.2).

2.5.1 High-Performance Liquid Chromatography

HPLC was developed, as its name shows, to enhance the performance of the classical LC on open columns. This was possible because of improvement in equipment, detectors, column packing and, perhaps more importantly, a thorough understanding of the separation theory. Many basic studies have been carried out on most of these aspects. They have included, for example, new types of stationary phases, small-diameter particles of controlled porosity, improved pump systems able to achieve high inlet pressures at very constant flow rates, as well as a new understanding of reversed-phase and gradient-evolution chromatography.[25]

The 'high-performance' behaviour of this type of LC primarily depends on the stationary phase characteristics. Due to the small-size column packings (usually 5–10 μm) the pressure in the chromatographic column may rise to 20 MPa (3000 psi). Therefore, HPLC is also a 'high-pressure' LC. In the first years of its use HPLC was of course also a 'high-price' LC, but today the price of the equipment seems to be very modest taking into account the results which may be obtained with this technique. Therefore, the HPLC method ('High-Performance Liquid Chromatography' is today the only name used) is now widely employed for the analysis of very different compounds which cannot be separated by GC. HPLC, with efficiencies approaching those of GC, offers many potential advantages for the separation of polymer additives because most of them are non-volatile or thermally labile compounds.

The power in modern LC is the wide variety of modes available to the analyst. The LC technique is defined by the nature of the predominant interaction that occurs between the sample molecules and the stationary phase. Thus, according to the separation mechanism, there are three basic HPLC techniques (adsorption, partition and ion-exchange LC) and the unique LC technique based on the size-exclusion mechanism, i.e. size-exclusion chromatography (Table 2).

(A) *Liquid–Solid (adsorption) Chromatography* (LSC) uses a stationary phase with polar active sites, usually —OH groups in silica gel or alumina, which interact with the polar portions of the sample molecules.

(B) *Liquid–Liquid (partition) Chromatography* (LLC) utilises a mechanically held liquid stationary phase immiscible with the mobile phase. Although this HPLC technique was used for the analysis of stabilisers it is today replaced, as in other applications, by bonded-phase chromatography.

(C) *Bonded-Phase Chromatography* (BPC) resulted from the modification of LLC, and it is the most popular form of HPLC today. The stationary phase consists of various polar or non-polar compounds chemically bonded to a silica gel base. The separation mechanism is mainly partition, when sample molecules actually penetrate the bulk of a thick bonded phase. However, adsorption may also occur when the polar or non-polar sample molecules are attracted to the polar or non-polar functional groups of the bonded phase.

(D) *Ion-Exchange Chromatography* (IEC) and *Ion-Pair Chromatography* (IPC) are the other LC techniques which resulted from the modification of LLC and may be regarded as a combination of LLC or BPC with IEC. They allow the separation of compounds with ionic or multiple ionisable groups. These LC techniques have so far found no significant application in the analysis of antioxidants and light stabilisers.

Normal-Phase (NP) and *Reversed-Phase* (RP) *Liquid Chromatography* are simple divisions of the LC techniques based on the relative polarities of the mobile and stationary phases. 'Normal-phase' LC is referred to when the stationary phase is more polar than the predominant solvents of the mobile phase. The opposite mode, where the stationary phase is less polar than the mobile phase, was termed 'reversed-phase' LC. Except for SEC and IEC, the division 'normal' versus 'reversed' is applied to all LC techniques, i.e. traditional column LC, paper chromatography, TLC and HPLC. Although this division is somewhat arbitrary it is widely accepted. Therefore, the analysis of antioxidants, light stabilisers and other polymer additives will be presented as NP-HPLC (Section 4.1) and RP-HPLC (Section 4.2) separations. Both NP- and RP-HPLC analysis make use of either of the *isocratic* or *gradient-elution* modes of separation (i.e. constant or variable composition of the mobile phase, respectively).

2.5.2 Size-Exclusion Chromatography (SEC)

The other technique of modern LC, formerly known as gel permeation chromatography (GPC), is the most important tool for the analytical separation of macromolecules. SEC separates molecules according to their effective size in solution with practically no interaction between the sample molecules and the stationary phase. Therefore, SEC is of less value in the separation of low-molecular-weight compounds. However, SEC has sometimes been utilised for the analysis of additives extracted from polymers (Section 5.1). SEC is the only LC technique that allows the direct analysis of additives in polymers, without the preliminary extraction step (Section 5.2).

Papers concerned with the analysis of antioxidants, light stabilisers and other chemical additives in polyolefins and other polymers by HPLC and SEC have been reviewed.[27-91]

3 SAMPLING AND STANDARD SAMPLE PREPARATION

Except in the case of some SEC analyses (Section 5.2), the initial step of all the chromatographic procedures to determine antioxidants and light stabilisers in polymers involves the removal of these additives from the polymer matrix. In some cases, the polymer matrix is more or less soluble in the extracting solvents. However, extraction is usually required because no other method has been developed that is suitable for subsequent chromatographic analysis.

Stabilisers cannot be easily extracted from polymers. Besides the partially soluble nature of the polymer itself, other problems arise. Thus, due to the low concentration of stabilisers in polymers, i.e. usually less than 0·5%, any loss in sample handling is significant. Chemical additives, especially antioxidants, may suffer degradation during the extraction step. Therefore, a successful analysis depends primarily on extracting the stabilisers without decomposition at the low concentration encountered.

The extraction procedures as a preliminary step for various analyses of additives in polyolefins were intensively studied in the 1960s[92-110] and have been reviewed.[8,14,92] The focus of work dealing with analyses of additives in polymers by modern LC is on the chromatographic method. However, much emphasis must be placed upon additive extraction followed by LC analysis. The extraction of additives should not be considered as an isolated step because it may strongly influence the subsequent chromatographic separation. The success of an analysis may very often depend more on the extraction procedure than on the chromatographic separation. Therefore, tailor-made extraction procedures have to be developed to assure compatibility between the extraction and chromatographic systems. However, systematic investigation of the problems of quantitative extraction of additives in connection with the chromatographic analysis have only occasionally been performed.

The extraction method should meet the following requirements: (1) high extraction rate for a minimum recovery time; (2) minimum solubility for compounds other than stabilisers present in the sample; (3) no physical loss or degradation of stabilisers; (4) procedural

MODERN LIQUID-CHROMATOGRAPHIC ANALYTICAL METHODS   229

simplicity; and (5) compatibility with the subsequent chromatographic analysis.

Both solid–liquid (Section 3.1) and liquid–liquid extraction (Section 3.2) have been used to remove the additives from the polymer matrix. Table 3 reviews the main experimental conditions used in various extraction procedures for the chromatographic analysis of additives in polymers.

3.1 Solid–Liquid Extractions

Most of the separations of additives from polyolefins that have been reported make use of solid–liquid extractions, usually at the boiling point of the solvent in Soxhlet extractors. Related extractor types such as Wiley[105,106] and Kumagawa[78] extractors, refluxing of the polymer sample soaked in the extraction solvent[39–41,47,150] and extractions below the boiling point of the solvent[49a,84,89,104] have been also utilised.

Regardless of the extraction procedure, for any solid–liquid extraction it is necessary to use a solvent system where the maximum amount of organic additives and the minimum amount of the polymer are extracted, i.e. it must be a good solvent for the additives and a poor one for the polymer. The extraction efficiency basically depends on many factors, the following being the most important: solvent type, additive solubility, characteristics of the polymer sample (type, particle size and shape), extraction temperature and time.

3.1.1 Additive Solubility in the Extraction Solvents

The additive package of commercial polymers may contain antioxidants, light stabilisers and other additives. Some commercial stabilisers contain reaction by-products. Complex decomposition products of stabilisers and other additives may be also present in the sample of polymer to be analysed. Therefore, a complex mixture of chemical compounds is subjected to extraction but not all of them are soluble in the organic solvents employed for extractions. Thus, many inorganic, metallo-organic (e.g. calcium stearate, the widely used antacid agent in HDPE and PP) and even some organic additives are insoluble. However, the vast majority of polyolefin additives can be satisfactorily extracted from the polymer matrix. For solid–liquid extractions the most used extractants are $CHCl_3$, CH_2Cl_2, CCl_4, n-hexane, diethyl ether and tetrahydrofuran. The use of low-boiling-point solvents is advantageous in order to minimise stabiliser degradation during extraction and for the subsequent concentration of the

Table 3
Extraction techniques for additives from polymers prior to chromatographic analysis

Polymer	Additives	Extraction conditions	Chromatography	Ref.
(A) Solid–liquid extraction in Soxhlet or related extractors				
Powdered PE (10 g)	Cyasorb 531	Triple Soxhlet extraction with 500 ml n-hexane or $CHCl_3$ for 12 h, filtration, concentration, wax precipitation with methanol, evaporation to dryness, solution in $CHCl_3$	NP-HPLC, TLC, GC	52
PE or PP ground to 10-mesh (50 g)	DLTDP; DLTDS; Armostat 310, 410; Atmul 84	Soxhlet extraction with 250 ml $CHCl_3$ for 2 h; evaporation to 10 ml on a steam bath; dilution to 25 ml with $CHCl_3$	NP-HPLC	77
PP pellets or powder (10 g)	Tinuvin 770; Hostavin VP N-20	Kumagawa extraction with $CHCl_3$ for 16 h; extract concentration to 20 ml under a flow of nitrogen; wax precipitation with 80 ml acetone; filtration; precipitate washing with hot acetone; filtrate concentration under a flow of nitrogen	RP-HPLC with guard column	78
PP pellets (50 g)	DLTDP; DSTDP: TNPP; Goodrite 3114; Weston 618; Topanol CA; Irganox 1076; Cyasorb UV 531	Soxhlet extraction with 250 ml CH_2Cl_2 for 48 h; evaporation to dryness; redissolving in 5 ml THF; filtration through Millipore Teflon filter of 0.5 μm pore size	RP-HPLC	47
PP pellets (10 g)	Irganox 1010; Irgafos 168; Tinuvin 770; erucamide	Soxhlet extraction with diethyl ether for 15 h; evaporation to dryness; wax precipitation by refluxing with 5 ml ethanol; cooling; filtration	Capillary SFC	159 160
PE ground	Irganox 259, 1010, 1035, 1076; Santonox; Tinuvin P; Ionol; TNPP	Extraction with isopropanol/cyclohexane	RP-HPLC, RP-HSLC	36

Sample	Additives	Procedure	Method	Ref
HDPE pellets (10g)	Irganox 1010; Irgafos 168	Three consecutive Soxhlet extractions each with 60 ml CHCl$_3$ for 2 h; evaporation in rotary vacuum evaporator at 30°C under nitrogen stream until 5 ml extract; wax precipitation with 10 ml MeCN; filtration through a filtration syringe with Millipore FH filter paper of 0·5 μm pore size	RP-HPLC	58, 59
Polyolefins	BHT; Topanol CA; DLTDP; Cyasorb UV 531; Irganox 1010, 1076; stearamide; erucamide	Overnight Soxhlet extraction with isopropanol	RP-HPLC	81
LDPE foils	Crodinstab 1986/003, 007, 010, 014	Soxhlet extraction with CHCl$_3$, CH$_2$Cl$_2$ or hexane for 1–6 h	NP-HPLC	51
PE thin film cut into strips (10·2 g)	Phenolic antioxidants; DLTDP; benzophenones	Double Soxhlet extraction each with 150 ml acetone for 4 h in an inert atmosphere with light excluded; extract concentration in vacuum evaporator; wax precipitation with methanol; filtration; evaporation to dryness; redissolution in THF	SEC	67
LDPE film (5 g)	DSTDP; Irganox 1035; Santonox R; peroxide initiator Vulcup	Soxhlet extraction with 100 ml THF, extract concentration to 5 ml	SEC	56
PVC film or finely ground PVC particles (1g)	Tinuvin 320; Cyasorb UV-9; Uvinul N-539	Soxhlet extraction with diethyl ether for 16 h; evaporation to dryness; solution in THF; filtration through Millipore Teflon filter of 0·5 μm pore size	NP-HPLC, RP-HPLC	64

(continued)

Table 3—contd.

Polymer	Additives	Extraction conditions	Chromatography	Ref.
Dashboard foils (unspecified polymer) out into pieces of 1 mm² area	Irganox 259, 1076; epoxy compounds; unidentified additives	Soxhlet extraction with diethyl ether or hexane for 24 h; evaporation to dryness; solution in THF; filtration through regenerated cellulose filters SM 11606 from Sartorius	Preparative NP-HPLC and RP-HPLC	80a
(B) Solid–liquid extraction from polymers soaked in boiling solvents				
PP copolymer frozen and pulverised (0·5–5·0 g)	Irganox 1330	Refluxing for 40 min under nitrogen purging in 25–100 ml decalin, hexane, CHCl₃, THF in the presence of 0·5–1·5 mg BHT; cooling; filtration through Whatman GF/A microfibre filter	NP-HPLC	41
PP pellets (3g)	Irganox 1010, 1076; BHT; Topanol CA	Refluxing in 45 ml CHCl₃ for 6 h; decantation; concentration to 10 ml; wax precipitation with 10 ml acetone; filtration	RP-HLC	39
(a) LDPE (2g) (b) PP (2g)	BHT; Irganox 1076; Cyasorb UV-531; Tinuvin 327	(a) Refluxing with 10 ml CCl₄; wax precipitation with 10 ml acetone; filtration through 0·5 µm Millipore filter; evaporation to dryness; redissolution in THF (b) Refluxing with 10 ml THF; decantation of the extract; filtration through 0·5 µm Millipore filter; concentration to 1 ml	RP-HPLC	40
Powdered PE (5g)	Tinuvin P, 326, 327, 770; Chimassorb 81; Sandovur EPU; Irganox 1076; Irgafos 168; fatty amides	Refluxing with 200 ml CH₂Cl₂/CH₃OH (9:1) for 3 h; evaporation to dryness on a water bath; solution in CHCl₃	Capillary GC	150

Sample	Additives	Procedure	Method	Ref
Pressed PP foils of max. 0·005-inch thickness	DLTDP; DSTDP; TNPP; Goodrite 3114; Weston 618; Topanol CA, Irganox 1076; Cyasorb UV-531	Refluxing with boiling CH_2Cl_2 for 2 h; evaporation to dryness; solution in 5 ml THF; filtration through 0·5 μm Millipore Teflon filter	RP-HPLC	47

(C) Solid–liquid extraction from polymers soaked in solvents below their boiling point

PP beads or piece cut into small shavings (1 g)	BHT; Topanol CA; Irganox 1076	Overnight extraction with 5 ml MeCN at ambient temperature in a sealed amber vial with constant stirring, filtration	RP-HPLC with precolumn filter	84
Talc-filled PP (4g of 8-mesh pellets)	DSTDP; BHT; Topanol CA; Santowhite powder	Shaking in a 50 ml screw-capped, darkened glass vial with (a) 20 ml THF or (b) CH_2Cl_2 for 24 h at room temperature	(a) SEC (b) NP-HPLC	89

(D) Liquid–liquid extraction

PE or PP pellets (2 g)	BHT; Irganox 1010, 1076; Santonox R; Ethanox 330; Goodrite 3114; Topanol CA; Tinuvin 144	Stirring at 110°C in 50 ml decalin for 30 min; polymer precipitation by cooling; polymer removal with a microspatula; solution drawn into a syringe through a porous metal filter inserted into the solution	NP-HPLC with RP-HPLC guard column	74 75 76
PE piece	Santonox R	Polymer solution in hot toluene; wax precipitation with cold methanol; filtration	NP-HPLC	82
Calendered PVC films	Di-2-(ethylhexyl) phthalate; epoxidized soya-bean oil; tris(nonyl phenyl) phosphite	Polymer solution in THF; precipitation with methanol; filtration; evaporation to dryness in a vacuum oven; solution in CCl_4/CH_2Cl_2 (65:35, v/v) used as the mobile phase	NP-HPLC	29
PVC film, unground or finely ground PVC particles (1 g)	Tinuvin 320; Cyasorb UV-9; Uvinul N-539;	Polymer solution in 20 ml THF; precipitation with 300 ml hexane; filtration; evaporation to dryness; solution in 10 ml THF	NP-HPLC, RP-HPLC	64

(continued)

Table 3—contd.

Polymer	Additives	Extraction conditions	Chromatography	Ref.
PVC films	Tris(2-ethylhexyl) trimellitate and phthalate plasticisers	Polymer solution in THF; filtration through Millipore filters	SEC	29a
Powdered polyurethane (with dry ice)	BHT; Irganox 245; Tinuvin 328	Polymer solution in hot dimethylformamide (DMF)	High-temperature SEC (DMF mobile phase)	54

extract. Chloroform is widely used because of its high volatility and the high solubility of most polyolefin additives.

Some widely used stabilisers have low solubility in typical extraction solvents, for example, the solubility of Irganox 1010 at 20°C in hexane is 0·3 wt% compared with 71 wt% in chloroform. The low extractability of polymeric hindered-amine light stabilisers (HALS) is of advantage in their technical application, e.g. for the stabilisation of PP fibres subjected subsequently to dry cleaning. However, such stabilisers with molecular weights higher than 2000 are more difficult to extract than the usual low-molecular-weight ones. Thus, Soxhlet extraction with chloroform for 16 h led to only 22% recovery of Chimassorb 944 from PP and a complete recovery was achieved only by a liquid–liquid extraction with decalin.[111]

Selective extraction of some antioxidants is basically not possible because of the similar solubility of additives other than antioxidants present in the sample. Therefore the extract generally contains most of the chemical additives of the sample and that complicates the subsequent chromatographic separation. However, commercial polymers usually contain up to five components in the additive package which can be separated relatively easily by modern LC techniques. Hence, the goal of an ideal extraction would be the complete extraction of all additives from the polymer for subsequent chromatographic separation, identification and quantification. Indeed, most published work concerning the analysis of antioxidants and light stabilisers in polymers determines simultaneously the other additives present in the samples investigated.

3.1.2 The Permeability of Polyolefins to the Extraction Solvent

Polyolefins are not soluble at moderate temperatures in extraction solvents but they do contain a fraction of low-molecular-weight polymer (wax) which is more soluble than the parent polyolefin. Therefore, most solvents extract a certain amount of polyolefin wax with subsequent contamination of the extract. Efforts have been made to minimise the quantity of the extracted wax, e.g. by extraction with selective solvents or at lower temperature. Thus, CH_2Cl_2 is a particularly good solvent for PP extractions because of the small amount of atactic material extracted from the polymer compared with other solvents.[8] Wax-free additive extracts from PE and PP have been obtained by extraction with n-hexane at low temperature (0°C).[103] However, typical extraction procedures produce extracts that generally contain a small amount of wax. Such extracts should not be directly injected into the chromatographic column in order to avoid the clogging of its inlet. Therefore, the removal of the wax from the extract is necessary; this is often a painstaking operation which may involve mechanical removal and/or filtration or centrifuging.

The dissolved wax may be precipitated by adding cold non-solvents such as acetone or methanol to the extract. Incomplete removal of the wax from the extract solution prior to injection into the column may cause very serious problems. Thus, wax may precipitate upon injection of the extract, plugging up the injector and even clogging the analytical column. Column clogging leads to a significant pressure increase and this strongly affects the separation. These phenomena may appear especially in RP-HPLC analysis when extracts in typical polyolefin extraction solvents (such as chloroform), when not subjected to wax removal by precipitation, are injected in columns using polar mobile phases which consist of non-solvents for the polyolefin wax, e.g. MeOH and MeCN. The opposite situation occurs in most SEC separations, where the presence of the wax does not usually disturb the separation (Section 5.1).

Wax removal is a very important step in order to improve the separation and to lengthen the column life. Following wax removal by precipitation, prior to injection the extract is usually filtered through very fine filters, e.g. Millipore with average pore size 0·5 μm. Best results have been obtained with guard columns, consisting of short protection columns placed between the injector and analytical column in order to retain sample impurities. Guard columns are usually packed with larger particles of the same type as the microparticles of the analytical columns.[54,78] Thus, a guard column (2 cm × 0·4 cm i.d.) packed with LiChroprep-NH$_2$ (25–40 μm) was connected to an analytical column (25 cm × 0·4 cm i.d.) packed with LiChrosorb-NH$_2$ (10 μm), i.e. both columns used an amino-bonded phase on silica gel.[78] However, Schabron et al.[74–76] protected the normal-phase analytical column (30 cm × 0·39 cm i.d., μ-Porasil, 10 μm porous silica) with a reversed-phase guard column (3 cm × 0·4 cm i.d., C$_{18}$ Corasil, 37–50 μm) which was able to retain the wax because of the strong interaction between the non-polar stationary phase and the polyolefin wax, which had a large aliphatic carbon number. After 20–30 injections of the decalin extract from PE or PP, the guard column had to be changed in order to avoid the pressure increase in the system. Precolumn filters were also used to remove particulate material from the injected sample.[84,85]

Despite all precautions, some additive may remain entrapped in wax so that the solution to be injected no longer contains the total amount of additives extracted from polymer. When non-solvents are used to precipitate the wax they may drive additives back into the polymer matrix.[74] However, it has also been reported that precipitation with acetone of the PP wax in the chloroform extract causes no losses of additives.[78]

The efficiency of solid–liquid extractions depends very much on the crystallinity, shape and size of the polyolefin sample. Polyolefins subjected to quality control or used for the analysis of unknown or competitors' products are usually powders or pellets. Finished products such as films[51,64,67] or moulded parts[82,84,85,89] have been also analysed.

Most polyolefins are semicrystalline polymers. In such polymers the low-molecular-weight stabilisers are concentrated in the amorphous phase, especially at the crystalline boundaries and at the defect centres in the spherulites. The extraction solvents permeate the amorphous phase but the crystallites may not be penetrated completely. Therefore, polyolefins with a lower degree of crystallinity, e.g. LDPE, are more permeable to solvents, allowing more complete recovery of additives from the matrix.[99,101]

Extraction efficiency improves with decreasing particle size. Thus, the time required for the extraction at room temperature of phenolic antioxidants from PEs varies linearly with polymer density and particle size but also with the nature of the extraction solvent.[104] However, when the polymer particles are small enough, the difference in the rate of extraction from PP at ambient temperature with hexane, chloroform or THF is very small despite the wide range of polarity of these solvents.[41] For PE, powdered to 50-mesh, only 3 h of shaking with chloroform at room temperature was found sufficient to remove 98% of the common additives.[104] The large-size samples such as pellets and moulded parts have sometimes been subjected to various procedures in order to increase the surface area/weight ratio and thus speed up the rate of solid–liquid extraction. Thus, polymer samples were cut into small shavings,[84] hot-pressed into thin films,[47] or ground into more or less fine particles.[36,64,77] Ball mills and Willey cutting mills,[77,94,102,105,106] microtomes[95] and grinding following polymer freezing below the glass transition temperature, with liquid nitrogen or solid carbon dioxide,[41,103] have been reported. During these procedures macroradicals may appear in the polymer by thermo-oxidative or mechano-chemical processes. The subsequent reaction of stabilisers with these macroradicals leads to a decrease of their effective concentration in the sample before the chromatographic analysis.

3.1.3 Extraction Temperature and Time

Long extraction times (up to 24 h or even more) have been used to achieve a high efficiency of the solid–liquid extraction (Table 2). In most of these extractions the polymer sample is permeated by the solvent at its boiling point. Therefore, it is advantageous to use volatile solvents such as chloroform in order to minimise stabiliser losses by volatilisation and degradation during extraction. The inherent volatility of some antioxidants such as BHT is so high that it provided the basis of a direct determination in PE by vacuum sublimation.[109] However, only limited attempts appear to have been made to study and to minimise the decomposition and physical loss of antioxidants during extraction.[41,89,104] As antioxidants are designed to be sensitive to oxidation, they obviously have to be protected against degradation during extraction. Extraction at lower temperatures and/or in the absence of oxygen and in the presence of an additional protecting antioxidant may retard or stop the degradation of the antioxidants of interest.[40,41,89,104] Extraction at room temperature requires very long extraction times to achieve a complete recovery of additives.[89,104] Thus, Wims & Swarin[89] found that 24 h of extraction in THF was necessary to recover completely a number of antioxidants from PP using a wrist-action shaker. With CH_2Cl_2, however, only 50% of the additives were extracted in 24 h. Low temperatures during solid–liquid extraction basically avoid antioxidant degradation, as studies using volatile solvents such as hexane, chloroform and THF show.[40,41,89] However, THF, a frequently used extraction solvent, has the tendency to react with oxygen to form peroxides which oxidise the antioxidants to be analysed when no additional protection is provided.[41] Very small amounts of metal and even trace catalyst residues in polyolefins may catalyse the antioxidant degradation during extraction in the presence of oxygen, especially at high temperatures.[41]

3.2 Liquid–Liquid Extraction

To overcome the disadvantage of the slow rate of solid–liquid extractions, some procedures for the separation of additives from polyolefins make use of liquid–liquid extractions which involve the complete dissolution of the polymer in the appropriate solvent, such as toluene and decalin, followed by solution cooling or by addition of a non-solvent such as methanol, ethanol or acetone. After cooling and/or non-solvent addition the polymer precipitates and it is separated by filtration or centrifuging. The filtrate contains all the additives soluble in the solvent or solvent/non-solvent mixture. In these extractions polyolefins have been dissolved by stirring in hot solvents[74–76] or by refluxing.[33,80,82,93] Temperatures over about 100°C are necessary to assure the complete dissolution of the polymer. An extraction procedure involving dissolution of the polymer in boiling toluene under reflux followed by precipitation with ethanol was standardised[93] and used by some workers.[33,80,82] However, Schabron et al.[74–76] found that the use of non-solvents such as ethanol or butyl cellosolve may drive the additives from solution back into the precipitated polymer matrix and only cooling of solution under stirring following dissolution in decalin at 110°C was found to lead to a good recovery of additives from polyolefins.[74–76] The insufficient recovery of Chimassorb 944 from PP pellets with n-decane and tetrachloroethane was attributed to the coprecipitation of this polymeric light stabiliser with the polyolefin. Only decalin gave an efficient recovery (96%) for this system.[111]

Although liquid–liquid extractions are less time-consuming than solid–liquid, the stabilisers are subjected to more losses due to volatilisation and/or degradation because of the higher temperatures employed. Gasslander & Jaegfeld[41] showed that the unstable antioxidant Irganox 1330 may be protected against degradation during reflux of PP samples in decalin solution only by the combined effect of continuous nitrogen purging and a 100-fold molar excess of BHT (Fig. 1). Freitag[111] prevented the autoxidation of decalin by adding Irganox 1010 before heating PP pellets in decalin at 150°C for the extraction of Chimassorb 944. Significant amounts of polar aromatic impurities in decalins from various commercial sources influenced the HPLC separation of stabilisers.[76]

3.3 Compatibility of the Extract with the Chromatographic System

The choice of a general method for the extraction of additives from polymer depends on its simplicity and efficiency. However, an ideal extraction method for subsequent LC analysis requires the compatibility of the extraction solvent with the mobile phase used in the chromatographic separation of additives. Figure 2 shows the incompatibility of chloroform and THF in an NP-HPLC analysis using

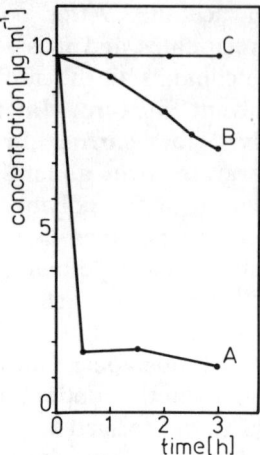
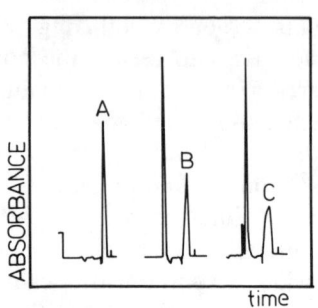

Fig. 1. Degradation of Irganox 1330 during reflux in decalin at c. 180°C: A, oxygen-saturated solution; B, continuous nitrogen purging; C, continuous nitrogen purging and 100-fold molar excess of BHT in solution.[41]

Fig. 2. Chromatograms from Irganox 1330 solutions in various solvents: A, hexane; B, 20:80 $CHCl_3$/hexane; C, 20:80 THF/hexane.[41] Chromatographic conditions: normal-phase column (15 cm × 0·4 cm i.d., Nucleosil-5 NO_2, 5 μm); mobile phase hexane/THF, 99:1; UV detection at 230 nm; injected sample 20 μl, 20 μg/ml.

hexane as mobile phase. Only a hexane solution of antioxidant gives a narrow peak with minimal baseline disturbance. Although the chloroform and THF extracts are diluted five times with hexane, the antioxidant peak becomes lower and broader, the incompatibility being more pronounced for THF, where the solution must be diluted 10 times in order to obtain acceptable peak shapes. The combined effects of dilution and peak broadening lowered the sensitivity by a factor of 10 or 20 for chloroform or THF, respectively.[41]

In some RP-HPLC analyses the injection of compounds such as hindered phenol-type antioxidants in a typical solid–liquid extraction solvent such as chloroform may result in distorted or split peaks.[72,73] Injection of BHT and Irganox 1010 in decalin solution, i.e. in a typical liquid–liquid extraction solvent, resulted also in a distorted BHT peak and a broadened Irganox 1010 peak.[74]

In some cases the same solvent is used for extraction and as the mobile phase or as the major component of the mobile phase in HPLC but especially in SEC separations. Sometimes, the extraction solvent is compatible with the mobile phase, e.g. the system decalin/heptane: CH_2Cl_2.[74–76]

When the extraction solvent is not compatible with the mobile phase the additive-containing extract has to be concentrated almost or even to dryness for subsequent dilution or redissolving in an appropriate solvent, usually the major component of the mobile phase. Extract concentration is also required, because of the low additive content in the polymer samples and the high ratio of extraction solvent to polymer employed. Many extraction procedures obtain extracts of too low an additive concentration. Therefore, the extracts have to be concentrated prior to injection in order to achieve an additive concentration higher than the detection limit of the detector employed (see Section 4.3).

The concentration of the extract (and especially concentration to dryness) is also a critical step which may result in stabiliser losses due to volatilisation and degradation. Therefore, the extracts have to be concentrated under an inert atmosphere and/or reduced pressure, e.g. using a rotary vacuum evaporator under a stream of nitrogen.[58,59] Stabiliser extracts in highly volatile solvents such as $CHCl_3$ and CH_2Cl_2 are again preferred. To avoid further risk of degradation, extracts are better analysed without delay or with only a short delay. If this is not possible the storage in actinic glassware under nitrogen in a refrigerator minimises the risk of stabiliser degradation.[8]

3.4 Standard Sample Preparation

For the study of extraction procedures and chromatographic separation, known amounts of additives are incorporated into the additive-free polymer. The amount of additive found by the chromatographic analysis of the solvent extract from these 'standard' samples is compared with the added amount in order to evaluate the percentage recovery of the additive. Therefore, the preparation of the polymer standards containing an exactly known amount of additive uniformly dispersed in the polymer matrix is also of great importance. These standards quantify the reproducibility of the analyses. However, the vast majority of papers dealing with chromatographic analysis of additives in polymers do not mention the preparation of the standard samples.

Stabilisers are generally totally miscible with the molten polymer, but at room temperature their solubilities are very limited. Stabilisers may be lost during the preparation of standards by volatilisation and degradation, or they may not be uniformly dispersed in the polymer matrix. Various approaches have been used for the preparation of standard samples. Thus, known amounts of additive have been incorporated by milling into the unstabilised polymer to produce a masterbatch which was then milled with known amounts of virgin polymer.[112] Significant loss of volatile antioxidants such as BHT occurred during standard sample preparation.[94]

The best method of obtaining a very uniform dispersion of stabilisers in polymers seems to be melt-mixing following solvent-blending. An additive-free polymer powder is mixed with a solution of the stabilisers in a volatile solvent (usually CH_2Cl_2). Polymer pellets or flakes may also be used. The solvent is then removed by evaporation, preferably at low temperature (max. 50–80°C) under an inert atmosphere and with stirring to improve the dispersion (e.g. in a rotary vacuum evaporator under a stream of nitrogen). Hot-pressing is then used for the preparation of standard films.[30,94,113] This method appears to give highly irreproducible results.[114] Melt-pressing in a vacuum press was always accompanied by significant loss of some antioxidants.[114] Immersion of preformed commercial PP films in iso-octane solutions of stabilisers for 50–200 h was also used to obtain standard samples.[30] Melt-mixing following solvent-blending was performed in small-size extruders (0·25–0·75-in (6·4–19·1 mm) screw diameter) and the extrudate was chopped into granules. Twin-screw extrusion assures homogeneous dispersion of stabilisers in the polymer.[47] When single-screw extruders were used two extrusions were necessary by recycling the granules through the extruder.[114] In HPLC analysis of a stabiliser mixture (BHT + Cyasorb 531 + Tinuvin 327 + Irganox 1076) extracted from PE and PP standard samples, Francis et al.[40] found 98–100% recovery for the solvent-blended standards and only 88–97% recovery for the extruded standard samples. No study of the homogeneity of additive concentration throughout the standard sample seems to have been published.

4 HIGH-PERFORMANCE LIQUID CHROMATOGRAPHY

The additives extracted from polymers may be separated by normal- or reversed-phase HPLC. Both methods may be applied either in the isocratic mode, i.e. the mobile phase has constant composition, or with gradient elution, i.e. the composition of the mobile phase varies during the separation. Normal- versus reversed-phase and isocratic versus gradient-elution HPLC separation will be discussed in the following sections. Selection from these four available separation techniques depends on many variables but basically on the number and chemical structure of the compounds to be separated and on the purpose of the analysis. Hence, although some general rules may be developed, the chromatographer must devise a tailor-made separation for each additive mixture.

4.1 Normal-Phase HPLC Separations

When the predominant functional group of the stationary phase is more polar than the commonly used mobile phases, the separation technique is termed 'normal-phase' HPLC. The term 'normal' has a historical meaning because early LC separations were primarily carried out either in the adsorption mode (LSC) using silica gel with polar silanol groups or in the partition mode (LLC) using supported polar coatings and relatively non-polar mobile phases.

Majors[55] in 1970 was perhaps the first to show the utility of modern LC for separating polymer additives such as antioxidants. Thus, aromatic amines and hindered phenolic antioxidants were separated by LSC, LLC and BPC. At that time pellicular packings were used, i.e. particles of average size 40 μm which consisted of a solid glass bead with a thin (1–2 μm) porous outer shell of silica (for LSC) or a silica layer to which a 'liquid' phase was mechanically held (for LLC) or chemically bonded (for BPC). Thus, Majors[55] utilised the following column packings: 37–50-μm silica particles (for LSC), 0·5% β,β'-oxidipropionitrile mechanically held on 37–50-μm silica particles (for LLC) and 3·7% OC_4H_4CN bonded to 36–75-μm silica particles (for BPC).

Today the HPLC analysis of antioxidants, light stabilisers and other polymer additives are performed only by LSC and BPC. The pellicular packings have been entirely replaced by the microparticulate packings, i.e. microparticles of 5 and 10 μm sizes which have considerably increased the column efficiency and speed of analysis. Therefore, shorter column lengths (25–30 cm vs 100–120 cm) achieved the same separation as did pellicular columns. The pellicular packings still find use as a packing material for guard columns[54,74–76,78] for the protection of analytical columns since they can be conveniently dry-packed.

The advent of stable, reproducible, chemically bonded phases has made BPC the most widely used separation technique for the analysis of stabilisers too. The functionality of the most widely used bonded phases in NP-HPLC are amino (μ-Bondapak NH_2, LiChrosorb NH_2) and cyanonitrile (μ-Bondapak CN). Other bonded-phase functional groups such as nitro (Nucleosil-5 NO_2) have sometimes been used.[41] Stationary phases with bonded polar functional groups are continuously replacing the unmodified silica gel in NP-HPLC because they are milder, e.g. they give rise to fewer chemisorption and tailing problems, and respond much more rapidly to mobile phase composition changes. However, many separations of polymer additives have been formed on unmodified silica packings.

The mobile phase of NP-HPLC is less polar than the stationary phase. The mobile phase consists of a non-polar solvent such as a hydrocarbon (the weak solvent) which may be mixed with a more polar solvent (the strong solvent). Weak solvents are those which interact very little with the stationary phase. Therefore, in NP-HPLC the weak solvents are non-polar hydrocarbons such as n-hexane, n-heptane, iso-octane and cyclohexane. Strong solvents are those which may interact very strongly with the stationary phase, thus replacing the compounds strongly retained on the stationary phase. Therefore, in NP-HPLC the strong solvents are more polar compounds such as CH_2Cl_2, $CHCl_3$ and CH_2Cl—CH_2Cl. The strength of the mobile phase may be varied to achieve the optimum separation by mixing strong solvents with weak ones.

In the NP-HPLC separation mode the more polar stabilisers are more strongly retained on the polar stationary phase and, consequently, the less polar stabilisers elute first at shorter retention times. Therefore, to elute a polar stabiliser faster, an increase of the mobile phase strength is usually performed. The elution of each compound is measured by the peak retention time (volume) or by the capacity factor

$$k' = (V - V_0)/V_0 = (t - t_0)/t_0$$

where $t(V)$ is the retention time (volume) of the eluted compound and $t_0(V_0)$ is the retention time (volume) of a substance that has not been retained in the column (usually a non-polar solvent such as hexane for NP-HPLC and a polar solvent such as MeCN for RP-HPLC).

Isocratic separations are generally preferred because of better precision and simplicity compared with the gradient-elution mode of separation. However, the range of polarity of polyolefin additives is too great to elute all of the compounds with a mobile phase of constant composition while maintaining sufficient resolution. The isocratic separation of a stabiliser mixture containing many compounds and with a wide range of polarity would result in a chromatogram with many agglomerated peaks at the beginning and broad peaks at very long retention times. Such separations do not allow accurate identification and quantification and, consequently, the gradient-elution mode has to be used. However, normal gradient-elution separation is excluded for the analysis of non-UV-absorbing additives which require the use of the refractive index (RI) detector and, therefore, an isocratic separation.

Although NP-HPLC has good resolution, the retention times are not reproducible over a long period of time because of the inevitable accumulation of polar material on the silica gel, thus decreasing its adsorption ability.

4.1.1 Isocratic Separations

Most of the NP-HPLC analyses of polymer additives have been performed in the isocratic mode.[41,43,50–52,55,64,70,71,75–78,80,82,83,89] Separation basically depends on the backing type, polarity of stabilisers and mobile phase strength. Table 4 exemplifies these dependences in the isocratic NP-HPLC separation of some stabilisers of different polarity, performed by LSC (silica gel packing) and BPC (CN and NH_2 bonded-phase packings) with mobile phases of increasing strength.[52] The more polar stabilisers such as Cyasorb UV-531 are more strongly retained on the polar stationary phase, i.e. they have higher capacity factors, than the less polar stabilisers such as BHT. Increasing the mobile phase strength by increasing the chloroform content in cyclohexane resulted in a faster elution of stabilisers. Mobile phases of different strengths were also obtained from mixtures of various weak (cyclohexane, n-heptane) and strong ($CHCl_3$, CH_2Cl_2) solvents. The bonded-phase packings are milder than the silica gel packing. Thus, except for Cyasorb UV-531, all stabilisers are more strongly retained on the silica column than on the CN bonded-phase column, for the same composition of the mobile phase (cyclohexane $CHCl_3$, 90:10).

When the mixture of stabilisers to be separated contains compounds strongly retained on the stationary phase, an increase of the mobile phase strength is usually performed in order to reduce the analysis time.[89] However, too great an increase of the mobile phase strength may result in loss of resolution for the less polar stabilisers, leading to

Table 4
Capacity factors (k') of some stabiliser standards and extracts with different statinary and mobile phases[52]

Stationary phase	Lichrosorb Si 60			CN bonded phase				NH$_2$ bonded phase		
Mobile phase[a]	CHx:CHCl$_3$				CHx:CHCl$_3$		CHx	CHx:CH$_2$Cl$_2$	CHx:CHCl$_3$	Hp:CHCl$_3$
Sample	90:10	70:30	0:100	100:0	95:5	90:10	100	95:5	98:2	98:2
Standards										
Tinuvin P	0.26	0.24	0.05	0.62	0.41	0.00				
Tinuvin 120	0.10	0.00	0.00	2.71	1.01	0.00				
Tinuvin 326	0.10	0.00	0.00	0.01	0.00	0.00	2.80	0.70	1.20	1.31
Tinuvin 327	0.12	0.00	0.00	0.01	0.00	0.00	2.80	0.68	1.30	1.35
Cyasorb UV-531	0.58	0.22	0.04	1.95	0.95	1.30	7.45	0.90	4.27	4.37
BHT	0.11	0.09	0.00	0.53	0.20	0.00	0.42	0.07	0.21	0.24
Extracts from PE										
Hexane extract	0.50	0.22	0.03	1.96	1.01	1.37	7.35	0.93	4.30	4.34
	0.09[b]	0.06[b]	0.00	0.50[b]	0.21[b]	0.00	0.40[b]	0.03[b]	0.18[b]	0.22[b]
CHCl$_3$ extract	0.58	0.18	0.03	1.90	0.98	1.01	7.40	0.90	4.33	4.40

[a] CHx, cycloxehane; Hp, heptane.
[b] Only in trace concentration.

many agglomerated or even overlapping peaks at the beginning of the chromatogram. Therefore, the optimum strength of the mobile phase has to be found. For the separation of a few stabilisers of similar polarity isocratic analysis is always preferred. Analysis of a single stabiliser in the polymer represents the easiest application of this technique. Such analyses are carried out typically for quality-assurance tests and to verify the efficiency of the extraction procedure.[41,75,77,82] Thus, the level of the antioxidant Santonox R in a PE batch[82] and of the light stabiliser Tinuvin 144 in PP[75] were established by the analysis of the solvent extract. The analysis of chloroform extracts from PE and PP standard samples, each containing a single additive (DLTDP, DSTDP, mono- and di-glycerides, tertiary $C_{12}-C_{16}$ alkyldiethanolamines) showed a practically complete recovery of all additives after 2 h of Soxhlet extraction.[77]

The development of hindered-amine light stabilisers (HALS), a new class of widely used polyolefin light stabilisers, required suitable procedures for their identification and quantification in polymers. Some chromatographic and spectrometric methods for the analysis of HALS in polymers have been reviewed.[75,78] The NP-HPLC analysis of amine-type stabilisers is preferred because their separation by RP-HPLC presents some difficulties due to tailing effects. For the separation of Tinuvin 770 and Hostavin TMN-20, Sevini & Marcato[78] used an amino bonded-phase column and the MeCN H_2O mixture, a non-typically mobile phase for NP-HPLC separations where water is the strong solvent because it is more polar than MeCN. The strength of the mobile phase was modified in order to optimise the separation (Fig. 3). With a stronger mobile phase (i.e. higher water content), the peaks were very symmetrical but the retention times were considerably reduced and approached each other as well as the solvent front with some interferences [Fig. 3(C)]. With a weaker mobile phase the retention times increased but a very bad tailing effect was evident affecting the resolution and quantitative determination [Fig. 3(A)]. The best results were obtained with MeCN/H_2O 99·5:0·5 where the peaks of the two stabilisers were symmetrical, well resolved and their retention times were reproducible [Fig. 3(B)]. For the analysis of Tinuvin 144 in PP, Schabron & Bradfield[75] developed an isocratic separation on a porous silica column using as mobile phase $CHCl_3/C_2H_5OH$/ammonia, 95:5:0·05, similar to the mobile phases used for silica TLC of amines.

Usually more than one additive is used in polyolefin formulation.

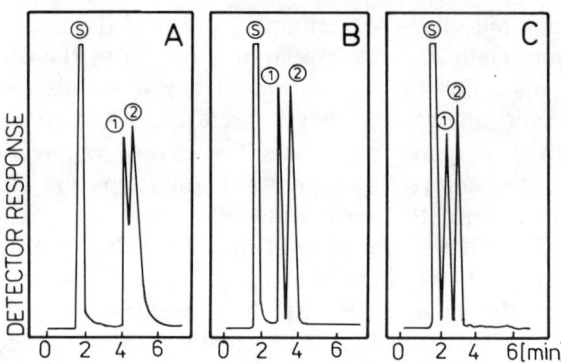

Fig. 3. Isocratic NP-HPLC separation of two hindered-amine light stabilisers at different mobile phase strengths.[78] Chromatographic conditions: analytical column (25 cm × 0·4 cm i.d., LiChrosorb-NH_2, 10 µm); guard column (2 cm × 0·4 cm i.d., Lichroprep-NH_2, 25–40 µm); mobile phase composition MeCN/H_2O (A) 99·9:0·1, (B) 99·5:0·5, (C) 98:2; flow rate 2 ml/min; UV detection at 280 nm; injected sample 10 µl. Peaks: S, solvent; 1, Tinuvin 770; 2, Hostavin TMN20.

Thus, in determining a stabiliser it is important to check possible interferences with the other stabilisers and additives which could be present in the polymer. In general, interference studies should be done with any other known additive present, and many of the papers concerned with HPLC analysis of polymer additives have reported such studies. For example, the chromatographic conditions established for the separation of Tinuvin 770 and Hostavin TMN-20 allowed both identification and quantification of these two HALSs with no interference with the following commercial additives: BHT; Irganox 1010, 1076; Irgafos 168; Ethanox 330; Cyasorb UV-531; Tinuvin 120, 144, 326, 327; fatty acids and amides all of which eluted at or near to the solvent front.[78]

Although the isocratic mode of HPLC separation is not ideally suited to the analysis of complex mixtures containing stabilisers with a wide range of polarity it may be used in some cases, at least for qualitative purposes. The optimisation of separation is achieved by changing the mobile phase strength. Thus, Perlstein[64] analysed a complex mixture of 15 light stabilisers on a silica column with n-hexane/$CHCl_3$ 90:10 mixture as mobile phase [Fig. 4(A)]. These conditions do not allow the separation of stabilisers with very close

MODERN LIQUID-CHROMATOGRAPHIC ANALYTICAL METHODS   249

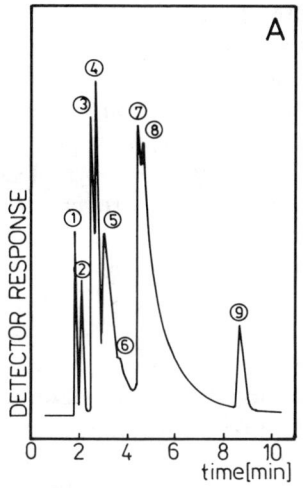
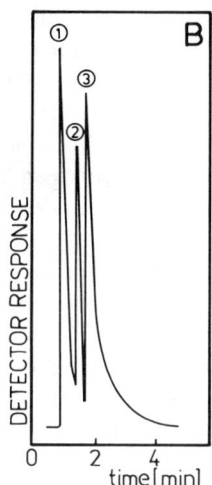

Fig. 4. Isocratic NP-HPLC separation of a synthetic mixture of 15 light stabilisers.[64] Chromatographic conditions: column (30 cm × 0·78 cm i.d., μ-Porasil, 10 μm); UV detection at 254 nm.
A, Mobile phase n-hexane/CHCl$_3$, 90:10, at 1 ml/min; injected sample 5 μl, 0·2% solution in CHCl$_3$. Peaks: 1, Cyasorb 1084; 2, Tinuvin 320 + 326 + 327; 3, Tinuvin P; 4, Sunkem MS + Salol; 5, Cyasorb UV-531 + Uvinul M-410; 6, impurity in Uvinul M-410; 7, Cyasorb UV-9; 8, Uvinul N-539; 9, Uvinul N-35.
B, Mobile phase CHCl$_3$ containing 0·6% ethanol at 3·5 ml/min; injected sample 4 μl; 1% solution in CHCl$_3$. Peaks: 1, Cyasorb UV-24; 2, Eastman RMB; 3, Uvinul 400.

chemical structures, i.e. Tinuvin 320, 326, 327 (benzotriazoles), Sunkem MS and Salol (methyl and phenyl salicylates) and Cyasorb UV-531 and Uvinul M-410 (benzophenones). Each of these three groups of compounds gave a single peak in the chromatogram. Two benzophenone-type stabilisers showed extremely long retention times, being strongly retained on the silica column because of the presence of two hydroxyl groups in their molecule (Uvinul 400 and Cyasorb UV-24). Resorcinol monobenzoate, a compound with a single OH group, showed the same behaviour. An increase of the mobile phase strength to n-hexane/CHCl$_3$ 50:50 reduced the retention times of these three stabilisers but the two benzophenones showed very strong tailing. A further increase of the mobile phase strength to 100% CHCl$_3$ produced a good separation with no tailing for benzophenones

[Fig. 4(B)]. However, under these conditions all the other light stabilisers eluted together at the solvent front.[64]

Light stabilisers of the benzophenone and benzotriazole type have been determined by acetylation and NP-HPLC on a silica column.[48] However, acetylation is an additional step of analysis and it is possible only if an active hydrogen atom is present in the molecule; many important light stabilisers do not contain such hydrogen atoms.

Synthetic mixtures of antioxidants and solvent extracts from polyolefins containing mixtures of antioxidants and other chemical additives have also been separated by isocratic NP-HPLC.[43,50,51,74,77,80,83,89] Dicumyl peroxide, trimethylolpropane trimethacrylate and polymerised 2,2,4-trimethyl-1,2-dihydroquinoline in compounded, uncured EPDM elastomers have been determined using a silica column and the mobile phase cyclohexane/THF, 98:2.[70] Mixtures of di(2-ethylhexyl) phthalate, epoxidised soya-bean oil and tris(nonylphenyl) phosphite extracted from PVC films have been separated on a μ-Porasil column with CCl_4/CH_2Cl_2 (65:35) as the mobile phase.[29]

4.1.2 Gradient-Elution Separations

The gradient-elution mode of HPLC separation is used when complex mixtures of stabilisers with very different polarity have to be analysed. The composition of the mobile phase is varied during the separation to obtain an increase of their strength. Compared with the isocratic mode, additional parameters are thus available to influence the separation of stabilisers, notably the mobile phase strength at the beginning and at the end of separation and the gradient time and shape. By proper choice of these parameters, the separation may be drastically improved in terms of peak resolution, detection and analysis time. However, the gradient-elution mode of separation does not allow the use of the universal RI detector and, consequently, non-UV-absorbing additives are not amenable to analysis by this technique.

Various synthetic mixtures of polymer additives and solvent extracts from polymers have been separated by gradient-elution NP-HPLC making use of n-hexane or n-heptane as weak solvents and CH_2Cl_2 as strong solvent.[43,53,66,74,76,89] In an isocratic separation of three antioxidants with different polarity using the n-heptane/CH_2Cl_2 mixture as mobile phase, Schabron & Fenska[74] demonstrated that BHT showed weak retention while Irganox 1076 and 1010 showed stronger

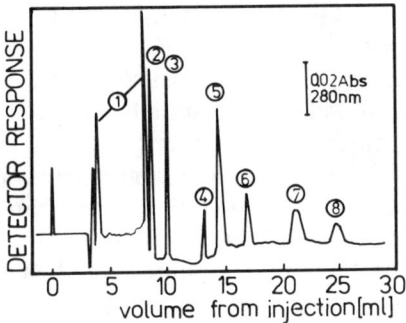

Fig. 5. Gradient-elution NP-HPLC separation of a synthetic mixture of seven antioxidants.[76] Chromatographic conditions: analytical column μ-Porasil (30 cm × 0·39 cm i.d., 10 μm); reversed-phase guard column (3 cm × 0·4 cm i.d., C_{18}-Corasil, 37 μm); linear gradient from 100% n-heptane to 100% CH_2Cl_2 in 5 min at 2 ml/min; UV detection at 280 nm; injected sample 25 μl in decalin. Peaks: 1, blank decalin; 2, 0·59 μg BHT; 3, 0·75 μg Ethanox 330; 4, 0·61 μg Irganox 1076; 5, 0·84 μg Santonox R; 6, 0·74 μg Goodrite 3114; 7, 0·64 μg Irganox 1010; 8, 0·65 μg Topanol CA.

retention. A linear gradient-elution from 100% n-heptane to 100% CH_2Cl_2 in 5 min improved the separation. The same chromatographic conditions allowed the separation of a wider range of polyolefin antioxidants[76] (Fig. 5).

Wims & Swarin[89] identified the additive package of two commercial PP samples by comparing the chromatograms from the CH_2Cl_2 extracts with the chromatogram from the synthetic mixture of seven stabilisers of widely varying polarity. The elution gradient from 0·9% to 70% CH_2Cl_2 in hexane at 10% per minute allowed the complete separation of the stabilisers in the following elution order: BHT; Ethanox 330; Antioxidant 425; Santonox R; Irganox 1010; Santowhite; and Topanol CA.

4.2 Reversed-Phase HPLC Separations

In RP-HPLC the stationary phase is usually a hydrophobic bonded phase such as octadecylsilane or octylsilane. The development of stable, reproducible, chemically bonded phases in the early 1970s has made this type of packing the most widely used. The vast majority of RP-HPLC separations are carried out today on C_{18} or C_8 hydrocarbon phases bonded to silica microparticles. Thus, most of the RP-HPLC analyses of polymer additives make use of the following columns: Lichrosorb RP-18, μ-Bondapak C_{18}, HC-ODS/Sil-X. Other non-polar substrates such as phenyl-bonded phases are sometimes used.[54]

In RP-HPLC the mobile phase usually consists of polar solvents such as water and water-soluble organic solvents such as methanol, MeCN and THF. Less polar sample components are more strongly retained on the non-polar stationary phase than the more polar components which present little interaction with the stationary phase. Therefore, to increase the rate of elution of the less polar components, a mobile phase of low polarity is necessary. Thus, in RP-HPLC the solvent strength decreases with increasing polarity so that water, the most polar compound, is the weakest solvent. In mixtures of water and organic solvents the latter are strong solvents which are used as modifiers of the mobile phase strength. In such mixtures the relative strength of the most used modifiers increases with decreasing polarity in the following order: methanol, ethanol, acetonitrile, dioxan, isopropanol, tetrahydrofuran. The same rule applies for mobile phases without water. For example, in a mixture of two organic solvents such as MeCN + THF, the more polar MeCN is the weak solvent and the less polar THF is the strong solvent.

At present RP-HPLC dominates the application of HPLC, including the analysis of antioxidants, light stabilisers and other polymer additives, because of several important advantages. Thus, the elution order is often predictable on the basis of the degree of hydrophobicity of the sample molecule. The less polar stabilisers are more strongly retained on the non-polar stationary phase. The more polar stabilisers prefer the polar mobile phase and elute before non-polar components, which are forced into the hydrocarbon bonded phase. The bonded-phase columns are fairly reproducible, relatively stable and retention times are more reproducible than in NP-HPLC. Mobile phases typically employed use solvents of lower refractive index such as water, MeOH and MeCN. Thus, in the isocratic separation of non-UV-absorbing additives, the RI detector can be operated at higher sensitivity.

4.2.1 Isocratic Separations

Many RP-HPLC analyses of polymer additives have been performed in the isocratic mode.[28,30,31,36,37,39,43,47,57–59,63,86,91] Isocratic RP-HPLC has been used to separate not only simple mixtures of polymer additives but also very complex ones, containing 10, 20 or more

Table 5
Capacity factors (k') and UV/RI response ratios of some common polyolefin additives[47]

Additive	k'	UV/RI
BHT	0·29	5·6
TNPP	0·33	4·1
Topanol CA	0·43	6·1
Cyasorb UV-531	0·43	21·9
Goodrite 3114	0·57	2·0
Oleamide	0·53	0·0
Irganox 1010	0·83	3·0
Stearamide	0·90	0·0
Weston 618	1·12	0·0
Ethanox 330	1·19	3·5
Erucamide	1·26	0·0
DLTDP	1·64	0·0
Irganox 1010	2·43	1·9
DSTDP	8·04	0·0

Chromatographic conditions: Column C_{18} μ-Bondapak; Mobile phase MeCN/THF/acetic acid, 83:15:2 at 1·3 ml/min; Injected sample volume 10 μl in THf; UV detection at 280 nm (0·5 AUFS sensitivity) and RI detection (4 × sensitivity).

compounds of different polarity. The main problem involved in these analyses was to find the optimum strength of the mobile phase in order to separate the majority of mixture components. However, in very complex mixtures not all the additives may be separated. Thus, in an analysis of a synthetic mixture of 14 polyolefin additives, Haney & Dark[47] found that the spread of capacity factors of various additives was good enough to separate most combinations of additives but there are several possibilities of overlap, e.g. BHT/TNPP (Table 5).

To emphasise the interactions between various compounds, Munteanu et al.[59] performed successive injections of 22 stabilisers starting with the most polar compound, namely the antioxidant Ethanox 754 containing two hydroxyl groups, and adding the other stabilisers one by one. The simple mobile phase employed, i.e. the strong eluent MeCN, does not permit the complete separation of stabilisers and the first peaks, corresponding to the more polar compounds, overlap. However, these conditions allow the elution of the strongly retained non-polar antioxidants DLTDP, Irgafos 168 and Irganox 1076 at not excessively long retention times [Fig. 6(a)]. Using a weaker mobile phase, (MeCN/H_2O, 90:10), many stabilisers are separated well enough to allow qualitative analysis but the peaks of some polar stabilisers are still overlapping [Fig. 6(b)]. Using a still weaker mobile phase, i.e. MeCN/H_2O, 80:20, a satisfactory separation of the stabilisers was achieved. However, the retention times of some stabilisers are close and Goodrite 3114 presents a very long retention time [Fig. 6(c)]. In such a separation strategy that decreases the strength of the mobile phase, the retention times of the non-polar stabilisers would be longer the weaker is the mobile phase and, consequently, they were not injected (Table 6; see for example Goodrite 3114).

Inversions of the elution order of polar stabilisers may occur (Table 6). Changes in the order of the eluted solutes on varying the composition of the mobile phase have been observed with polyalkylbenzenes and -phenols, notably when the alkyl substitutents were branched or contained more than three carbon atoms.[65] This behaviour may be explained by the opposite influence of the following two factors on the separation mechanism: (i) the hydrophobic nature of the molecule, and (ii) the higher affinity for the mobile phase by solvation. In the separation of some phenolic antioxidants with different composition of the MeOH/H_2O mobile phase (90:10, 80:20 and 70:30) the retention times decreased with the MeOH/H_2O ratio.[31]

Such separation studies[31,58,59] with mobile phases of different strengths show that suitable conditions for the separation may be found. The strength of the mobile phase can be looked upon as an adjustable parameter. Thus, if the non-polar additives strongly retained on the stationary phase are not of interest, the analysis time is reduced considerably and weaker mobile phases may be used to improve the separation of polar additives. If such non-polar additives are of particular interest they can be eluted faster using stronger mobile phases. This will result, of course, in loss of resolution of the quickly eluting components. However, an optimum value of the mobile phase strength may be found for general-purpose work. It should be pointed out that the additive package of most polyolefins usually does not contain more than three or four additives. Conse-

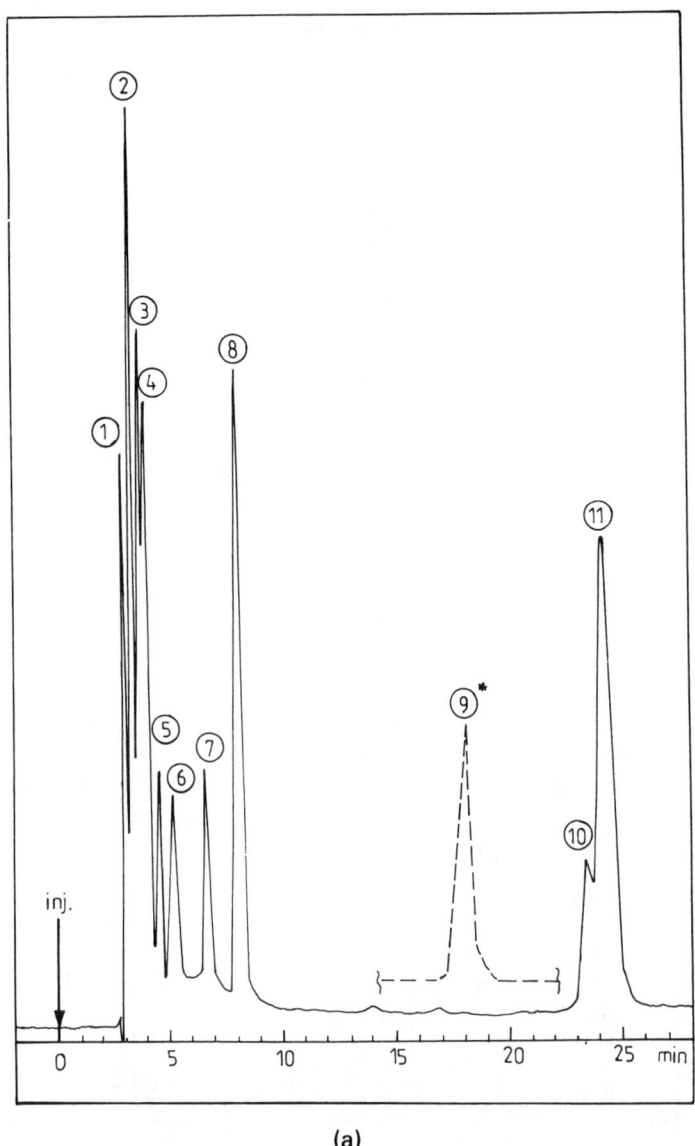

(a)

Fig. 6. Isocratic RP-HPLC separation of a synthetic mixture of 21 antioxidants and light stabilisers at different strengths of the mobile phase: (a) 100% MeCN; (b) 90:10 MeCn/H_2O; (c) 80:20 MeCN/H_2O.[59] Chromatographic conditions: column (25 cm × 0·4 cm i.d., LiChrosorb RP-18, 5 μm); flow rate 1 ml/min; UV detection at 280 nm (DLTDP monitored with the RI detector). Peaks: see Table 6.

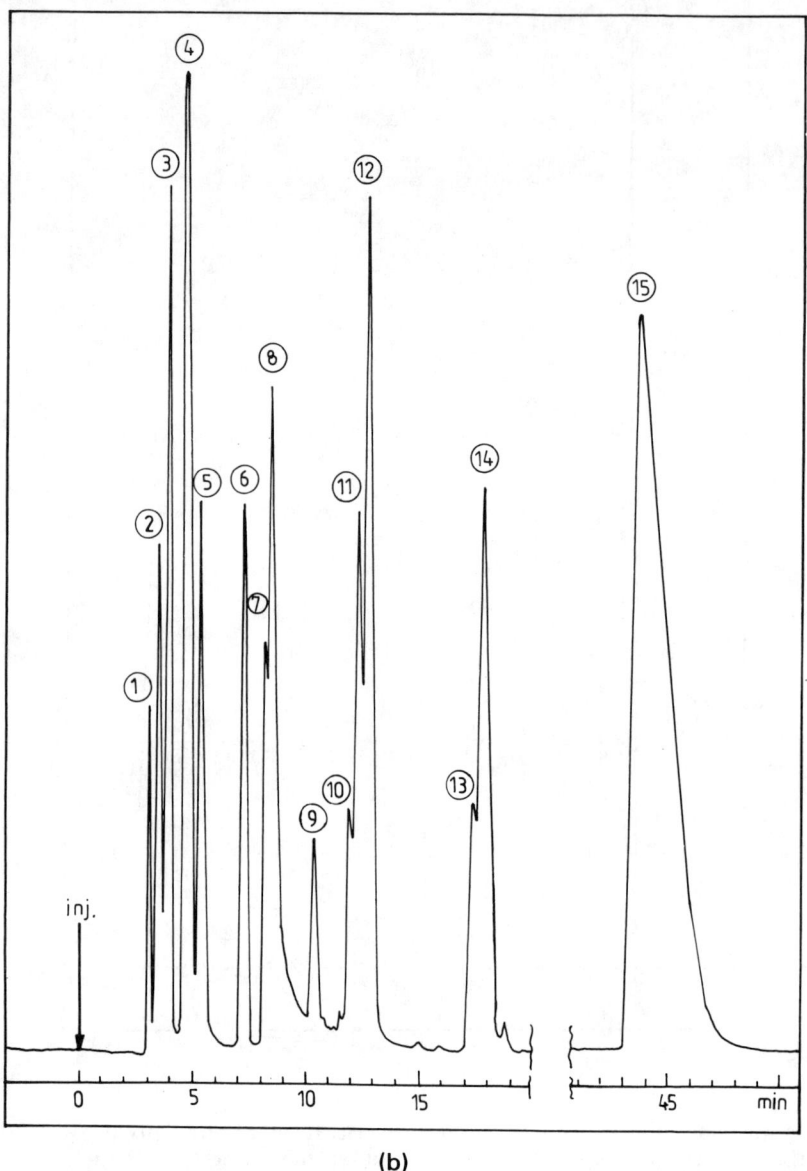

(b)

Fig. 6—*contd.*

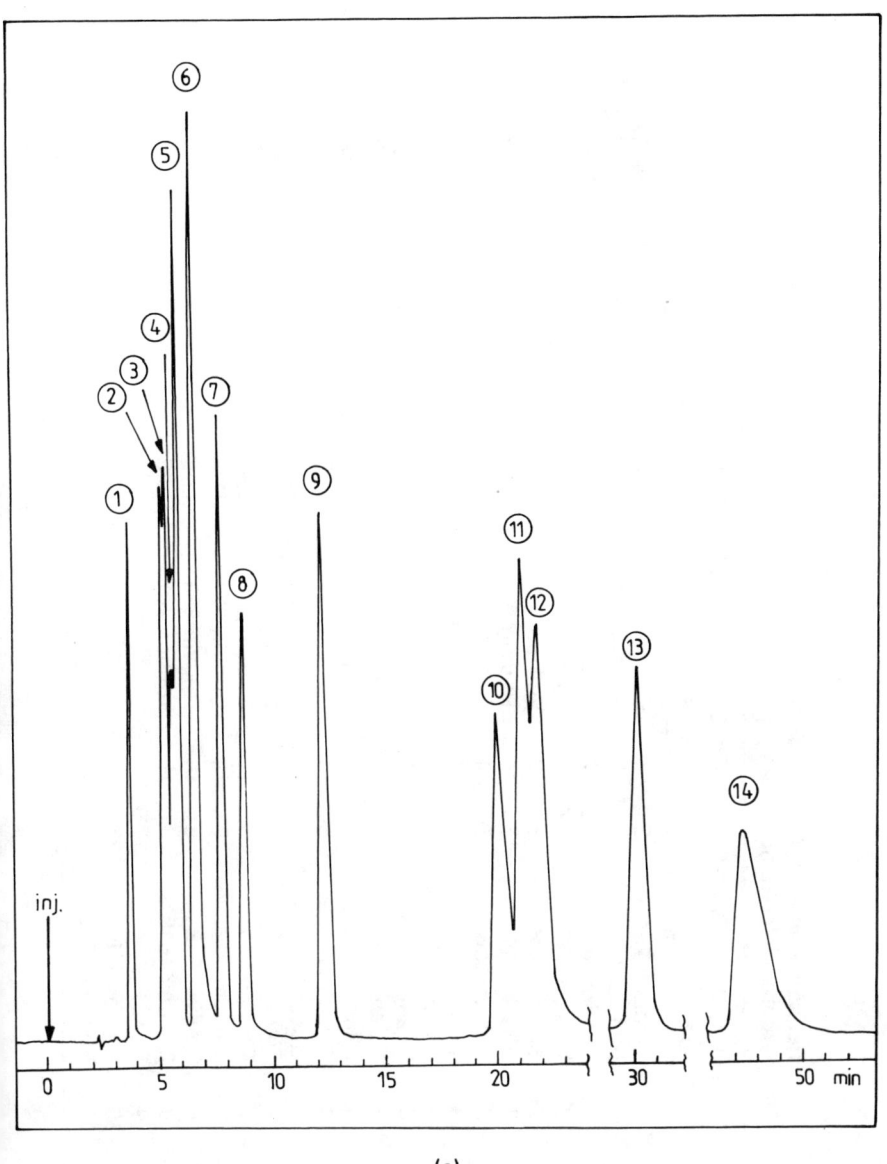

(c)

Fig. 6—contd.

Table 6
Retention times t_R (min) of stabilisers at different mobile phase compositions[59]
Chromatographic conditions: see Fig. 6.

Fig. 6(A): 100% NeCN			Fig. 6(B): MeCN/H_2O, 90:10			Fig. 6(C): MeCN/H_2O, 80:20		
Peak	Stabiliser	t_R	Peak	Stabiliser	t_R	Peak	Stabiliser	t_R
1	Ethanox 754	3·0	1	Ethanox 754	3·1	1	Ethanox 754	3·7
1	Irganox 245	3·0	2	Irganox 245	3·4	2	Irganox 245	5·2
1	Irganox 1024 MD	3·2	2	Irganox 1024 MD	3·6	3	Irganox 1024 MD	5·4
2	Vulkanox NKF	3·4	3	Santonox R	3·8	4	Santonox R	5·7
2	Santonox R	3·4	3	Vulkanox NKF	4·0	5	Vulkanox NKF	5·8
3	BHT	3·7	4	Tinuvin P	4·6	6	Tinuvin P	6·6
3	Vulkanox BKF	3·7	4	BHT	4·8	7	BHT	7·8
3	Tinuvin P	3·7	4	Vulkanox BKF	4·8	8	Vulkanox BKF	8·8
4	Topanol CA	4·0	5	Topanol CA	5·4	9	Topanol CA	12·3
4	Irganox 1035	4·0	6	Irganox 1035	7·4	10	Chimassorb 81	20·0
5	Ethanox 702	4·5	7	Ethanox 702	8·3	11	Irganox 1035	21·2
6	Chimassorb 81	5·0	8	Chimassorb 81	8·5	12	Ethanox 702	21·8
6	Tinuvin 120	5·2	9	Tinuvin 120	10·5	13	Tinuvin 120	30·2
6	Goodrite 3114	5·2	10	Goodrite 3114	12·1	14	Goodrite 3114	47·4
7	Tinuvin 326	6·6	11	Tinuvin 326	12·4			
7	Tinuvin 320	6·6	12	Tinuvin 320	12·8			
8	Tinuvin 327	8·0	13	Tinuvin 327	17·6			
8	Tinuvin 328	8·0	14	Tinuvin 328	17·9			
8	Irganox 1010	8·0	15	Irganox 1010	35·8			
9	DLTDP	17·9						
10	Irgafos 168	23·4						
11	Irganox 1076	24·2						

quently, in the particular analysis of a certain additive package the optimum strength of the mobile phase may be more easily selected than for general-purpose separations, and this allows both identification and quantification. Thus, antioxidants, light stabilisers and other additives present in commercial samples or finished products have been determined by isocratic RP-HPLC separations of the additive mixtures extracted from polyolefins.[28,36,39,47,58,59,63,86]

Some polyolefins contain amide-type additives, e.g. slip agents such as stearamide, erucamide and oleamide. In such cases the addition of an acid such as acetic[47] or propionic acid[58,59] is usually employed to improve the peak shape of these compounds, i.e. to eliminate the tailing effect. This so-called ion suppression technique, which requires buffering the aqueous mobile phase at a particular pH to suppress solute ionisation, is often used to reduce band-tailing in RP-HPLC.[25] Munteanu et al.[59] showed that the addition of 0·5% propionic acid to the mobile phases of different strengths (MeCN/H_2O, 100–80:0–20) does not significantly change either the separation behaviour or the response of the detectors. If such compounds are present in the additive package but are not of interest in analysis, the acid may be omitted from the mobile phase. The strong retention of the HALSs Hostavin TMN 20, Tinuvin® 144 and 770 to the stationary phase was suppressed and the effectiveness of the isocratic separation was improved by the addition of an ionogenic compound such as $Na_2B_4O_7$, NaCl, NaH_2PO_4 and H_3PO_4 or cetyl trimethylammonium bromide to the MeOH/H_2O mobile phase.[86]

4.2.2 Gradient-Elution Separations

Gradient-elution RP-HPLC was also used for the analysis of antioxidants, light stabilisers and other polymer additives.[27,31,36,44,46,54,64,65,81,84,85] Compared with isocratic analysis the gradient-elution mode of separation of complex mixtures with many additives of very different polarity improves the peak resolution and detection and reduces the analysis times. If the analysed mixture of polymer additives is not very complex, complete separations may be accomplished. The main parameter to control these separations is also the variation of the strength of the mobile phase. This is achieved by the use of various solvents of different polarity and by the shape of the elution gradient. The flow rate of the mobile phase may also be modified during the separation. Thus, Perlstein[64] used a gradient programme in which both mobile phase composition and flow rate

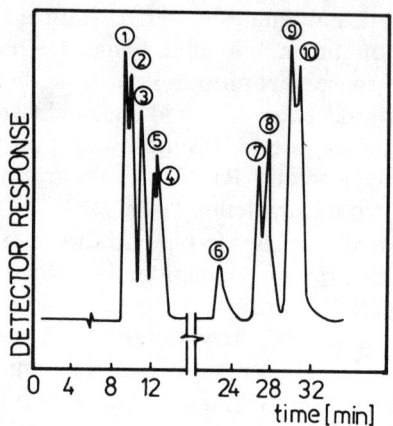

Fig. 7. Gradient-elution RP-HPLC separation of a synthetic mixture of 16 light stabilisers.[64] Chromatographic conditions: column (30 cm × 0·78 cm i.d., μ-Bondapak C_{18}), mobile phase 15 min 55:45 THF/H_2O at 0·8 ml/min, then the ratio was changed to 65:35 THF/H_2O and the flow rate to 1·2 ml/min during 9 min by using the steepest concave curve (No. 2) on the Waters Model 720 programmer; UV detection at 225 nm. Peaks: 1, Uvinul 400 + Cyasorb UV-24 + inhibitor RMB; 2, Sunkem MS; 3, Uvinul N-35 + Cyasorb UV-9; 4, Salol; 5, Tinuvin P; 6, Uvinul N-539; 7, Tinuvin 320 + Cyasorb UV-1084 + Cyasorb UV-531; 8, Tinuvin 326; 9, Uvinul M-410; 10, Tinuvin 327.

were changed in order to separate the majority of the stabilisers contained in a complex mixture of 16 compounds. However, some of these stabilisers could not be separated (Fig. 7).

Temperature control is essential for achieving reliable and reproducible HPLC separations. Increasing the effective operating temperature generally decreases retention, improves the separation efficiency (because of the decrease in the mobile-phase viscosity), and may alter the separation selectivity. RP-HPLC separations generally use aqueous mobile phases that are more viscous than typical NP-HPLC mobile phases. Therefore, in RP-HPLC separations it is advantageous to operate at higher temperatures. However, the same temperature change will not affect each sample component to the same degree and therefore temperature control may sometimes be used as an adjustable parameter, i.e. an additional variable that can be used to improve the analysis. Thus, Stoveken[81] performed a gradient-elution separation of eight polymer additives at different temperatures of the reversed-phase column. The selectivity of the column drastically depended on temperature [Fig. 8(A)] because the partition coefficient between the mobile phase and stationary phase has a different thermal dependency for each individual additive. Consequently, an optimal value of the temperature (80°C) was selected that allowed the complete separation of the additive mixture and shortened the analysis time [Fig. 8(B)]. However, the direction and extent of the temperature effect are not easily predictable and varying temperature is not generally used as the initial means of adjusting selectivity.[25] Indeed, almost all HPLC analyses of polymer additives have been performed at room temperature. Isocratic and gradient-elution RP-HPLC of polyolefin additives at 40°C, 50°C and 60°C have been reported, however.[36]

Gradient-elution RP-HPLC separation of the solvent extract has been used for quali-quantitative analyses of additive packages in

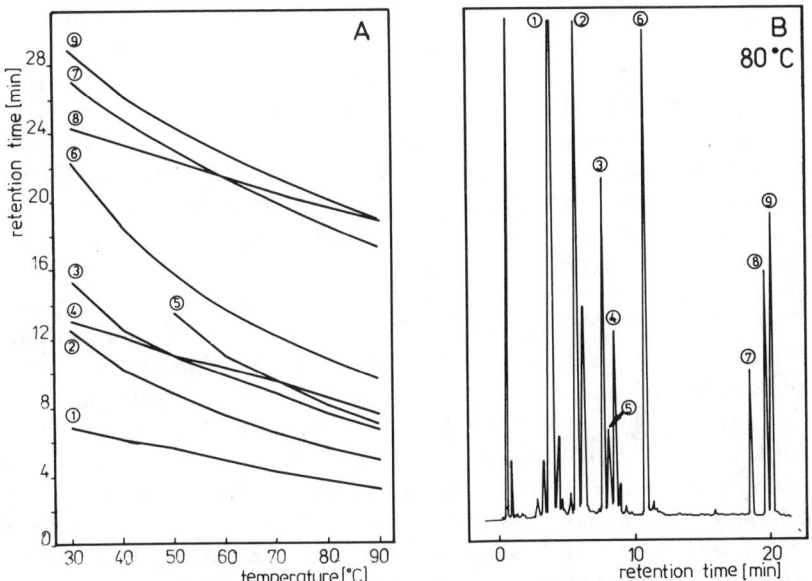

Fig. 8. The influence of temperature on the separation of polymer additives by gradient-elution RP-HPLC.[81] Chromatographic conditions: column HC-ODS/Sil-X; linear gradient from 50:50 H_2O/MeCN to 100% MeCN in 25 min at 1·5 ml/min; UV detection at 200 nm. Peaks: 1, BHT; 2, oleamide; 3, Topanol CA; 4, Cyasorb UV-531; 5, stearamide; 6, erucamide; 7, DLTDP; 8, Irganox 1010; 9, Irganox 1076.

polyolefins[36,81,84,85] and other thermoplastics such as PVC[64] in polyurethane[57] or in rubber articles.[44]

4.3 Detection and Identification

The vast majority of antioxidants and light stabilisers have chemical structures possessing strong chromophores above 254 nm. Thus, the absorption maxima of most phenol-type antioxidants is near 280 nm.[73] Therefore, most of the separations of these compounds have been monitored with UV detectors operated at 280 nm[28,36,47,51,54,58,59,63,74–76,84] or at 254 nm.[37,52,58,59,64,70,82,86] In this region the chromatographic-gradient solvents of usually mobile phases do not absorb UV radiation. Besides these two standard wavelengths, related wavelengths have been also employed, e.g. 270 nm,[39,86] 242 nm,[53] 240 nm,[85] and 230 nm.[41] Mixtures of light stabilisers without antioxidants have been monitored at 225 nm[64] or even at 208 nm[78] in the case of the HALSs Tinuvin 770 and Hostavin TMN.

Some polyolefin additives, e.g. antioxidants such as dialkyl thiodipropionates and slip agents such as fatty acid amides, do not absorb at the usual wavelength range employed for UV detection (i.e. 254–280 nm). Such additives may be monitored with the universal RI detector. However, the RI detector is less sensitive than the UV detector and excludes the gradient-elution mode of separation. Therefore, when possible, the elution of these additives was monitored at lower wavelengths where they show acceptable UV absorption. However, some of these additives can be detected only with an RI detector. Thus, Schabren[77] used UV detection at 230 nm for the isocratic NP-HPLC separation of DLTDP and DSTDP antioxidants with 1,2-dichloroethane as mobile phase. The detection of three other additives (mono- and di-glycerides, tertiary C_{12}–C_{16} alkyldiethanolamines) was possible only with the RI detector. Simultaneously, detection with UV and RI detectors allowed the analysis of mixtures containing additives without UV absorption at the usual wavelengths, e.g. fatty acid amides and dialkyl thiodipropionates.[30,47,58,59]

RI detection and UV detection at wavelengths lower than 254 nm may present some difficulties. The refractive indices of typically mobile phases for NP-HPLC are fairly high so that the RI detection of additives is not sensitive enough. Antioxidants such as DLTDP and DSTDP possess moderate UV absorption at 230 nm. Consequently the concentration of these additives in the injected solution should be about two orders of magnitude greater than in the case of UV detection at 254–280 nm. In the separation of such additives with RI or UV at 230 nm detection, Schabron[77] used concentrations 1–10 mg/ml greater than that for the separation of phenol-type antioxidants with detection at 280 nm (0·02–0·2 mg/ml).[74]

Mixtures of phenolic antioxidants and fatty acid amides have been monitored with UV detection to allow the gradient-elution mode of separation. The detectors were operated at 200 nm to give the required sensitivity for the amides.[36,81] For the HPLC analysis of UV absorbers in paints a fluorescence detector was used.[62] Electrochemical detection was employed in the analysis of food antioxidants (see Section 4.4). Masoud & Cha[57] monitored the isocratic RP-HPLC separation of phenolic antioxidants with electrochemical detection, fluorescence detection at 370 nm, and UV detection at 230 and 280 nm.

For the identification of the compounds present in real samples, most analyses use the comparison of retention times or capacity factors (k') with those of known standards (see for example Tables 4–6). Identification by this method is very time-consuming and requires one to have all of the suspected compounds on hand so that identification can be attempted. Although retention times are more reproducible in RP-HPLC than in NP-HPLC, erroneous identification can sometimes be made due to various causes, e.g. the chemical similarity of many compounds; the relatively poor separating capability of HPLC (compared with capillary GC) which results in overlapping peaks; changes in the order of eluted additives on varying the composition of the mobile phase (see for example Table 6). Consequently, the identification of compounds in additive mixtures should not be based only on peaks' retention times or capacity factors.

Retention times are not the only data available for identifying the additives. Haney & Dark[47] made use of the ratio of UV absorbance to refractive index both as a guide to identification of additives and to detect superimposed peaks (Table 5). Munteanu et al.[59] obtained more valuable data by simultaneously monitoring the elution of 22 stabilisers with three detectors: UV at 254 nm and 280 nm, and RI. From this was obtained the ratio of the peak heights/areas recorded at 254 nm and 280 nm. Although this ratio is frequently used in qualitative analyses, publications on HPLC separation of polymer additives do not mention it. Both UV/RI and especially UV_{254}/UV_{280} response ratios depend on the chemical structure of stabilisers and may be used together with retention time values for identifying the stabilisers and to detect overlapping peaks. Thus, the UV_{254}/UV_{280} ratio of sterically hindered phenol-type antioxidants (BHT; Ethanox 754 and 702; Goodrite 3114; Irganox 1076, 1010, 1035, 1024MD and 245; Topanol CA; Vulkanox BKF and NKF) varies between 0·2 and 0·4 because the maximum absorption of most of these compounds is near 280 nm. The UV_{254}/UV_{280} ratios of the alkyl-substituted benzotriazoles (Tinuvin P, 320 and 328) are 0·5–0·8 and the substitution of the aromatic ring with a chlorine atom (Tinuvin® 326 and 327) increases this ratio to 1·5. Consequently, the ratio UV_{254}/UV_{280} provides very useful information about the chemical structure of the analysed stabiliser.[59] Although this ratio is an intrinsic characteristic of each compound and depends on its UV spectrum, it can be very sensitive to the environment in which it was measured, e.g. the pH and polarity of the mobile phase. Therefore this ratio has to be calculated for each stabiliser and separation condition. It was found, for example, that the UV_{254}/UV_{280} ratio does not sensibly depend on the composition of a $MeCN/H_2O$ mobile phase containing 0–20% water.[59]

The possibility of detecting superimposed peaks is exemplified in Fig. 9. The two antioxidants may not be emphasised at 280 nm because of the very low UV_{280} absorption of Santonox R as compared with Vulkanox NKF: the UV_{254}/UV_{280} ratios are 2·1 and 0·2, respectively. Consequently, the detection at 254 nm emphasised both antioxidants.[59] The oxidation products of Irganox 1076 were monitored with a multiple UV detector. UV scanning assisted the identification of the oxidation products separated by RP-HPLC.[49a]

Stand[80a] used preparative HPLC for the characterisation of low-molecular-weight compounds in dashboard films. The additives were extracted with diethyl ether or hexane and separated by gradient-elution HPLC on preparative columns (250 mm × 20 mm i.d., 7-µm packing) combined with precolumns (50 mm × 20 mm i.d., 5–20 µm packing). Both NP-HPLC (LiChrosorb SI 60 packing) and RP-HPLC (LiChrosorb RP-8 packing) separations were used and the choice of appropriate gradients was selected on analytical columns (250 mm × 4·6 mm i.d.) on the same packing. Following evaporation in vacuum, IR-transmission spectra of the fractions were recorded in KBr and interpreted using reference spectra. The results showed that it is possible to gain essential information about the unknown additives from a polymer sample. It was possible to recognise at least the type of substances from their IR spectra and, sometimes, even to determine the exact species. Thus, Irganox 259 and 1076 antioxidants were

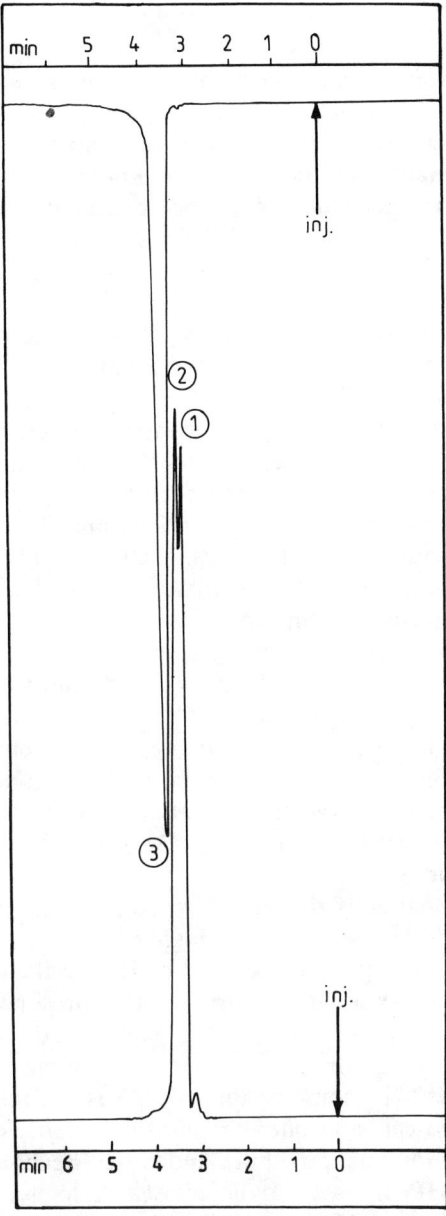

Fig. 9. Isocratic RP-HPLC separation of a synthetic mixture of two antioxidants with UV detection at two wavelengths.[59] Chromatographic conditions: see Fig. 6(A). Peaks: 1, Vulkanox NKF and 2, Santonox R at 254 nm; 3, Vulkanox NKF + Santonox R at 280 nm.

identified. However, the method is time-consuming and does not allow the determination of higher-molecular-weight additives which are not extracted from the polymer matrix by the boiling solvent extraction method. More exact identification of unknown additives would be achieved by NMR spectrometry of the separated fractions, but this method requires larger quantities of sample than IR spectroscopy.

For the qualitative analysis of solvent extracts containing unknown polymer additives TLC may be used as a rapid screening method, to show the mixture's complexity and the chemical composition of the individual components. TLC gives approximate data about the presence of some compounds in the additive package and the results can be used as an elimination step. Thus, if the measured retention factors of some standards are not approximately the same as those of unknown additives in the extract, the presence of these standards may be eliminated from further HPLC identification. TLC can also assist HPLC in the investigation of the best mobile phase for complete separation of the unknown additive mixture, since changing the mobile phase in HPLC is more complex and time-consuming than in TLC. The use of TLC as a complementary technique that can be employed effectively in conjunction with HPLC was reported for the analysis of antioxidants and light stabilisers in polyolefins[52] and polyurethane.[54] However, even with TLC as a preliminary rapid screening method, the identification of unknown additives in the HPLC chromatogram on the base of known standards is time-consuming. Therefore efforts have been made to use mass spectrometry (MS) for the identification of unknown compounds.

A mass spectrometer is the ideal detector for all chromatographic separation techniques because of its higher sensitivity compared with other detectors, and especially because of its specificity in identifying unknown compounds or for confirming the presence of suspected compounds. The combination GC/MS has already accomplished a great success because the effluent is compatible with gas-phase ionisation methods. The combination LC/MS is becoming increasingly important in analytical work but its realisation is more difficult because the operating conditions for these individual techniques are quite different. Thus, HPLC uses mobile phases of higher density and at higher pressure than GC, and MS operates under high vacuum accepting liquid flow rates of only 10–20 μl/min. For higher flow rates it is necessary to modify the vacuum system (thermospray interface), to remove the solvent before entry into the ion source (moving belt interface) or to split the effluent of the column (direct liquid introduction interface) so that only a small fraction (10–20 μl/min) of the total effluent is introduced into the ion source where the mobile phase provides for chemical ionisation of the sample.

Vargo & Olson[84,85] demonstrated the feasibility of using the HPLC/MS combination to analyse additives in polyolefins. A quadrupole mass spectrometer was equipped with a moving belt LC/MS interface where solutes were dissolved at 230°C followed by methane chemical ionisation at 120°C, 0·3 torr and 70 eV. The removal of the mobile phase before the entry into the ion source was facilitated by use of a narrow-bore column (2 mm i.d.) which was operated at a lower flow rate (0·2 ml/min) than a normal-bore column (see Section 6.3). UV absorbance and MS detectors were operated in series, without any significant loss of chromatographic resolution, thus providing unique quali-quantitative information on polymer additives separated by gradient-elution RP-HPLC (Fig. 10). Nanogram quan-

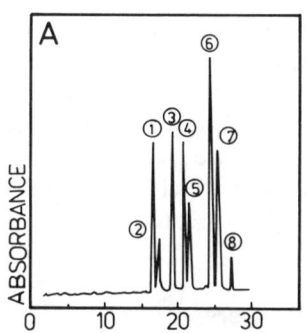
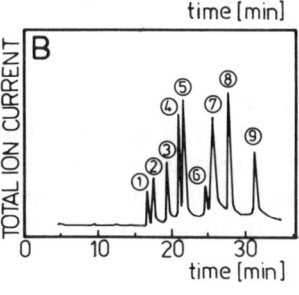

Fig. 10. Gradient-elution RP-HPLC separation of a synthetic mixture of antioxidants and light stabilisers. (A) Absorbance chromatogram at 280 nm. (B) Total ion-current chromatogram.[84] Chromatographic conditions: column (25 cm × 0·21 cm i.d., ODS, 5 μm); eluent A, 75:25 MeCN/H$_2$O; eluent B, 50:50 THF/MeCN.

Gradient elution scheme:

Time (min)	0	10	20	30	32
Eluent A (parts by vol.)	100	60	0	0	100
Eluent B (parts by vol.)	0	40	100	100	0

Flow rate 0·2 ml/min; sample of 10 μl was injected when the gradient controller was 7 min into the gradient-elution programme. Peaks: 1, BHT, 3·1 μg; 2, Santowhite powder, 0·9 μg; 3, Topanol CA, 2·7 μg; 4, Cyasorb UV-531, 0·66 μg; 5, Cyasorb UV-5411, 0·68 μg; 6, Irganox 1010, 4·5 μg; 7, Ethanox 330, 2·4 μg; 8, Irganox 1076, 1·2 μg; 9, DSTDP, 1·7 μg (not detected at 280 nm).

tities of additives were detected by MS following HPLC separation. Lower detection limits approaching those of UV absorption at 280 nm were obtained from reconstructed mass chromatograms of selected ions instead of total ion current chromatograms. The detection of both UV and MS detectors simplified the identification of the additives present in an automotive component moulded from PP.[84]

Recent developments have facilitated the on-line combination of HPLC with MS, using microcolumns which enable the introduction of the total effluent into the ion source (Section 6.3).

Quantitative evaluation of each separated additive was generally carried out by the external standard method, i.e. by comparison of peak heights/areas obtained with the sample and the standard solutions using average results from duplicate or more injections.

The precision, accuracy and limits of detection in the HPLC analysis of polymer additives depends on many factors such as extraction procedure, polymer sample size, additive type and concentration in polymer, isocratic vs gradient-elution separation and detection. Spiking experiments using standard polymer examples (see Section 3.4) have often been performed in order to evaluate the analysis performance. The limits of additive detection in polyolefins varies between 6 and 160 ppm, but for the vast majority of stabilisers they are in the range 10–60 ppm. These limits are quite sufficient for the analysis at additive levels of about 500 ppm which are typical for most of the commercial polyolefins. The relative standard deviations are usually 1–2%.

4.4 Miscellaneous Analysis of Stabilisers

Most papers concerning the analysis of antioxidants, light stabilisers and other additives refer to their identification and/or quantification in polymers. However, related analyses have also been performed by modern LC and they may interest both polymer researchers and chromatographers. Therefore, this section briefly presents the use of modern LC for other purposes such as the study of chemical transformation of stabilisers in polymers, the analysis of the composition of stabilisers and the analysis of stabilisers in non-polymeric materials such as monomers, fuel and foods.

Various chemical transformations of stabilisers and other additives during the polymer processing or the long-term life of the finished articles have been studied by HPLC analyses of solvent extracts.[30,47,64,86] Thus, Haney & Dark[47] emphasised the chemical transformation of antioxidants and antacid agents during their incorporation into the polymer matrix by melt mixing. The HPLC separation of the CH_2Cl_2 extract from standard PP samples (Fig. 11) showed only 60% recovery of the antioxidant Goodrite 3114 due to its thermal degradation. The two small extraneous peaks on either side of the Goodrite 3114 peak were attributed to the degraded antioxidant. If these two peaks were added to the real peak of Goodrite 3114, the recovery approached 100%. Another two unexpected peaks were identified as stearic and palmitic acids, respectively. The metal stearates used as antacid agents in the PP standard samples are not soluble in the extraction solvent and their presence in the chromatograms was unexpected. However, they react with acidic components such as HCl in the PP forming the free fatty acids (metal stearates are obtained from stearine—a mixture of palmitic and stearic acid). These acids are soluble in CH_2Cl_2 so that they are present in the extract and, consequently, in the chromatogram (Fig. 11).

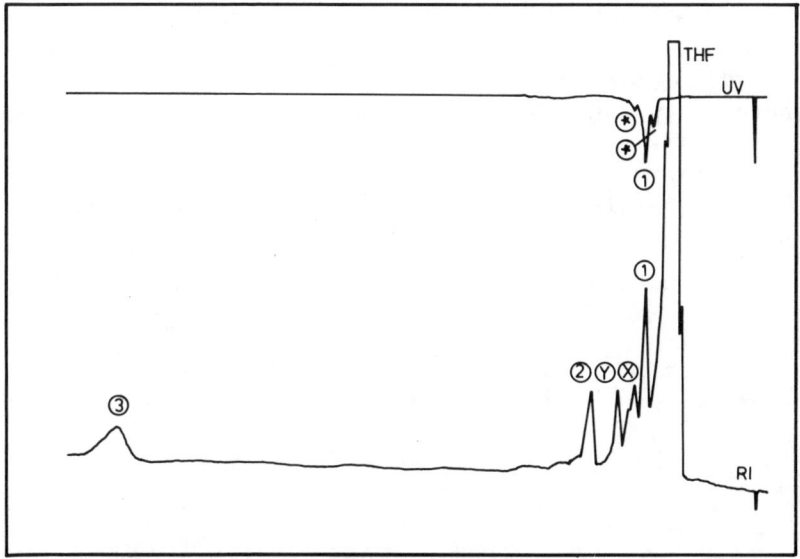

Fig. 11. Isocratic RP-HPLC separation of the CH_2Cl_2 extract of a polypropylene reference sample containing 0·10% Goodrite 3114 (peak 1), 0·25% Weston 618 (peak 2), 0·10% DSTDP (peak 3) and 0·10% calcium stearate.[47] Chromatographic conditions: see Table 5. Other peaks: *, thermal degradation products of Goodrite 3114; X, Y, stearic and palmitic acid, respectively.

Following an isocratic RP-HPLC separation, the detection and identification was reported of the antioxidant Goodrite 3114 and its degradation products in radiation and ethylene oxide treatments of polyolefins.[63] The product of the thermal degradation of Irgafos 168 was investigated by the isocratic RP-HPLC analysis of the solvent extract from PP.[86] Carlsson et al.[30] studied the decomposition of some light stabilisers during the photodegradation of PP films. Thus, 2,2,6,6-tetramethyl-4-hydroxypiperidine, 2,2,6,6-tetramethyl-4-hydroxypiperidine-N-oxyl and their transformation products were extracted with CH_2Cl_2 from the irradiated PP films and then separated by isocratic RP-HPLC.

Jonas et al.[49a] studied the oxidation of Irganox 1076 using three different oxidation methods for the antioxidant itself and two methods of antioxidant oxidation in polyethylene (processing at 200°C in a Brabender Plasticorder and oven-ageing at 110°C). Following extraction, the oxidation products were separated by RP-HPLC and monitored with a multidiode UV detector. During processing and ageing of PE stabilised with 0·1% Irganox 1076 (**a**), various large molecules are formed, the most important ones being: (**b**) cinnamate, (**c**) biscinnamate, (**d**) di- and *meso*-diastereoisomers of unconjugated bisquinonemethide, (**e**) conjugated bisquinonemethide and (**f**) phenol/quinone

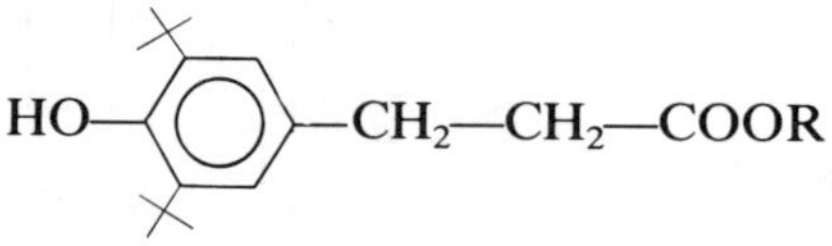

(**a**) R = $C_{18}H_{37}$

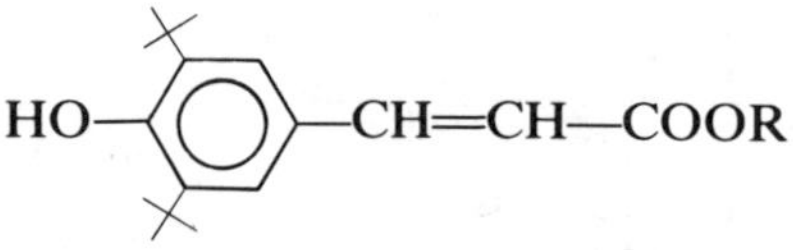

(**b**)

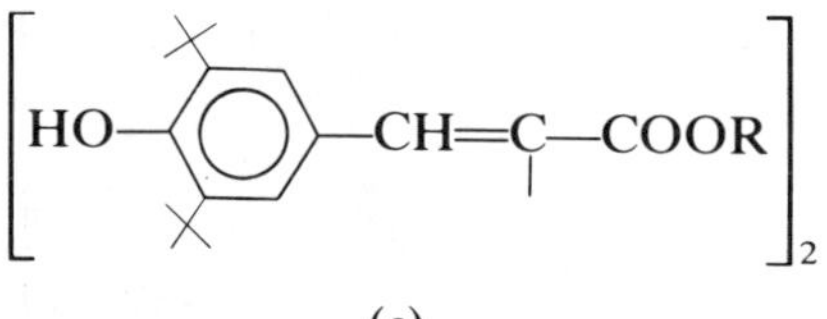

(**c**)

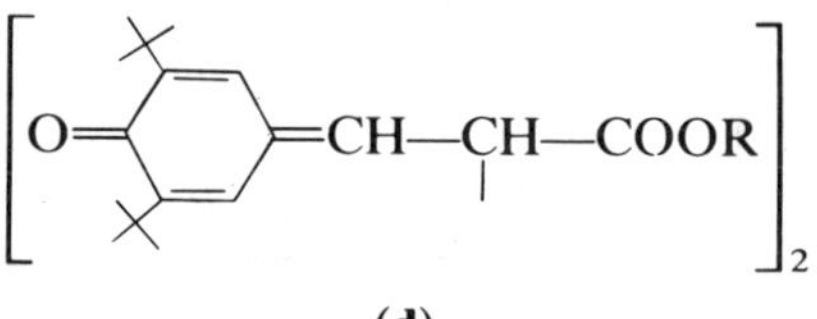

(**d**)

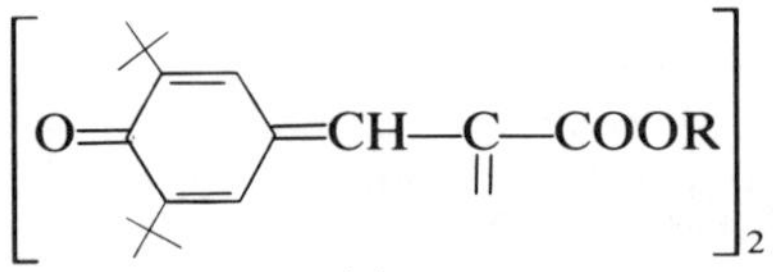

(**e**)

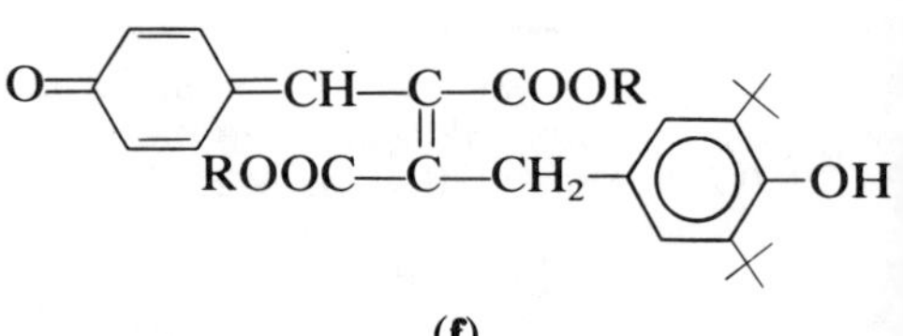

(**f**)

dimer. Each method of oxidation yielded all the compounds in varying degrees except the oven-ageing where compounds **c** and **e** were not detected. The level of Irganox 1076 in PE fell progressively during oxidation but it was only partially replaced by the oxidation products. The most probable cause is the loss by evaporation of some lower-molecular-weight products such as benzoquinone. Moreover, the formation of some polymer-bound species may occur but these compounds were found to be not very effective as antioxidants.

HPLC alone, but especially when combined with other analytical methods such as the spectroscopic ones, has been shown a very effective technique for the characterisation of polymer additives themselves. However, little information has been published on this topic. It is probable that there is a large amount of unpublished work in this area because of the proprietary nature of many in-house methods.

The monomeric antioxidant 2,2,4-trimethyl-1,2-dihydroquinoline (TMDQ) may be polymerised to form a resinous product which is a typical commercial rubber antioxidant (AgeRite Resin D). The

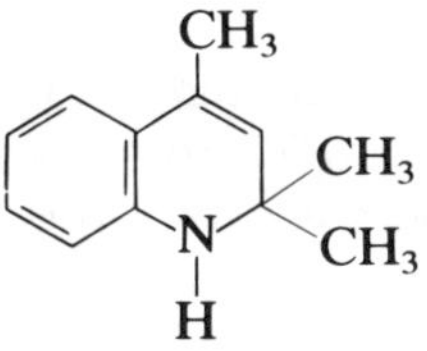

TMDQ

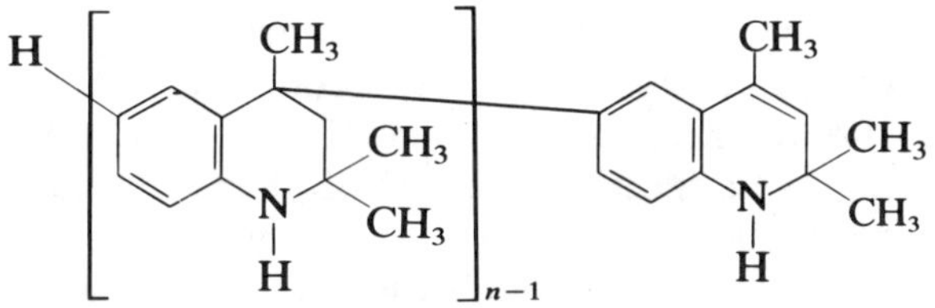

Poly-TMDQ

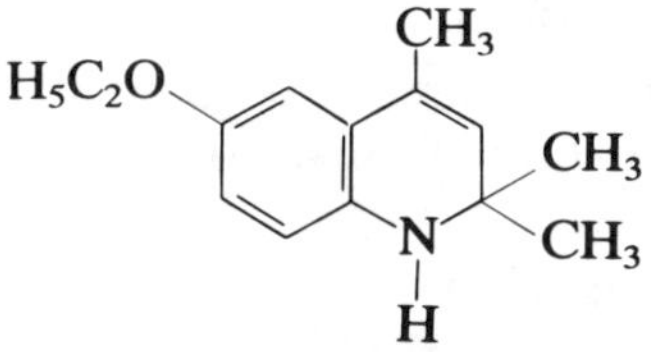

ETMDQ

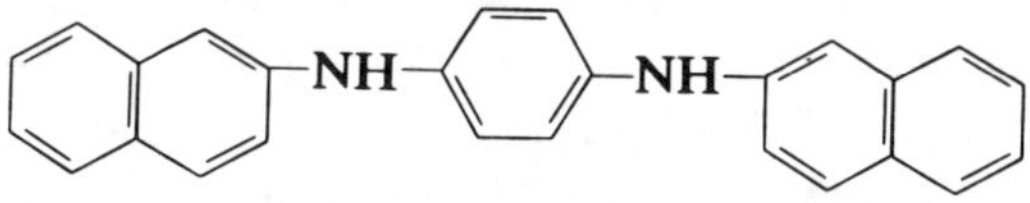

AgeRite White

detailed composition of this polymeric antioxidant was generally unknown before modern LC became available. The antioxidant is a mixture of low-molecular-weight oligomers (poly-TMDQ), as SEC separation showed.[115,116] Pausch[117] reviewed the analytical approach utilised in the B. F. Goodrich laboratories[118-122] for determining the antioxidant composition. The off-line coupling of HPLC with special spectroscopic techniques has proved to be a very effective combination for the characterisation of complex oligomer mixtures. The HPLC separation procedure utilised a reversed-phase column and a very long linear gradient-elution programme (from 50:50 THF/H_2O to 100% THF in 60 min) with UV detection at 254 nm. The HPLC effluent peaks were trapped and then analysed by special spectroscopic techniques able to provide positive and remarkably clear identification of the separated components. The conventional IR and iron-magnet NMR instrumentation does not possess the necessary sensitivity to investigate the very small quantities (up to 10 μg) of the separated components. A superconducting NMR magnet was used to obtain useful proton spectra after averaging a number of pulsed scans. Micro-ATR–IR spectroscopy was also used because it allowed the enrichment of the sample relative to the solution. The trapped effluent of the HPLC separation was added dropwise to the ATR crystal where the chromatographic solvent evaporated. Field desorption mass spectroscopy (FDMS) has been shown to be a very effective technique for determining molecular weights of thermally labile and non-volatile compounds such as polymer additives which do not give good molecular ion spectra during electron impact or chemical ionisation. Moreover, FDMS may analyse oligomers and the molecular weight values thus obtained assist the identification of individual components.[118-120] Earlier researches of Latimer et al.[121] to characterise the poly-TMDQ mixture of oligomers by off-line combination of HPLC and FDMS emphasised the presence of a series of eight TMDQ oligomers, the main components of the antioxidant resin (see poly-TMDQ formula). The off-line coupling of HPLC with the special spectroscopic microtechniques later permitted identification of the antioxidant components.[122] Thus, a total of eight oligomeric series containing 42 components have been identified.[117] The same analytical approach was used to determine the composition of the tar by-product from the synthesis of AgeRite White, an aromatic amine antioxidant produced from p-phenylenediamine and β-naphtol.[117]

Kato & Kanoshita[123] studied the autoxidation of the antioxidant

6-ethoxy-1,2-dihydro-2,2,4-trimethylquinoline (ETMDQ) and found that the major product was a dimer which, however, was converted to a range of unidentified products on exposure to light. The dimer was separated by RP-HPLC on a silica gel column with CH_3OH/H_2O, 90:10, isocratic elution and UV detection at 254 nm and was then identified by MS and NMR. An RP-HPLC separation using H_2O dioxan as mobile phase was employed to analyse technological mixtures of phenolic antioxidants, i.e. to determine the transesterification products of methyl β-(3,5-di-*tert*-butyl-4-hydroxyphenyl)-propionate with pentaerythritol, diethylene glycol or thiodiethylene glycol.[124]

An isocratic RP-HPLC separation was developed for the direct determination of the content of free phenolic compounds in the aryl esters of the phosphoric acid used as plasticisers and flame retardants for polymers.[124a] The mixture of phenols formed in the synthesis of Agidol-123 antioxidant was analysed by RP-HPLC.[124b]

Such analyses of stabilisers are very useful for various purposes: (a) for determining material balances on production processes; (b) to verify the purity of production batches; (c) to study the decomposition pathways: (d) for FDA approval and, generally, for a better understanding of the behaviour and performance of these additives.

Mixtures of inhibitors and antioxidants in acrylic monomers such as acrylic and methacrylic monoesters, tri- and tetra-ethyleneglycol dimethacrylates have been determined by isocratic NP-HPLC and RP-HPLC analyses.[125,126] Thus, a mixture of hydroquinone, methyl ether hydroquinone and Santonox was determined in a commercial sample of tetraethyleneglycol dimethacrylate used as crosslinking agent in some polymer formulations.[125b] Other inhibitors such as p-methoxyphenol and phenothiazine have been also determined.[126b] Usually, UV detection at 295 nm was used because the molar absorptivity of stabilisers is maximum at this wavelength and that of the monomer is low. The UV detection allowed a sensitivity up to 10^{-3} wt%. However, due to the 'refractometric' effect, i.e. a low signal caused by the presence of a substance in the detector which should not absorb light at that wavelength, there is a possibility of peaks of the monomer or its impurities overlapping with the peaks of the antioxidants and inhibitors. Therefore, for such analyses, where the monomer/inhibitor ratio can reach 10^7-10^6, the use of an electrochemical detector increased the sensitivity to about 10^{-5} wt%.[126b]

Aviation turbine fuel contains low levels (17–24 mg/l) of antioxidants such as 2,4-dimethyl-6-*tert*-butylphenol and 2,6-di-*tert*-butyl-4-methylphenol (BHT). A 'direct' HPLC analysis with UV detection was able to determine the level of 2,4-dimethyl-6-*tert*-butylphenol in the fuel.[127] Similar attempts to determine BHT by NP- or RP-HPLC have failed due to the more hindered structure of this antioxidant which causes it to elute with the hydrocarbon components of the fuel. However, the direct quantitative determination of BHT in fuel at the 2–30 ppm level was possible by an isocratic RP-HPLC separation with 90:10 CH_3OH/acetate buffer (0·05 M, pH 4·8) as the mobile phase and electrochemical detection. The Zorbax 10 μm C_{18} packing gave better results than Waters Radial-Pak 5 μm C_{18} where the BHT peak exhibited a shoulder due to an unresolved component which could not be removed by changing the mobile phase composition. The less hindered phenolic antioxidant 2,4-dimethyl-6-*tert*-butylphenol was lost in the major fuel peak and could not therefore be determined.[128] Additives in jet fuels have also been determined by SEC.[128a]

HPLC analysis is frequently used to find solutions for analytical and technological problems in the chemistry and technology of foodstuffs. Thus, besides other ingredients and additives, various antioxidants are incorporated in foods. The most used antioxidants in the food industry are phenolic-type compounds such as BHT, BHA (a mixture of two positional isomers, 3- and 2-*tert*-butyl-4-hydroxyanisole in an approximately 85:15 ratio); TBHQ (*tert*-butylhydroquinone); THBP (2,4,5-trihydroxybutyrophenone); *n*-alkyl gallates such as ethyl, propyl, octyl and dodecyl gallates and NDGA (nordihydroguaiaretic acid). Besides these synthetic antioxidants some 'natural' antioxidants such as tocopherols and ascorbyl palmitate are also used. Some of these antioxidants such as BHT are widely used as polymer antioxidants. Antioxidants based on α-tocopherol are also used as polymer antioxidants, e.g. the Sicostab AO-3 (BASF) processing antioxidant.

HPLC analysis was used for the identification and quantification of these antioxidants in edible oils and fats, e.g. tallow and vegetable oils, lards and shortenings.[129–146] The analytical problem resembles the determination of antioxidants in polymers. Thus, generally, the antioxidants are extracted with organic solvents from foods; they are then separated, identified and eventually quantified. Normal-phase[129,132,135,140] but more frequently reversed-phase, HPLC[132,137,138,141,143,144] has been used for the separation of the antioxidants. The isocratic mode of separation[129,132,135,137,138,141,144] was

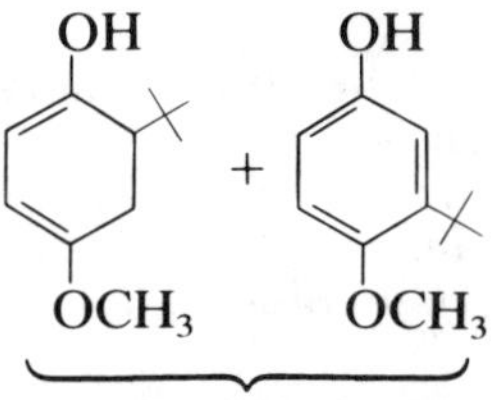

[Structures of BHA, TBHQ, THBP, Collates (n = 2, 3, 8, 12), NDGA, Vitamin C (L(+)-Ascorbic acid), Vitamin E (DL-α-Tocopherol)]

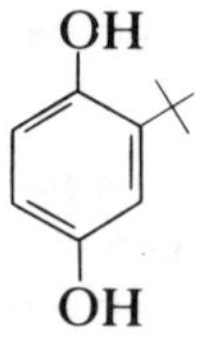

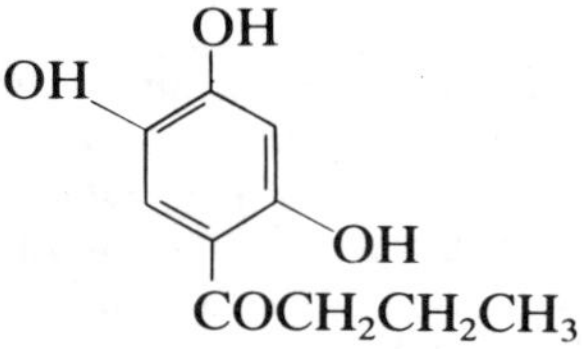

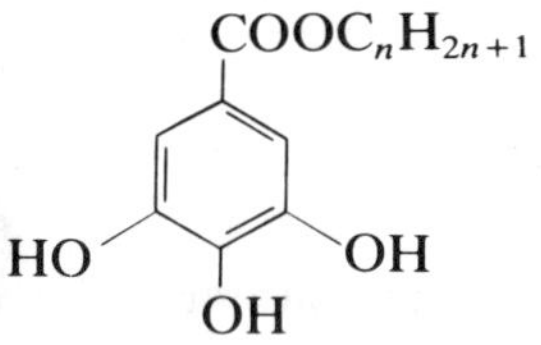

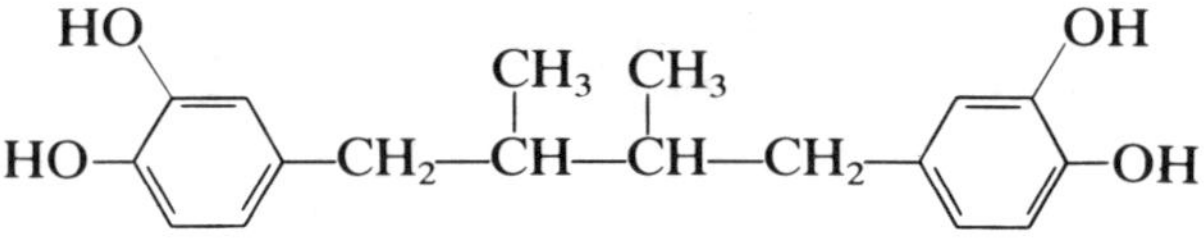

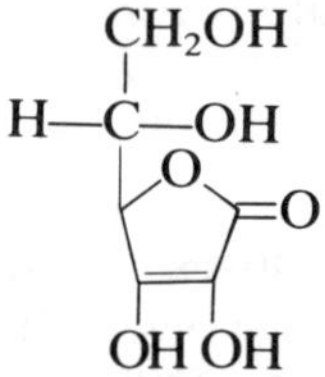

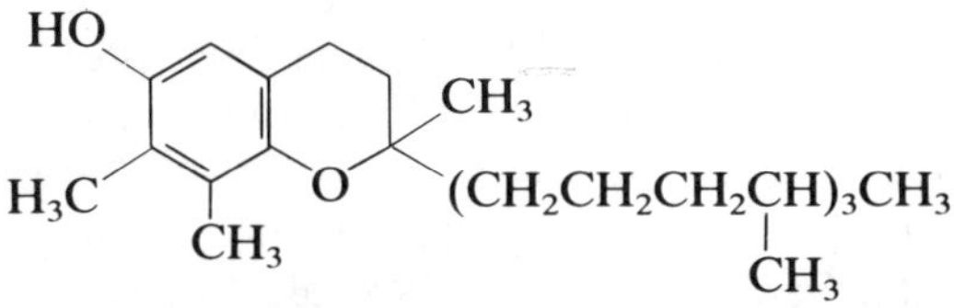

preferred for simplicity in the analysis of food extracts containing a single antioxidant or a relatively simple mixture of antioxidants, e.g. the BHA + BHT + TBMQ mixture.[137] The separation of a larger number of antioxidants of very different polarity has been generally achieved by gradient-elution HPLC,[132,138,143] e.g. the separation of BHT + BHA + NDGA + ascorbyl palmitate + DL-α-tocopherol mixture.[138] Gradient-elution is also required to obtain total separation of the antioxidants and co-extracted sample components. Ansari[129] was able to resolve the separation of the two positional isomers of BHA only on a column used for the separation of optical isomers and by adding isopropanol to the hexane mobile phase. The separation of BHA isomers is probably based on the interaction of the phenolic group of BHA with the stationary phase. The phenolic group of the major isomer (3-*tert*-butyl-4-hydroxyanisole) is sterically hindered and therefore cannot interact effectively with the stationary phase; this

Table 7
Isocratic NP-HPLC separation of 2- and 3-*tert*-butyl-4-hydroxyanisole[129] Chromatographic conditions: column Pirkly Type I–A (25 cm × 0·46 cm i.d.): spherical 5-μm γ-aminopropyl bonded-phase packing modified with *N*-dinitrobenzoyl derivative of D-phenylglycine; flow rate 1 ml/min; UV detection at 288 nm; injected sample volume 20 μl BHA in *n*-hexane (10 mg/ml).

2-*propanol in n-hexane* (%)	0	1	2	3	5	7	10
Retention time of 2-BHA (min)	—	96·0	30·4	17·2	10·7	8·2	6·8
Retention time of 3-BHA (min)	—	47·8	21·6	13·3	9·0	7·0	6·0

results in a faster elution than for the other isomer. Retention times of both isomers rapidly decrease with increasing isopropanol content of the mobile phase (Table 7).

The eluted antioxidants may be monitored both with UV detection[129–131,133–135,138,141] generally at 280 nm and with electrochemical detection.[136–138,143,146] The NP-HPLC separation of α, β, γ and δ-tocopherol and α-tocotrienol in seed oils was monitored with a mass spectrometric detector.[144a] None of these analyses requires a prior derivatisation of the antioxidants. However, Galensa[132] showed that after benzoylation it is possible to analyse together food additives of varying chemical structures such as phenolic antioxidants, alcohols, saccharides and sugar alcohols, and esters of *p*-hydroxybenzoic acid. The derivatised food additives are water-soluble and have a strong UV absorption maximum at 230 nm. A synthetic mixture of 22 benzoylated standard compounds including antioxidants such as BHA, NDGA and six alkyl gallates was separated by RP-HPLC making use of a sophisticated gradient-elution programme. However, isocratic separations are also able to determine these chemicals in foods.[132]

'Direct' analysis of food antioxidants in fats and oils have been also reported.[135,135a] The oils and fats were dissolved in the mixtures used as mobile phases in isocratic NP-HPLC separation and injected, without further clean-up, into the columns. Thus, BHT, BHA and TBHQ were separated on μ-Porasil or Rad-Pak Cyano columns with hexane/CH_2Cl_2/MeCN as the mobile phase[135] and for the analysis of BHA, TBHQ, NDGA and alkyl gallates a LiChrosorb DIOL column with hexane/dioxan/MeCN as the mobile phase was used.[135a] Such 'direct' analyses avoid the manipulation requirements involved in solvent extraction and the consequent losses.

5 SIZE-EXCLUSION CHROMATOGRAPHY

SEC, unlike various HPLC techniques, does not involve the interaction between the sample molecules and the stationary phase, which is a porous, three-dimensional matrix. The most popular stationary phase consists of small semi-rigid organic gel particles made from heavily crosslinked styrene–divinylbenzene copolymers but in such a way that small pores of controlled size exist within it.

The separation mechanism is based on the diffusion of sample molecules into and out of the pores in the stationary-phase particles. The molecules to be separated may or may not enter these small pores, depending on the size of the molecule which is defined by its hydrodynamic radius. All molecules larger than the pores are excluded and elute in the exclusion volume. All molecules smaller than the smallest pore elute at the total permeation volume. Molecules with sizes between these two extremes can be separated because they permeate part of the pores but are excluded from others and elute in order of decreasing molecular size, i.e. between the exclusion volume and the permeation volume. SEC is also known as 'Gel Permeation Chromatography' (GPC), a term which suggests the 'permeation' of the sample molecules into the pores of the 'gel' particles when separations are carried out in organic solvents. The term 'Gel Filtration Chromatography' (GFC) is preferred when separations are performed in aqueous solutions.

The SEC packings are available in pore sizes that allow separation over a wide molecular weight range of about 10^2–10^7. Thus, packings with small pores (e.g. less than 100 Å) allow the separation of relatively small molecules, with molecular weights up to about 1000. Column selection involves matching the pore size of the packing to the molecular size of the sample molecules. Therefore, several columns with packings of varying pore sizes, e.g. 60, 100, 250 and 1000 Å, may be required to separate a sample consisting of species of widely different molecular sizes.

SEC is an extremely useful analytical tool for the characterisation of oligomers and high polymers, mainly for the determination of their molecular weight distribution. However, for the separation of low-molecular-weight compounds SEC is basically not a suitable technique. The separation mechanism is based almost entirely on the size-exclusion principle since practically no interaction appears between the analysed sample and both stationary and mobile phases.

Consequently, this chromatographic technique has very poor resolution so that it is not able to separate individual components from complex nixtures, for example many antioxidants, light stabilisers and other polymer additives which are of very similar molecular weight/size (Table 1). Although the retention times are reproducible, sensitivity is low due both to broad peaks and to the high refractive indices of the mobile phases typically employed. However, despite the two main disadvantages, i.e. very poor resolution and low sensitivity, SEC has been used for the analysis of polymer additives, although to a lesser degree than HPLC.[29a,32–35,42,45,49,54,56,60,61,67–69,74,79,79a,87,89,90] Compared with HPLC the main advantage of SEC lies in its simplicity. In contrast with the other LC techniques, all components usually elute between the excluded volume and the total permeation volume at a fixed time/volume interval. Thus, less operator experience in chromatography is required. The main parameter to be selected is the optimum pore size and that is simply achieved by knowing the molecular-weight operation range of the columns and matching this with the suspected molecular-weight range of the sample. For the analysis of polymer additives this problem is very simple because most of them have molecular weights up to 1000 (except HALSs with molecular weights over 2000–4000, e.g. Tinuvin 622, 765, Chimassorb 944). It must be remembered that SEC was introduced by Moore[147] in 1964, before the advent of HPLC, so that it is not surprising that attempts have been made to use this separation technique despite its disadvantages for the analysis of polymer additives extracted from polymers (Section 5.1). However, SEC may perform analyses that are not amenable to HPLC, e.g. direct determination of additives in polymers, sample clean-up, and analysis of higher-molecular-weight and polymer-bound stabilisers (Section 5.2).

5.1. Size-Exclusion Chromatographic Analysis of Additive Extracts

SEC analysis of additive extracts is performed basically in the same manner as HPLC analysis. The preliminary step is the extraction of additives from the polymer matrix (see Section 3). The extracted additives are then separated by SEC according to their molecular size using columns of properly selected pore size. Following the identification of the separated peaks. quantitative analysis may be performed.

In the SEC analysis of additive extracts from polymers, the effect of the extraction solvent on the mobile phase is less critical than in HPLC analysis. The extraction solvents typically employed generally do not interfere with the SEC mobile phases. Moreover, the same solvents are often used both as extraction solvent and as mobile phase, e.g. THF[56,89] and chloroform.[60,61] Therefore, there is no need to evaporate the extract to dryness prior to analysis and then to redissolve it in a suitable solvent. Also, the presence of small amounts of polyolefin wax do not affect the SEC column; provided that the molecular weight of the wax is higher than that of the analysed compounds, no interference between wax and stabilisers will occur.

Earlier SEC analyses of polymer additives[32-35,49,60,61,67-69,89,90] were performed by low-pressure SEC (Table 2) using columns packed with gel of high particle size (40–80 μm), e.g. Styragel (crosslinked styrene–divinylbenzene copolymer). To achieve acceptable resolution, long columns (120 cm) and low linear velocities at low flow rates were employed. However, even in those conditions, the SEC method allowed some analyses of additives in polymers. Howard's work[49] seems to have been the first to show how useful SEC can be for the analysis of polyolefin additives but it did not adequately cover the separation of complex mixtures into individual components.

The most important contribution to the SEC separation of stabilisers belongs to Coupek, Protitova and Pospisil.[32-35,67-69] They generally studied the separation of synthetic mixtures of additives, especially amine and phenolic antioxidants, and some model substances such as phenols, amines and aromatic hydrocarbons. Qualitative analysis of rubber extracts[33] and the determination of stabilisers in polyolefins and PVC[67] were also reported.

Since SEC does not allow the direct determination of the molecular weight or molar volumes of the samples investigated, a calibration with standard compounds has to be made by plotting their molar versus elution volume. This method is used to calculate the molecular weight distribution of polymers but does not allow the identification of polymer additives other than by molecular weight. Thus, Protitova et al.[67-69] used normal hydrocarbons (C_{15}–C_{18}) whose molecular volumes could be calculated from known atomic volumes and structural characteristics. The results of SEC measurements on the elution volumes of some amine stabilisers, their molar volumes calculated from molecular weights and the effective molar volumes observed and read from calibration curves, revealed deviations in the behaviour of all amine stabilisers investigated compared with similar standard model compounds. Solvation of the sample molecules via hydrogen bonding and interactions between aromatic solutes and the crosslinked polystyrene gel did not allow the prediction of elution volumes based on the chemical composition alone. Thus, phenols and amines of similar molecular weight could often not be separated by SEC.[67-69]

Wims & Swarin[89] used a set of four Styragel columns (250 Å, 100 Å, 2 × 60 Å) designed to separate samples with molecular weights of less than 1000. In addition to the usual universal calibration curve, they plotted a calibration curve using common stabilisers of known molecular weight as standards. Thus, the peak elution volumes could assist the identification of stabilisers in solvent extracts from PP samples. However, for PP samples in which the antioxidants were completely unknown, additional analytical methods such as TLC and MS were necessary to give a positive identification. Such SEC chromatograms were used as 'fingerprints' in the routine screening of polymer samples to determine whether the supplier had changed his formulation.[89] The quantitative determination of additives in the solvent extract was also performed by means of linear calibration curves of peak height versus stabiliser concentration in a standard solution.[67,89]

Besides the identification and quantification of polymer additives in solvent extracts, some earlier SEC analyses attempted to determine the purity of individual stabilisers or to show possible chemical transformation of stabilisers, e.g. oxidation and decomposition, during the processing and end-use of the stabilised polymers. Thus, Protitova et al.[67] found that although most technical phenolic stabilisers for polymers used as food wrappings are very pure compounds, some of them are typical admixtures or contain by-products and/or impurities. 'Fingerprint' SEC chromatograms showed that the concentrations of stabilisers in a moulded PP part subjected to heating at 120°C for 500 h are much lower than in a new moulded PP because the stabilisers are consumed in the prevention of polymer degradation. The chromatograms also showed the thermal decomposition of DSDTP to give stearyl propionate.[67]

To improve the resolution of SEC in the separation of polymer additives, Nakamura[60,61] used the technique of recycling the additive mixture extracted from PP pellets and fibres and PVC sheets by means of a recycle valve through the same column a number of times (up to five cycles). The recycle technique was also applied to the separation and identification of the components in commercial additives such as lauramide and lauryl stearyl thiodipropionate, and of a yellow oxidation product in the BHT antioxidant. Resolution increased with the number of cycles, allowing the separation of some unresolved adjacent peaks. However, despite the better resolution, the recycling technique was not suitable for routine analysis due to the increased separation time, i.e. 3 h/cycle vs 15 min/cycle for HPLC. The ratio of RI to UV absorbance was used as an additional guide to the identification of additives.[61]

Following the development of microparticulate columns for adsorption, partition and ion-exchange LC in the early 1970s, microparticles also became available for SEC, allowing more rapid separations. The SEC microparticles, e.g. μ-Styragel (8–10-μm diameter vs 40–80-μm for Styragel particles) resulted in shorter columns (30–50 cm vs over 120 cm) consuming less solvent and allowing more convenient thermostating. High-pressure or high-performance SEC (Table 2) achieved an improved separation of polymer additives by better resolution, higher sensitivity and shorter analysis time (Fig. 12).

Despite these improvements, SEC analysis of additive extracts gained no practical importance because of its much lower analytical power in comparison with HPLC. Thus, in a mixture of three antioxidants, Irganox 1010 and 1076 were not separated from each other, but were separated from BHT on one 500-Å followed by three 100-Å μ-Styragel columns with CH_2Cl_2 as mobile phase.[74] Some analyses have been reported, however.[56,79,87] For the identification of the separated peaks the comparison of their retention times/volumes with those of suspected additives was usually performed. The comparison was based both on the relative reproducibility of retention times and the dependence of peak location in the chromatogram on the molecular size of the corresponding additive (Table 8).

Today the SEC analysis of additive extracts seems to be virtually extinct. However, Gharfeh[42] developed a very interesting technique described as size-exclusion non-aqueous reversed-phase chromatography for the separation of the polymeric HALS Chimassorb® 944 from the other usual polyolefin additives. Chimassorb® 944 adsorbs strongly to silica and to bonded silica packings due to the interaction between the silanol groups present on the packing and the multiple amine functional groups of this HALS. Chimassorb® 944 could not be eluted from these columns, even by the addition of ammonia or triethylamine to the mobile phase. Chimassorb® 944 was also adsorbed on neutral polymer packing (polystyrene–divinylbenzene) when toluene was used as the mobile phase but could be eluted from

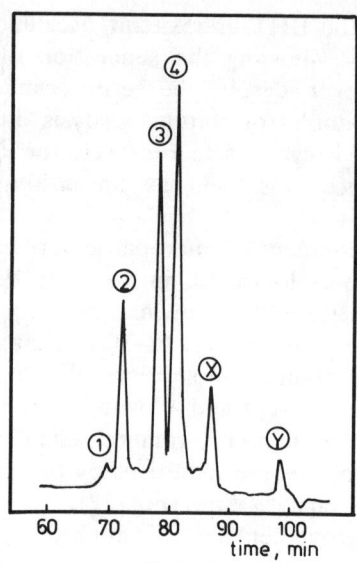

Fig. 12. SEC separation of additives extracted from polyethylene.[56] Extraction conditions: 5 g LDPE film extracted with 100 ml THF; extract concentration to 5 ml prior to injection. Chromatographic conditions: Three Mikropak TSK columns; 3000 H (50 cm × 0·8 cm i.d., 1500 Å) + 2000 H (50 cm × 0·8 cm i.d., 250 Å) + 1000 H (80 cm × 0·8 cm i.d., 40 Å); mobile phase THF at 0·5 ml/min; UV detection at 215 nm; injected sample 10 µl. Peaks: 1, DSTDP; 2, Irganox 1035; 3, Santonox R; 4, residual peroxide initiator α,α'(bis-*tert*-butylperoxy)di-isopropylbenzene; X,Y, presumably decomposition products of the peroxide, possibly phenols.

the column following addition of piperidine to the toluene. Piperidine competes with Chimassorb® 944 so that the retention time decreases with an increase of the concentration of piperidine in toluene up to 0·1 M. The concentration of piperidine required to elute Chimassorb 944 is related to the column type, shorter retention times being obtained on a short column (50 mm × 7·8 mm, Ultragel 100–1000 Å) than on a long one (300 mm × 7·8 mm i.d., µ-Styragel 100 Å). A step-gradient from toluene to 0·1M-piperidine in toluene was used for the analysis. First toluene eluted the polyethylene wax and any other low-molecular-weight additives that might be present in the PE sample (BHT, DLTDP, Irganox 1010 and 1076, Goodrite 3114, Ultranox 624,

Table 8
Separation of polymer additives by size[87]
Chromatographic conditions: four μ-Styragel columns of 100 Å pore size; mobile phase THF at 1 ml/min.

Additive	Relative location in chromatogram
Irganox 1010	0·818
1-Monostearin	0·882
DSTDP	0·889
Irgafos TPP	0·907
Topanol CA	0·938
Irganox 858	0·945
DLTDP	0·963
Erucamide	0·985
Stearamide	0·997
Oleamide	1·000
4,4'-Bis(2,6-di-*tert*-butylphenol)	1·000
4,4'-Methylenebis(2,6-di-*tert*-butylphenol)	1·016
Santonox	1·042
Cyasorb UV-1084	1·051
Dioctyl phosphite	1·145
Chimassorb 81	1·153
BHT	1·181

Ethanox 330, Lupersol 130 and triallyl cyanurate were investigated). Then piperidine/toluene eluted the strongly adsorbed HALS. Except Chimassorb 944 the elution of all the other additives was of a size-exclusion type so that the chromatogram contained only two peaks corresponding to PE wax plus additives, and Chimassorb 944.[42]

5.2 Direct Size-Exclusion Chromatographic Analysis of Stabilisers in Polymers

Despite its disadvantages in comparison with HPLC, SEC may perform analyses that are not amenable to HPLC and this will most probably be the field of further developments in the use of SEC for the analysis of stabilisers in polymers. Thus, SEC may be used for the direct determination of stabilisers in polymers and other materials such as lubricating and vegetable oils, provided that the matrix is of a sufficiently higher molecular weight to be almost completely resolved from the higher-molecular-weight stabiliser of the mixture. Thus, the antioxidant Irganox® 245 and the light stabiliser Tinuvin 328 were

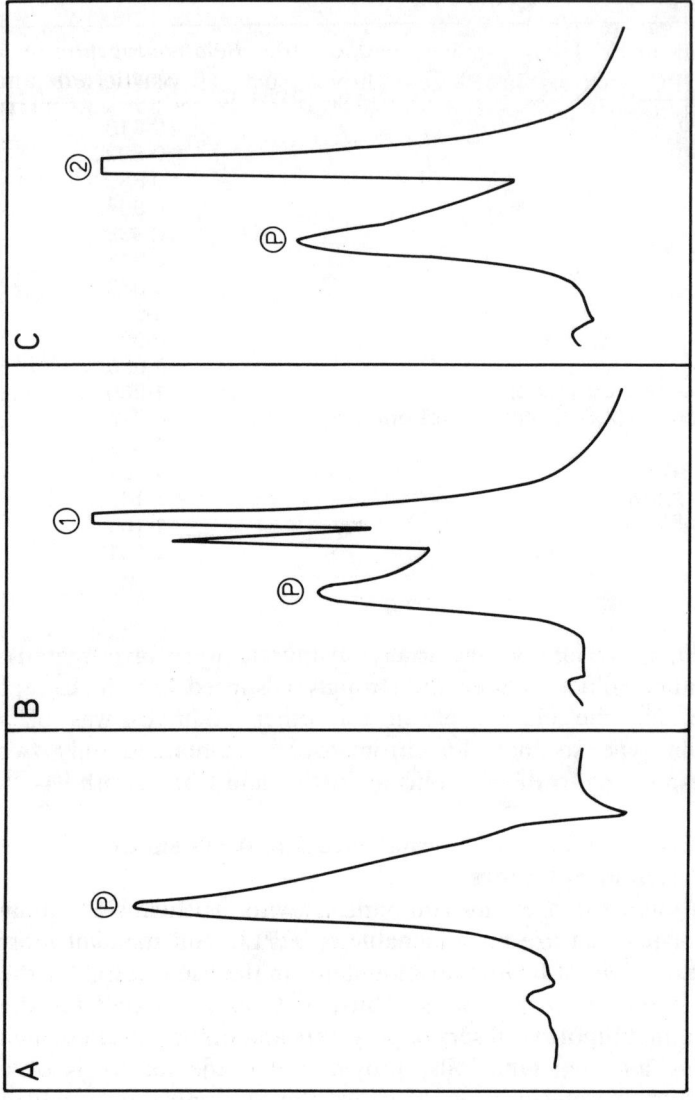

Fig. 13. Direct SEC analysis of stabilisers in a polyurethane. (A) Unstabilised; (B) stabilised with Tinuvin 328.[54] Chromatographic conditions: column PSM Bimodal Kit (Du Pont), 50 cm × 0·6 cm i.d., mobile phase dimethylformamide at 0·5 ml/min; temperature 100°C; RI detection. Peaks: P, polymer; 1, Irganox 245; 2, Tinuvin 328.

directly determined in a polyurethane by SEC separation (Fig. 13).[54] This direct analysis method eliminates the need for the time-consuming extraction step and hence the possibility of stabiliser loss. An extraction procedure of the same polyurethane sample followed by a gradient-elution RP-HPLC analysis yielded only 90% recovery.[54]

Direct GPC analysis of PVC additives such as plasticisers and thermal stabilisers have been also reported.[29a,79,79a] The use of THF, i.e. a good solvent for PVC, as the mobile phase allowed both RI and UV detection. In the analysis of Hatcol 200 plasticiser, i.e. tris(2-ethylhexyl) trimellitate, PVC is excluded from the column and eluted earlier than Hatcol 200, whereas the usual phthalate plasticisers do not interfere with the analysis.[29a]

For certain mixtures of additives, resolution is often such that they can be determined directly. Thus, in a stabilised lubricating oil, the antioxidants Ethanox 702 and tricresyl phosphate with molecular weights of 436 and 368, respectively, could be separated from each other by SEC due to their difference in molecular weight. The oil matrix was of a sufficiently higher molecular weight to be almost completely resolved from the Ethanox 702, the antioxidant of higher molecular weight.[56]

In some cases SEC can be used to provide a prefractionation or clean-up of the sample, with the actual separation being carried out with a secondary chromatographic method such as HPLC. For such work SEC has the advantage of low peak dilution. Thus, the food antioxidants BHT and BHA, with molecular weights of 220 and 180, respectively, were separated by SEC from a vegetable oil which consisted mainly of triglycerides with molecular weights of over 600, but they could not be separated from each other. The SEC effluent corresponding to the antioxidant peak was collected, concentrated and analysed by RP-HPLC which allowed the separation, identification and quantification of these two antioxidants.[56]

Gupta & Salovey[45] used SEC for the analysis of chemically bound stabilisers. They studied commercial copolymers of butadiene and acrylonitrile (BAN) (Hycar elastomers) using multiple detection, i.e. RI and UV absorption at 254 and 280 nm. These BAN copolymers contained variable amounts of an additive with a strong UV absorption at 280 nm, presumably an aromatic antioxidant. Some samples contained an extractable low-molecular-weight antioxidant so that the UV response at 254 nm was separated into a broad copolymer peak and a sharp antioxidant peak which eluted subsequently. Other

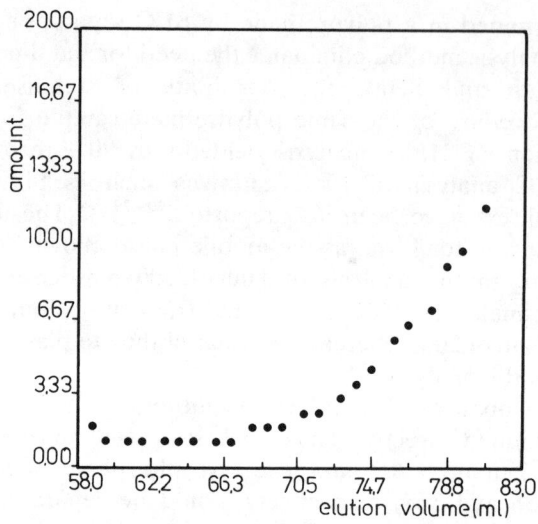

Fig. 14. Distribution of a bound antioxidant in a sample of butadiene–acrylonitrile copolymer.[45] SEC separation on five μ-Styragel columns in series. Mobile phase THF at 2 ml/min. Detection: RI, UV at 254 and 280 nm.

samples showed a much larger area response in UV at 254 nm than the previous ones and no separation of the antioxidant sharp peak. However, those samples contained an unextractable, bound antioxidant which was emphasised by SEC with UV detection at 280 nm where the UV absorption is due solely to antioxidant because the BAN elastomers do not absorb at this wavelength. The mass of the antioxidant bound to the polymer was calculated from the UV absorption at 280 nm and represented as a function of elution volume (Fig. 14). Because the concentration of antioxidant increases with elution volume (i.e. increases with a decrease in the molecular size/weight of the copolymer), it was suggested that the antioxidant may be bound to the chain ends.[45]

Consequently SEC may be successfully employed to study the distribution of stabilisers chemically bound to polymer chains by various methods and to relate this distribution to the effectiveness of stabilisers. If the distribution of stabilisers could be determined following the degradation of polymer, additional information about the mechanism of polymer stabilisation and degradation might be obtained.

6 RECENT DEVELOPMENTS IN CHROMATOGRAPHIC ANALYSIS OF STABILISERS IN POLYMERS

The rapid development of HPLC makes it today one of the most widely used and powerful analytical techniques. However, of all the chromatographic techniques (Table 2), GC provides basically the highest resolution because the diffusion coefficients are higher in the gaseous than in the liquid mobile phase. In order to approach the unsurpassed separation capability of high-resolution capillary GC (Section 6.1), new generations of chromatographic techniques having dramatic improvements in performance have been developed.

Thus, advances in packings, column technology and HPLC instrumentation have resulted in the development of high-speed LC (Section 6.2) and microcolumn LC (Section 6.3). Each of these new HPLC techniques uses one of the two possibilities to improve the column efficiency, i.e. the decrease of the packing particle size for high-speed LC and the decrease of the column internal diameter for microcolumn LC. Both these improvements resulted in the increase of the number of theoretical plates per unit column, i.e. higher efficiency due to the decrease of peak equivalent theoretical plates.

New insights into the mobile phase in chromatography resulted in the development of supercritical fluid chromatography (Section 6.4).

The new chromatographic techniques basically aimed at the improvement of: (a) the separation capability, i.e. higher resolution and efficiency; (b) the possibility of detection and identification, i.e. higher mass sensitivity and more specific detectors; and (c) the economics of the analysis, i.e. shorter analysis time and lower mobile phase consumption. All of these techniques have been successfully used for the analysis of antioxidants, light stabilisers and other polymer additives.

6.1 High-Resolution Capillary Gas Chromatography (HR-CGC) with Cold On-Column Injection

The unrivalled separation capability of high-resolution capillary gas chromatography (HR-CGC) combined with its compatibility with a wide range of sensitive detection systems such as MS and FT-IR spectroscopy makes it one of the most powerful tools for the analysis of complex mixtures of compounds. However, HR-CGC suffers from the same disadvantage as classical GC in the analysis of antioxidants and light stabilisers because most of them, and especially the modern ones, are non-volatile compounds. The advent of HR-CGC, with cold on-column injection,[148] resulted in improved GC analysis of polymer additives.[149-151] The solution of the additive mixture is injected directly into the cold end of the capillary column by means of a cold injector. Thus, sample discrimination, the instantaneous evaporation of the sample solvent, is avoided. The non-vaporising, on-column injection combined with the very high resolution of the capillary columns allows accurate separation, identification and quantification of additives in complex mixtures.

DiPasqualle et al.[149,150] chromatographed a solvent extract from a standard polyolefin sample containing 300 ppm of each additive from a mixture of three antioxidants, six light stabilisers and three amides. The components were completely separated on an FSOT capillary column (15 m × 0·32 mm i.d.) with temperature programming up to 280°C in 32 min allowing thus quali-quantitative analyses. However, the analysis was carried out for compounds with molecular weights up to 647, e.g. Irgafos 168, and many other stabilisers having higher molecular weights.

Blum & Damascenco[151] chromatographed a synthetic mixture of 21 antioxidants and light stabilisers with molecular weights from 323 to 1196 at maximum working temperatures of 420°C (Fig. 15). This high-temperature GC method used a specially prepared glass capillary column with OH-terminated polysiloxane stationary phases. However, such drastic separation conditions may give rise to some problems, e.g. working temperatures above 380–400°C reduce the durability of the usual capillary glass columns and the inertness of the coatings. Many other antioxidants and light stabilisers cannot be separated even by high-temperature GC since their vapour pressure is too high or their thermostability is insufficient. Therefore, even the authors[151] hesitate to consider high-temperature GC an alternative to LC and recommend it as a complementary method.

6.2 High-Speed Liquid Chromatography

The continuous tendency to enhance the separation capability of HPLC by decreasing the size of the stationary-phase particles resulted in the development of shorter columns packed with smaller particles (3–5 µm) and with a narrower size distribution than the conventional HPLC packings (5–10 µm). Therefore, higher separation efficiency per unit column length and a reduction in analysis time without sacrificing resolution have been obtained. This 'higher-performance' LC tech-

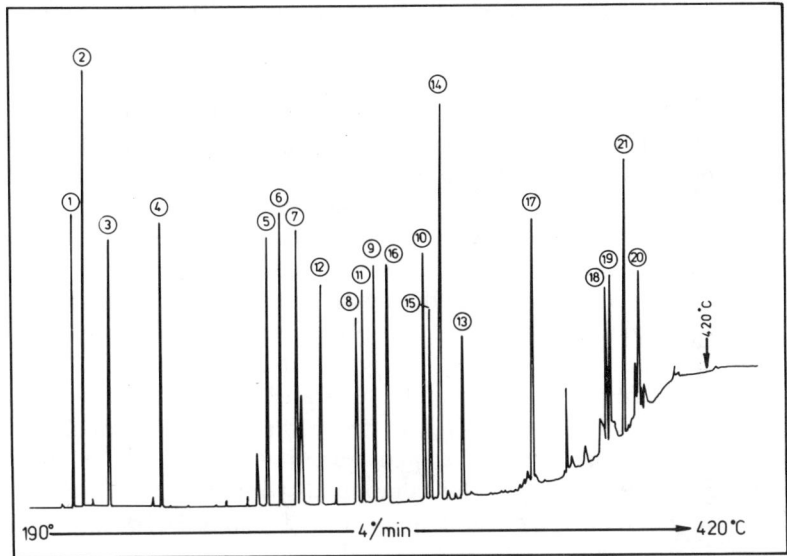

Fig. 15. High-resolution capillary gas chromatographic separation at high temperature with cold on-column injection of a synthetic mixture of 21 antioxidants and light stabilsiers.[151] Chromatographic conditions: capillary glass column (20 m × 0·33 mm i.d. coated with 0·2% OH-terminated polysiloxane stationary phase PS-089 (equivalent to SE-54 or SE-52) additionally crosslinked. Temperature programming at a rate of 8°C/min to 160°C and then at a rate of 4°C/min up to 420°C. Peaks: 1, Tinuvin 320; 2, Tinuvin 326; 3, Tinuvin 327; 4, Tinuvin 120; 5, Tinuvin 900; 6, Irganox 1076; 7, Irgaphos 21–40 (**X**); 8, Irganox 245; 9, Irganox 1019; 10, Irganox 1098; 11, Irganox 259; 12, Tinuvin 144; 13, Chimassorb 21–1231 (**XIXg**); 14, Ethanox 1330; 15, Irganox 3114; 16, Chimassorb 13–176 (**XIXb**); 17, Irganox 1010 triester (**VIIId**); 18, Irganox 3125; 19, Chimassorb 22–992 (**XIXh**); 20, Irganox 1010; 21, Polphrino (**XIXi**). For the development products Irgaphos 21–40, Chimassorb 21–1231, 13–176, 22–992, Polphrino and Irganox 1010 triester see formulae **X**, **XIXa, b, h, i** and **XIIId** respectively in the Appendix.

nique was therefore named High-Speed Liquid Chromatography (HSLC). In the early days of modern LC the term HSLC was sometimes used instead of HPLC.[25] However, HPLC is today the only name accepted, so that the term HSLC may be used in order to emphasise the higher performance of this HPLC variant. Table 9 presents the typical features of HSLC in comparison with those of

Table 9
Comparison of conventional high-performance liquid chromatography (HPLC) and high-speed liquid chromatography (HSLC)

	HPLC	HSLC
(A) Instrumentation		
Columns		
Internal diameter (mm)	4·6	4·6
Particle diameters (μm)	5–10	3–5
Length (mm)	250	100–125
Sample volume (μl)	Up to 20	Up to 6
UV detector		
Flow cell volume (μl)	8	2·4
Full-scale response (μs)	850	290
Typical instrumental bandwidth values (μl)	175	30–40
(B) Performance		
Column efficiency (plates/column)	3 000–10 000	8 000–15 000
Analysis time		
Isocratic separation (min)	5–15	1–3
Gradient-elution separation (min)	15–30	5–10
Detection limits	Nanograms	3–5 times lower
Solvent consumption (litres/100 samples)	1	0·5

conventional HPLC. The most important advantages are shorter analysis time, higher separation efficiency, greater analytical sensitivity and lower solvent consumption. These advantages make HSLC a very attractive analytical tool for the determination of additives in polymers. However, because of the small peak elution volumes (25–100 µl), specially designed chromatographs with miniaturised sample flow paths and minimal dispersive effects must be used.

Due to its advantages, the use of HSLC in quality assurance is becoming widespread in the plastics industry, including the analysis of polymer additives. In the isocratic mode many useful separations of these additives may be performed in 1–3 min. Thus, Dong & DiCesare[36] performed a rapid analysis of five common antioxidants in only 70 s (Fig. 16), obtaining excellent repeatability in both retention times and peak area in spite of the extremely fast separation, with the first four antioxidants eluting in 40 s. In the gradient-elution mode of HSLC separations of polymer additives, analysis times of 5–10 min are typical. Thus, Dong & DiCesare[36] reported a comparative separation

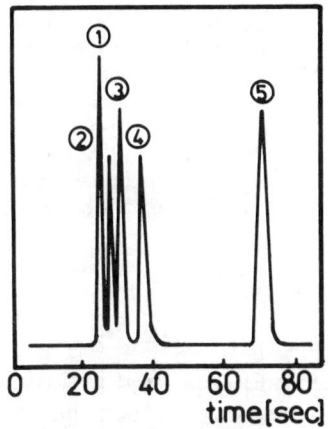

Fig. 16. Isocratic HSLC separation of antioxidants.[36] Chromatographic conditions: HS-3 C_{18} Perkin–Elmer reversed-phase column (10 cm × 0·46 cm i.d., C_{18} bonded phase, 3 µm); mobile phase MeCN at 3·5 ml/min; temperature 40°C; UV detection at 280 nm. Peaks; 1, Santonox R: 2, antioxidant CAO-5; 3, Irganox 1035; 4, Irganox 259; 5, Irganox 1010.

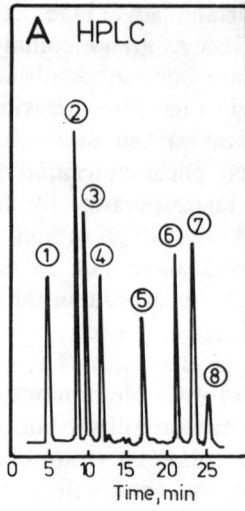 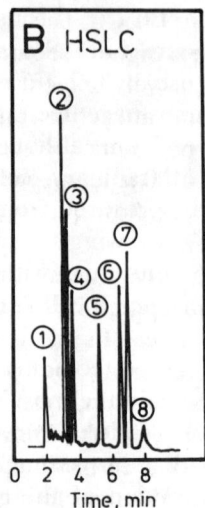

Fig. 17. Gradient-elution separation of polyethylene additives by conventional HPLC (A) and by HSLC (B).[36] Chromatographic HPLC conditions: HC-ODS Perkin–Elmer reversed-phase column (25 cm × 0·26 cm i.d.). Linear gradient from 50:50 MeCN/H_2O to 100% MeCN in 20 min; 6 min purge; 1 min ramp; 8 min equilibration; flow rate 1 ml/min; temperature 60°C; UV detection at 200 nm. Chromatographic HSLC conditions: HS-5 C_8 Perkin–Elmer reversed-phase column (12·5 cm × 0·46 cm i.d., C_8 bonded phase; 5 μm); linear gradient from 60:40 MeCN/H_2O to 100% MeCN in 6 min; 2·5 min purge; 1 min ramp; 2 min equilibration, flow rate 3 ml/min; temperature 50°C; UV detection at 200 nm. Peaks: 1, Tinuvin P; (2) BHT; (3) 2,6-di-tert-butyl-4-ethylphenol; (4) oleamide; (5) Kemamide E; (6) Irganox 1010; (7) Irganox 1076; (8) Trisnonyl phenylphosphite.

of eight PE additives in one-third of the analysis time compared with conventional HPLC (Fig. 17).

In the RP-HPLC separation of food antioxidants (see Section 4.4), Nissen & Kreysel[138] reduced very much the analysis time both in the isocratic and gradient-elution mode using an HSLC column (Ultrasphere ODS, 3 μm, 7·5 cm × 0·46 cm i.d.) instead of the usual HPLC column (Ultrasphere ODS, 5 μm, 25 cm × 0·46 cm i.d.).

6.3. Microcolumn Liquid Chromatography

The other approach to improving the efficiency of HPLC columns was to decrease their internal diameter. This resulted in the development of microcolumn LC.[152,153] Thus, narrow-bore (2·0–2·5 mm i.d.), microbore (0·5–1·5 mm i.d.) and slurry packed capillary columns (0·05–0·50 mm i.d., usually 0·2–0·3 mm i.d.) have been gradually developed. All these columns have been packed under high pressure with particles as small as 2 or 3 μm, although the typical particle size is 5 μm. Two other types of capillary columns, i.e. packed capillary columns (40–80 μm i.d., prepared from a prepacked glass tube by means of a glass capillary drawing machine) and the open-tubular capillary columns (1–50 μm i.d., without packing) have not so far gained practical importance in HPLC in view of the experimental difficulties involved.[152]

Although packed columns with large inner diameters (typically 4·6 mm i.d.) remain the most important tools in the practice of HPLC, the advantages of microcolumn HPLC are increasingly being realised: (a) easier fabrication of high-efficiency columns; (b) drastically decreased consumption of the mobile phase and column materials; (c) the possibility of using 'exotic' solvents; (d) dramatic increases in the mass sensitivity of concentration-sensitive detectors; (e) increased compatibility with certain types of detectors; (f) the use of very small samples. However, the development of widely applicable instrumentation for LC has been the bottleneck of overall acceptance. Microcolumn HPLC is feasible only with significant departure from the instrumental design of conventional HPLC, e.g. suitable detectors (small cell and connecting tube volumes, fast response times), injection systems and flow-controlled micropumps for the gradient-elution mode of separation. These difficulties increase with decreasing column inner diameter.[152]

The standard commercial HPLC columns used for the separation of polymer additives are normal-bore columns (4–5 mm i.d., usually 4·6 mm i.d.) which are usually operated at flow rates of 1–2 ml/min. However, some separations were performed on narrow-bore columns operated at lower flow rates. Thus, the use of a narrow-bore column (250 mm × 1·2 mm i.d.) packed with 5 μm ODS-silica was reported for a gradient elution separation of polyolefin additives at a low flow rate of the mobile phase, i.e. 0·2 ml/min.[84,85] The RP-HPLC separation of antioxidants in acrylic monomers was performed on a glass column (100 mm × 2·7 mm i.d.) packed with C_{10}-bonded Silasorb-600 (5 μm) silica gel, placed in a stainless steel cartridge and operated at a flow rate of 0·2–0·4 ml/min.[126b]

The microbore and packed capillary columns are operated at significantly lower flow rates, i.e. microlitres instead of millilitres.

The combination of HPLC separation with MS detection (Section 4.3) is facilitated by using microbore columns instead of normal-bore columns. The lower flow rates in microcolumn HPLC (10–50 μl/min) allow the introduction of the total effluent into the ion source. This increases the sensitivity of the method, compared with systems which require splitting of the effluent. Hirter et al.[154] have described the design of a capillary interface for microcolumn RP-HPLC separation/ MS detection and the performance of this combination in the analysis of antioxidant mixtures. The total effluent (35 μl/min) of the microcolumn (25 cm × 0·1 cm i.d., Nucleosil 5 C_{18}) was fed into the ion source of a quadruple MS via a desolvation chamber, whereby the mobile phase (MeCN/THF, 50:50) provides for chemical ionisation.

6.4 High-Resolution Capillary Supercritical Fluid Chromatography

Many of today's analytical separation problems cannot be solved using GC or HPLC. Microcolumn HPLC is still not able to approach the separation capabilities of HR-CGC because of the low solute diffusion characteristics of the HPLC mobile phases. However, research has provided the analyst with another chromatographic technique able to separate involatile or thermally labile compounds, with resolution approaching that of HR-CGC. The ability of highly compressed gases in the critical temperature range to dissolve large amounts of relatively involatile substances has resulted in the development of supercritical fluid chromatography (SFC) and supercritical fluid extraction (SFE). SFC bridges the gap between GC and LC while retaining many of the advantages of both techniques. Therefore, the role of this new chromatographic technique in analytical chemistry is continuously expanding.[155]

The supercritical fluids which may be used as the mobile phase in SFC (Table 10) have diffusion coefficients which are about 100 times greater than those of liquids, and viscosities which are gas-like. Therefore, efficiencies and analysis times are intermediate between those of GC and HPLC.[156] When open tubular columns are used in SFC, overall efficiencies of 10^5–10^6 theoretical plates are obtained in reasonable analysis time.[157] The physico-chemical properties of CO_2 (Table 10) make it especially suitable for SFC. The compressibilty of CO_2 above its critical temperature and pressure is considerable, and densities similar to liquids can be achieved at moderate temperatures. Unlike GC, the solvating power of the mobile phase can be changed

Table 10
Physico-chemical properties of some compounds that can be used as supercritical solvents in fluid extraction and/or mobile phases in supercritical fluid chromatography[156,157]

Compound	Boiling point (°C)	Critical temperature (°C)	Critical pressure (atm)	Critical density (g/cm^3)
CO_2	−78·5	31·04	72·85	0·468
NH_3	−33·35	132·4	112·5	0·235
H_2O	100·00	374·15	218·3	0·315
N_2O	−88·56	36·5	71·7	0·450
CH_3—CH_3	−88·63	32·28	48·16	0·203
CH_2=CH_2	−103·70	9·21	49·66	0·218
$CH_3CH_2CH_3$	−42·1	96·67	41·94	0·217
$CH_3(CH_2)_3CH_3$	36·1	196·5	33·25	0·237
$CCIF_3$	−81·2	28·9	38·7	0·579
SF_6		45·6	37·1	0·752
$CCIF_2$—$CCIF_2$		146·7	35·5	0·582
$CH_3CH(OH)CH_3$		253·3	47·0	0·273
$CH_3(CH_2)_3CH_3$		196·6	33·3	0·232

by altering the density of the supercritical fluid, i.e. by pressure programming. Thus, density programming can effectively control the solvating power of the supercritical fluid and thereby control separation. The mobile phase density can be programmed at low temperatures to elute large-molecular-weight, polar and thermally labile compounds. Other advantages of SFC are its potential for being coupled with a whole range of GC and HPLC detectors and the volatility of the typically mobile phases such as CO_2 on decompression; this eliminates interference from the mobile phase in the subsequent identification of the separated compounds by techniques such as FT-IR spectroscopy. Due to its advantages, SFC has been applied to the analysis of antioxidants, light stabilisers and other polymer additives.[158-160]

Doeht et al.[158] compared SFC separation with flame ionisation detection (FID) of some low-to-medium-polarity polymer additives on reversed-phase packed columns and open-tubular capillary columns using both carbon dioxide and nitrous oxide as mobile phase. On the two reversed-phase columns investigated, i.e. 150 mm × 1·2 mm i.d. 4-μm C_{18}-silica and 30 mm × 2·1 mm i.d. 5-μm cyano silica, the aliphatic amine Armostat 400 (antistatic agent) and the slip agents erucamide and oleamide were strongly adsorbed due to the interaction between their NH or NH_2 groups with the active, unreacted silanol groups in the bonded phase packing. The disulphide antioxidant Hostanox SE-10 as well as the aromatic stabilisers Irganox 1076 and Tinuvin 327 were readily eluted. The use of polar modifiers like methanol in order to deactivate the unreacted silanol groups in the stationary phase is not acceptable because they are not compatible with FID. To avoid the use of modifiers in the SFC separation of compounds with polar groups, open-tubular capillary columns were used because the stationary phase does not contain silanol groups. Indeed, all the additives were easily eluted from fused-silica capillary columns (20 m × 100 μm i.d. and 5·7 m × 48 μm i.d.) with non-polar methylsilicone or phenyl–methyl polysiloxane stationary phases. Simple mixtures of up to three additives could be separated under isobaric conditions but more complex mixtures required a pressure programme to shorten the analysis time. However, nitrous oxide was not suitable for pressure programming and FID due to a severe baseline drift caused by the high background level. Flame ionisation detection in SFC separation is well suited for the analysis of polymer additives without chromophores, which are usually monitored in HPLC separation by the less sensitive RI detector.

The fact that supercritical carbon dioxide and xenon are absorbing minimally in the mid-IR region has been utilized in the direct coupling to a high-pressure FT-IR cell and applied to the analysis of antioxidants.[157a] However, greater sensitivities can be achieved following the desolvation and deposition of a solute on a solid matrix prior to spectral acquisition.

Raynor et al.[159,160] showed that high-resolution capillary SFC is a very efficient technique for the separation and identification of additives in polymers. The separation capability of this technique was demonstrated with a synthetic mixture of 21 additives varying in chemical composition and with a molecular weight range from 225 to 1178.[160] The compounds were separated on a non-polar capillary column, using CO_2 as the mobile phase at 140°C and with density programming from 0·2 to 0·6 g/ml (Fig. 18). The linear relationship between the pressure and density of CO_2 at 140°C resulted in small variations in density for small variations in pressure. Thus, the solvating ability of the mobile phase and hence the solute retention times could be controlled. Compared with HPLC, the higher tempera-

MODERN LIQUID-CHROMATOGRAPHIC ANALYTICAL METHODS 297

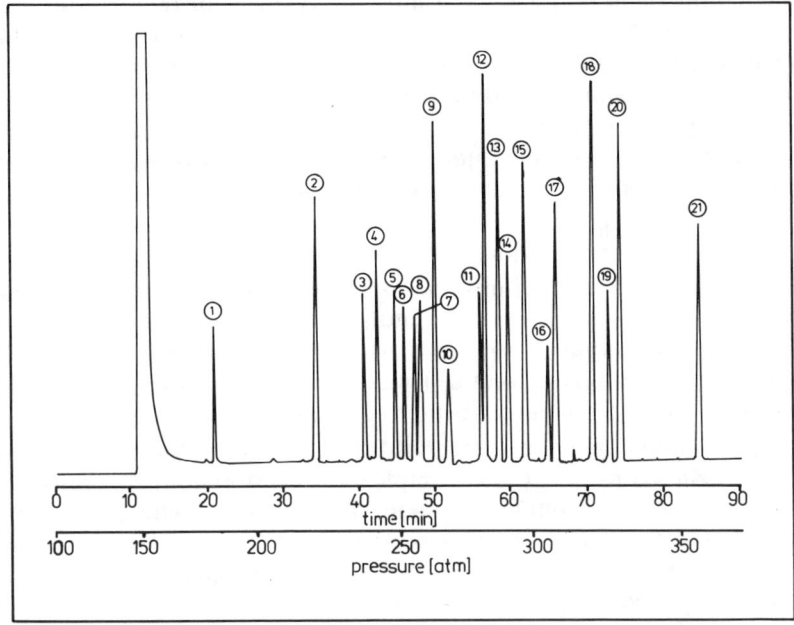

Fig. 18. High-resolution capillary supercritical fluid chromatographic separation of a synthetic mixture of 21 polymer additives.[160] Chromatographic conditions: fused silica capillary column (10 m × 0·05 mm i.d.) coated with 0·25-μm crosslinked methylpolysiloxane stationary phase; mobile phase CO_2 at 140°C; pressure programming 150 atm to 350 atm at 3 atm/min after an initial 12 min isobaric period; injected sample 0·2 μl of synthetic mixture containing 400 ng/μl of each additive, i.e. 20 ng/component introduced in to the column (assuming the uniform split ratio of 1:3); flame ionisation detector. Peaks: 1, Topanol OC: 2, Tinuvin P; 3, Tinuvin 292; 4, Tinuvin 320; 5, Tinuvin 326; 6, Tinuvin 328; 7, Chimassorb 81; 8, erucamide; 9, Tinuvin 770; 10, Irgafos 168; 11, Tinuvin 440; 12, Tinuvin 144; 13, DLTDP; 14, Irganox 1076; 15, Irganox 1024MD; 16 Irganox 245; 17, Irganox 1035; 18, Irganox 3114; 19, DSTDP; 20, Irganox 1330; 21, Irganox 1010.

ture of the mobile phase in SFC may result in thermal degradation of certain additives. Thus, some degradation of Tinuvin P in the column at 140°C occurred. Lowering the temperature lowered chromatographic efficiency.[160] Therefore, SFC analyses have to be performed at the highest temperature that samples can withstand without risk of degradation. Near the critical point the pressure–density relationship is non-linear and very slight variations in pressure or temperature have a significant effect on the fluid density, and hence on solvating ability, so that the separation may be highly irreproducible without precise control of temperature and pressure.

Peak retention times in the separation of additives at 140°C were found to be reproducible enough for identification purposes.[160] However, identification by this method is time-consuming, requires that all the suspected compounds are on hand and, sometimes, can also result in erroneous interpretation. Thus, for example, the peaks of Tinuvin 144 and the degradation product of Irgafos 168 have the same retention times. The accurate identification of unknown additives was achieved by coupling the capillary SFC separation with FT-IR detection using a microscope accessory and the solvent elimination interface. The Soxhlet extracts from standard and commercial PP samples containing unknown additives were separated. Polymer additives which eluted from the column were deposited on a KBr disc as spots of about 200 nm in diameter and spatially separated. Good-quality spectra which allowed additive identification by comparison with reference spectra were subsequently obtained by FT-IR microspectroscopy.[159,160]

The ability of supercritical fluids to dissolve large amounts of substances of low volatility is of considerable practical interest, for example in the SFE of caffeine, nicotine, hops, spices, drugs, etc.[156,161] Chalmers et al.[162] recently showed that SFE provides comparable and sometimes better extraction efficiencies than conventional Soxhlet extraction and with an order-of-magnitude increase in the rate of extraction. It is believed that on-line coupling of (a) the SFE with CO_2 of additives from polymers, (b) the high-resolution capillary SFC separation of the extract, and (c) the identification of the separated additives by FT-IR microspectroscopy following solvent elimination, may be possible.[160]

7 ANALYSIS OF ANTIOXIDANTS AND LIGHT STABILISERS IN POLYMERS: THE STATE OF THE ART

The determination of antioxidants, light stabilisers and other additives in polymers has always given rise to considerable interest in research laboratories in order to characterise the polymers better and to develop new applications. Analyses of practical interest have therefore been developed, mainly for quality assurance of polymer batches.

Antioxidants and light stabilisers are generally used either mixed together or with other additives, and many of these compounds show very little difference in their chemical structure. Therefore, only the chromatographic separation techniques following solvent extraction are generally able to achieve quali-quantitative analyses of additive packages in polymers. Despite all developments in GC, this technique cannot separate many antioxidants and light stabilisers and especially today's additives since their vapour pressures are often too high or since their thermostability is insufficient. Therefore, in spite of its lower efficiency in comparison with modern GC, HPLC is today the basic technique for the analysis of additives in polymers and has been predominantly used in recent years for such analyses.

Many applications of HPLC usually involve the development of specific analytical methods for each additive or class of additive of interest. This specific approach is often satisfactory for quality control or similar purposes involving routine analysis of a small number of additives. However, many situations arise where a general method for the analysis of additives is desired, e.g. the identification and quantification of unknown additives in polymer samples.

Isocratic HPLC separation of polymer additives is generally preferred because of better precision and simplicity compared with the gradient-elution mode of separation. Additive packages for commercial polyolefins usually do not contain more than a few components. Therefore, for routine work involving the analysis of known additives in polymer batches, it is nearly always possible to develop an isocratic separation which allows both identification and quantification of individual components. Optimised isocratic elution of components in the particular analysis of a certain additive package is generally achieved by varying the mobile phase strength. Sometimes, optimised isocratic analyses may be developed to separate complex mixtures containing many compounds of very different structure and polarity; however, the gradient-elution mode of separation is preferred for such purposes. Gradient-elution HPLC is also used for the rapid screening of unknown additives and for pilot work to find a suitable mobile phase for an isocratic separation.

New developments in HPLC such as HSLC and microcolumn HPLC and the advent of high-resolution capillary SFC are basically aimed at the improvement of separation capability, detection and identification and improved economics of analysis. However, despite their advantages, these chromatographic techniques are feasible only with significant departures from the conventional HPLC instrumentation, and therefore are not available to everyone. Conventional HPLC is still able today to solve satisfactorily almost all the problems encountered in the analysis of antioxidants, light stabilisers and other polymer additives. Routine analyses are carried out without difficulties by NP- or RP-HPLC separation, conventional detection (UV and/or RI detectors) and identfication by means of additive standards. This traditional approach often allows more complicated analyses such as identification and quantification of unknown additives in polymers. However, such problems may be better solved by off- and preferably on-line coupling of HPLC separation with powerful identification tools such as FT-IR spectroscopy and especially MS. The trend towards such interfacing is steadily increasing.

Due to its disadvantages in comparison with HPLC, the use of high-performance SEC in the analysis of additive extracts from polymers seems to be virtually extinct. However, SEC will most probably be used for some analyses that are not amenable to HPLC, e.g. the direct determination of additives in polymers without the preliminary extraction step, and the analysis of high-molecular-weight and polymer-bound stabilisers.

In view of the fact that monitoring of polymer additives has become increasingly important because of stricter laws and regulations, the need for fast and reliable analytical methods has increased. This is especially true for some very exigent fields such as analysis of additives in biocompatible and biomedical polymers and in food packaging. It is probable that there is a large amount of unpublished work concerning the analysis of polymer additives because of the proprietary nature of many in-house methods. However, the analyst has to remember Lord Kelvin's famous words: 'Measurement is the basis of all knowledge', and has to be prepared for the advent of new types of antioxidants and light stabilisers which are probably even now under development.

IN MEMORIAM

The author dedicates this contribution to the memory of his beloved friends and colleagues Anabela and Codruţ Işfan.

ACKNOWLEDGEMENT

The author is indebted to Dr L. Cotarcă of the Department of Organic Chemistry, Polytechnic Institute, Timişoara, for stimulating discussions on the subject.

REFERENCES

1. Mascia, L., *The Role of Additives in Plastics*. Edward Arnold, London, 1974.
2. Seymour, R. B., *Additives for Plastics*. Academic Press, New York, 1978.
3. Henman, T. J., *World Index of Polyolefin Stabilisers*. Kogan Page/Royal Society of Chemistry, London, 1982.
4. Stepek, J. & Daoust, H., *Additives for Plastics*, Vol. 6, Polymers Properties and Applications. Springer Verlag, Berlin, 1983.
5. Gächter, R. & Müller, H., *Plastics Additives*. Carl Hanser Verlag, München, 1984.
6. Hardy, B. W. In *Developments in Polymer Photochemistry—3*, ed. N. S. Allen. Applied Science Publishers, London, 1982, p. 287.
7. Gould, R. W., Henman, T. J. & Billingham, N. C., *Brit. Polym. J.*, **16** (1984) 284.
8. Crompton, T. R., *Chemical Analysis of Additives in Plastics*, Vol. 46. International Series of Monographs in Analytical Chemistry, 2nd edn. Pergamon Press, Oxford, 1977.
9. Haslam, J. H. & Willis, H. A., *Identification and Analysis of Plastics*. Van Nostrand, Princeton, NJ, 1965.
10. Wake, W. C., *The Analysis of Rubber and Rubber-Like polymers*, 2nd edn. Wiley Interscience, New York, 1969.
11. Braun, D., *Simple Methods for Identification of Plastics*. Carl Hanser Verlag, München, 1982.
12. Krause, R., Lange, A. & Ezrin, M., *Plastics Analysis Guide*. Carl Hanser Verlag, München, 1983.
13. Crompton, T. R., *The Analysis of Plastics*, Vol. 8. Pergamon Series in Analytical Chemistry. Pergamon Press, Oxford, 1984.
14. Wheeler, D. A., *Talanta*, **15** (1968) 1315.
15. Walter, R. B. & Johnson, J. F., *J. Polym. Sci., Macromol. Rev.*, **15** (1980) 29.
16. Vimalasiri, P. A. D. T., Haken, J. K. & Burford, R. P., *J. Chromatogr., Chromatogr. Rev.*, **300** (1984) 303.
17. Sreenivasan, K., *Chromatography*, **22** (1986) 199.
18. Airaudo, C. B., Gayte-Sorbier, A., Laurent, P. & Creusevau, R., *J. Chromatogr.*, **314** (1984) 349.
19. Airaudo, C. B., Gayte-Sorbier, A. & Creusevau, R., *J. Chromatogr.*, **392** (1987) 407; Airaudo, C. B., Gayte-Sorbier, A., Aujoulat, P. & Mercier, V., *J. Chromatogr.*, **437** (1988) 59.

20. Fahnrich, J., Popl, M. & Vyborna, E., *Sci. Pap. Prague Inst. Chem. Technol.*, **18** (1983) 105.
21. Poole, C. F., *Tr. A. C.*, **4** (1985) 209.
22. Brinkmann, U. A. Th., *Tr. A. C.*, **5** (1986) 178.
23. Witkiewicz, Z. & Bladek, J., *J. Chromatogr.*, **373** (1986) 111.
24. Sano, M., Abe, M., Yoshino, K., Matsuura, T., Sekino, T., Sato, S. I. & Tomita, I., *Chem. Pharm. Bull.*, **34** (1986) 174.
25. Snyder, L. R. & Kirkland, J. J., *Introduction to Modern LIquid Chromatography*, 2nd edn. John Wiley & Sons, New York, 1979.
26. Horvath, C. (ed.), *High-Performance Liquid Chromatography. Advances and Perspectives*, Vols 1 and 2 (1980), Vol. 3 (1985), Vol. 4 (1987). Academic Press, New York.
27. Archer, A. W., *Anal. Chim. Acta*, **128** (1981) 236.
28. Baylock, D., Majcherczyk, C. & Pellerin, F., *Ann. Pharm. Fr.*, **43** (1986) 329.
29. Sreenivasan, K., *J. Chromatogr.*, **357** (1986) 433.
29a. Sreenivasan, K., *J. Liq. Chromatogr.*, **9** (1986) 2425.
30. Carlsson, D. J., Grattan, D. W., Suprunchuk, T. & Wiles, D. M., *J. Appl. Polym. Sci.*, **22** (1978) 2217.
31. Constante, G. E., *Grases Aceites (Seville)*, **26** (1975) 150; *Chem. Abstr.*, **83** (1975) 130109.
32. Coupek, J., Kalovec, M., Krivakova, M. & Pospisil, J., *Angew. Makromol. Chem.*, **15** (1971) 137.
33. Coupek, J., Pokorny, S., Protitova, J., Holcik, J., Karvas, M. & Pospisil, J., *J. Chromatogr.*, **65** (1972) 279.
34. Coupek, J., Pokorny, S., Jirakova, L. & Pospisil, J., *J. Chromatogr.*, **75** (1973) 87.
35. Coupek, J., Pokorny, S. & Pospisil, J., *J. Chromatogr.*, **95** (1974) 103.
36. Dong, M. W. & DiCesare, J. L., *Plastic Engineering* (1983) 25.
36a. Dostal, A., *Chemicke Listy*, **81** (1987) 765.
37. Duval, M. & Giguere, Y., *J. Liq. Chromatogr.*, **5** (1982) 1347.
38. Fallick, G. J., Talarico, P. C. & McGough, R. R., *Proc. Soc. Plast. Eng, 34th Ann. Techn. Conf.*, 1976, p. 574.
39. Fähnrich, J., Vit, I., Popl, M. & Smejkal, F., *Chemicky Prymysl.*, **32/57** (1982) 487.
40. Francis, V. C., Sharma, Y. N. & Bhardwaj, I. S., *Angew. Makromol. Chem.*, **43** (1983) 219.
41. Gasslander, U. & Jaegfeldt, H., *Anal. Chim. Acta*, **166** (1984) 243.
42. Gharfeh, S. G., *J. Chromatogr.*, **389** (1987) 211.
43. Gross, D. & Strauss, K., *Kautsch. Gumm. Kunstst.*, **29** (1976) 741.
44. Guergens, U. & Doemling, H. J., *Deutsch. Lebensm. Rundsch.*, **78** (1982) 49.
45. Gupta, H. K. & Salovey, R., *Rubber Chem. Technol.*, **58** (1985) 295.
46. Hammond, K. J., *J. Assoc. Public. Anal.*, **16** (1978) 17.
47. Haney, M. A. & Dark, W. A., *J. Chromatogr. Sci.*, **18** (1980) 655.
48. Hodgeman, K. C., *J. Chromatogr.*, **214** (1981) 237.
49. Howard, J. M., *J. Chromatogr.*, **55** (1971) 15.
49a. Jonas, R. O., Parsons, B. A. G. & Wilkinson R., *Proc. Conf. Chemical*

and Physical Phenomena in the Ageing of Polymers, Prague, 11–14 July 1988, p. 84.
50. Kauffman, P., *J. Chromatogr.*, **132** (1977) 356.
51. Kovacic, T., Nardelli, T. & Klaric, I., *Hem. Ind.*, **37** (1983) 106.
52. Lehotay, J., Danecek, J., Liska, O., Lesko, O. J. & Brandsteterova, E., *J. Appl. Polym. Sci.*, **25** (1980) 1943.
53. Lichtenhalter, R. G. & Ranfelt, F., *J. Chromatogr.*, **149** (1978) 149.
54. Mady, N. H., Liu, R. & Wiczkus, J., *Proc. SPI Annu. Tech. Mark Conf.*, 1986, p. 332.
55. Majors, R. E., *J. Chromatogr. Sci.*, **8** (1970) 338.
56. Majors, R. E. & Johnson, E. L., *J. Chromatogr.*, **167** (1978) 17.
57. Masoud, A. N. & Cha, Y., *J. High Resolut. Chromatogr., Chromatogr. Commun.*, **5** (1982) 299.
58. Munteanu, D., Isfan, A., Isfan, C. & Tincul, I., *Mater. Plast. (Bucharest)*, **22** (1985) 173.
59. Munteanu, D., Isfan, A., Isfan, C. & Tincul, I., *Chromatography*, **23** (1987) 7.
60. Nakamura, S., *Kobunshi Kagaku*, **29** (1972) 372.
61. Nakamura, S., Ishiguro, S., Yamada, T. & Moriizumi, S., *J. Chromatogr.*, **83** (1973) 279.
62. Oguri, H., *Shikizai Kyokaishi*, **57** (1984) 441; *Chem. Abstr.*, **101** (1984) 173077.
62a. Pacco, J. M. & Mukherji, A. K., *J. Chromatogr.*, **144** (1977) 113.
63. Pellerin, F., Majcherczyk, C. & Baylocq, D., *Talanta*, **33** (1986) 85.
64. Perlstein, P., *Anal. Chim. Acta*, **149** (1983) 21.
65. Popl, M., Vit, I. & Smejkal, F., *J. Chromatogr.*, **213** (1981) 363.
66. Proseus, R. A., *J. Chromatogr.*, **97** (1974) 201.
67. Protitova, J., Pospisil, J. & Zikmund, L., *J. Polym. Sci., Polym. Symp.*, **40** (1973) 233.
68. Protitova, J. & Pospisil, J., *J. Chromatogr.*, **88** (1974) 99.
69. Protitova, J., Pospisil, J. & Holcik, J., *J. Chromatogr.*, **92** (1974) 361.
70. Quinn, M. E. & McGee, W. W., *J. Chromatogr.*, **350** (1985) 187.
71. Rotschova, J. & Pospisil, J., *J. Chromatogr.*, **211** (1981) 299.
72. Schabron, J. F., Hurtubise, R. J. & Silver, H. F., *Anal. Chem.*, **50** (1978) 1911.
73. Schabron, J. F., Hurtubise, R. J. & Silver, H. F., *Anal. Chem.*, **51** (1979) 1426.
74. Schabron, J. F. & Fenska, L. E., *Anal. Chem.*, **52** (1980) 1411.
75. Schabron, J. F. & Bradfield, C. Z., *J. Appl. Polym. Sci.*, **26** (1981) 2479.
76. Schabron, J. F., Smith, V. J. & Ware, J. L., *J. Liq. Chromatogr.*, **5** (1982) 613.
77. Schabron, J. F., *J. Liq. Chromatogr.*, **5** (1982) 1269.
78. Sevini, F. & Marcato, B., *J. Chromatogr.*, **260** (1983) 507.
79. Shepherd, M. J. & Gilbert, J., *J. Chromatogr.*, **218** (1981) 703.
79a. Shepherd, M. J. & Gilbert, J., *J. Chromatogr.*, **178** (1979) 435.
80. Smejkal, F., Popl, M. & Cihova, A., *J. Polym. Sci., Polym. Symp.*, **68** (1980) 145.
80a. Stand, U. D., *Angew. Makromol. Chem.*, **150** (1987) 13.

81. Stoveken, J. J., Perkin Elmer Technical Note 58, 1977.
82. Telepchak, J. J., Perkin Elmer Application Study 37, 1974.
83. Uchytil, B., *J. Chromatogr.*, **93** (1974) 447.
84. Vargo, J. D. & Olson, K. L., *Anal. Chem.*, **57** (1985) 672.
85. Vargo, J. D. & Olson, K. L., *J. Chromatogr.*, **353** (1986) 215.
86. Vit, I., Popl, M. & Fähnrich, J., *Chemicky Prymysl.*, **34** (1984) 642.
87. Waters Associates, *Application Highlight Polymer Additives*, AH-356, 1975.
88. Waters Associates, *Liquid Chromatography Procedures for Polyolefin Additives*, WAPP-100, 1978.
89. Wims, A. M. & Swarin, S. J., *J. Appl. Polym. Sci.*, **19** (1975) 1243.
90. Yoshikawa, T., Kimura, K. & Fujimura, S., *J. Appl. Polym. Sci.*, **15** (1971) 2513.
91. Zehner, J. M. & Simonaitis, R. A., *J. Chromatogr.*, **14** (1976) 326.
92. Schröder, E., *Pure Appl. Chem.*, **36** (1973) 233.
93. British Standard 2782, Part 4, Method 405 B and D (1965).
94. Campbell, R. H. & Wise, R. W., *J. Chromatogr.*, **12** (1963) 178.
95. Cook, C. D. & Woolworth, R. C., *J. Am. Chem. Soc.*, **75** (1953) 6242.
96. Cornish, P. J., *J. Appl. Polym. Sci.*, **7** (1963) 727.
97. Crompton, T. R., *J. Appl. Polym. Sci.*, **6** (1962) 538.
98. Crompton, T. R., *Europ. Polym. J.*, **4** (1968) 473.
99. Klute, C. H. & Franklin, P. J., *J. Polym. Sci.*, **6** (1962) 538.
100. Metcalf, K. & Tomlinson, R., *Plastics* (*London*), **25** (1960) 319.
101. Myers, A. D., Rogers, C. E., Stanett, V. & Szuare, M., *Mod. Plast.*, **34** (1957) 157.
102. Schröder, E. & Rudolph, G., *Plaste Kautsch.*, **10** (1963) 22.
103. Slonaker, D. F. & Sievers, D. C., *Anal. Chem.*, **36** (1964) 1130.
104. Spell, H. L. & Eddy, R. D., *Anal. Chem.*, **32** (1960) 1811.
105. Stafford, C., *Anal. Chem.*, **32** (1960) 1811.
106. Stafford, C., *Anal. Chem.*, **34** (1962) 794.
107. Van der Heide, R. F. & Wouters, O., *Lebensmitt. Untersuch. Forsch.*, **117** (1962) 129.
108. Woggon, H., Korn, O. & Jehle, D., *Die Nahrung*, **9** (1965) 495.
109. Yushkevichyute, S. S. & Shlyapnikov, Yu. A., *Plast. Massy*, No. 12. (1966) 62.
110. Yushkevichyute, S. S. & Shlyapnikov, Yu. A., *Plast. Massy*, No. 1. (1967) 54.
111. Freitag, W., *Fresenius Z. Anal. Chem.*, **316** (1983) 495.
112. Luongo, J. P., *Appl. Spectrosc.*, **19** (1965) 117.
113. Drushel, H. V. & Sommers, A. L., *Anal. Chem.*, **36** (1964) 836.
114. Billingham, N. C., Bott, D. C. & Manke, A. S. In *Developments in Polymer Degradation—3*, ed. N. Grassie. Applied Science Publishers, London, 1981, p. 63.
115. Krishen, A., *J. Chromatogr. Sci.*, **15** (1977) 434.
116. Krishen, A. & Tucker, R. G., *Anal. Chem.*, **49** (1977) 898.
117. Pausch, J. B., *Anal. Chem.*, **54** (1982) 89A.
118. Latimer, R. P., Hooser, E. R. & Zakriski, P. M., *Rubber Chem. Technol.*, **53** (1980) 346.

119. Latimer, R. P., Harmon, D. J. & Hansen, G. E., *Anal. Chem.*, **52** (1980) 1808.
120. Latimer, R. P. & Hansen, G. F., *Macromolecules*, **14** (1981) 776.
121. Latimer, R. P., Harmon, D. J. & Welch, K. R., *Anal. Chem.*, **51** (1979) 1293.
122. Latimer, R. P. & Welch, K. R., *Rubber Chem. Technol.*, **53** (1980) 151.
123. Kato, S. & Kanoshita, K., *J. Chromatogr.*, **324** (1985) 462.
124. Rubtsova, T. A., Melnikova, N. A., Cheresneva, A. F. & Glushkova, L. C., *Zh. Anal. Khim.*, **40** (1985) 721.
124a. Petro, M. & Bystricky, L., *Chem. Papers*, **40** (1986) 357.
125a. Eisenbeiss, F., Dumont, E. & Henke, H., *Angew. Makromol. Chem.*, **71** (1978) 67.
125b. Pasteur, G. A., *Anal. Chem.*, **49** (1977) 363.
125c. Mann, T. J., *J. Liq. Chromatogr.*, **3** (1980) 210.
126a. Lazaris, A. Ya. & Beloded, L. N., *Zh. Anal. Khim.*, **39** (1984) 1700.
126b. Lazaris, A. Ya., Beloded, L. N. & Kalinin, A. I., *J. Chromatogr.*, **365** (1986) 333.
127. Cunningham, A. F. & Hillman, D. E., *J. Chromatogr.*, **148** (1978) 528.
128. Hayes, G. E. & Hillman, D. E., *J. Chromatogr.*, **322** (1985) 376.
128a. Black, B. H., Wechter, M. A. & Hardy, D. R., *J. Chromatogr.*, **437** (1988) 203.
129. Ansari, G. A. S., *J. Chromatogr.*, **262** (1983) 393.
129a. Beker, D. & Lovrec, V., *J. Chromatogr.*, **393** (1987) 459.
130. Coustard, J. M. & Sudraud, G., *J. Chromatogr.*, **219** (1981) 338.
131. Dolan, J. W., Grant, J. R., Tanaka, R., Giese, W. & Karger, B. L., *J. Chromatogr. Sci.*, **16** (1978) 616.
132. Galensa, R., *Z. Lebensm. Unters. Forsch.*, **178** (1984) 475.
133. Gross, D. & Strauss, K., *Kautsch. Gumm. Kunstst.*, **32** (1979) 18.
134. Huong, M. & Bui-Nguyen, *J. Chromatogr.*, **196** (1980) 163.
135. Indyk, H. & Woolard, D. C., *J. Chromatogr.*, **356** (1986) 401.
135a. Anderson, J. & Niekerk, P. J., *J. Chromatogr.*, **394** (1987) 400.
136. King, W. P., Joseph, K. T. & Kissinger, P. T., *J. Assoc. Off. Anal. Chem.*, **63** (1980) 137.
137. Kitada, Y., Ueda, Y., Yamamota, M., Shynomiya, K. & Nakazawa, H., *J. Liq. Chromatogr.*, **8** (1985) 47.
138. Nissen, H. P. & Kreysel, H. W., *GIT Suppl.*, **2** (1986) 41.
139. Nilsson, B., Johannson, B., Janson, L. & Holmberg, L., *J. Chromatogr.*, **145** (1978) 169.
140. Okamoto, T. & Suezawa, Y., *Kagawa-Ken Hakko Shokuhin Shikenjo Hokoku*, **76** (1983) 79; *Chem. Abstr.*, **102** (1985) 22959.
141. Orsi, F. & Abraham-Szabo, A., *Elelmiszervizsgalati Kozl.*, **31** (1985) 78; *Chem. Abstr.*, **103** (1985) 177056.
142. Pachla, L. A. & Kissinger, P. T., *Methods Enzymol.*, **62** (1979) 15.
143. Page, D. B., *J. Assoc. Off. Anal. Chem.*, **62** (1979) 1239.
143a. Pellerin, F., Dumitrescu, D. & Bayloco, D., *Ann. Pharm. Fr.*, **38** (1980) 7.
144. Pujol-Forn, M., Lopez, S. & Maria, C., *Circ. Farm.*, **42** (1984) 3.

144a. Van de Greef, J., Speek, A. J., Tas, A. C. & Schriver, J., *LG-GC Magazine*, **4** (1986) 636.
145. Vatassery, G. T., Maynard, V. R. & Hagen, D. F., *J. Chromatogr.*, **161** (1978) 299.
146. Wagner, E. S., Lindley, B. & Coffin, R. D., *J. Chromatogr.*, **163** (1979) 225.
147. Moore, J. C., *J. Polym. Sci., Part A-2*, **2** (1964) 835.
148. Grob, K., *On-column Injection of Capillary Gas Chromatography*. Hüthig, Heidelberg, 1987.
149. DiPasqualle, G. & Galli, M., *J. High Resolut. Chromatogr., Chromatogr. Commun.*, **7** (1984) 484.
150. DiPasqualle, G., Giambelli, L., Soffientini, A. & Paiello, R., *J. High Resolut. Chromatogr., Chromatogr. Commun.*, **8** (1985) 618.
151. Blum, W. & Damascenco, L., *J. High Resolut. Chromatogr.*, **10** (1987) 472.
152. Novotny, M., *J. High Resolut. Chromatogr., Chromatogr. Commun.*, **10** (1987) 242.
153. Kucera, P., *Microcolumn High Performance Liquid Chromatography*. Elsevier, Amsterdam, 1984.
154. Hirter, P., Walther, H. J. & Daetwyler, P., *J. Chromatogr.*, **323** (1985) 89.
155. Chester, T. L., *J. Chromatogr. Sci.*, **24** (1986) 226.
156. Van Wasen, U., Swaid, I. & Schneider, G. M., *Angew. Chem.*, **19** (1980) 575.
157. Novotny, M., Springston, S. R., Peaden, P. A., Fjelsted, J. C. & Lee, M. L., *Anal. Chem.*, **53** (1981) 407A.
157a. French, S. B. & Novotny, M., *Anal. Chem.*, **58** (1986) 164.
158. Doeht, J., Farbrot, A., Greibokk, T. & Iversen, B. J., *J. Chromatogr.*, **392** (1987) 175.
159. Raynor, M. W., Davies, I. L., Bartle, K. D., Williams, A., Chalmers, J. M. & Cook, B. W., *Europ. Chromatogr. News.*, **1** (1987) 18.
160. Raynor, M. W., Bartle, K. D., Davies, I. L., Williams, A., Clifford, A. A., Chalmers, J. M. & Cook, B. W., *Anal. Chem.*, **60** (1988) 427.
161. Penninger, J. M. L., Radosz, M., McHugh, M. A. & Krukonis, W. J., *Supercritical Fluid Technology*. Elsevier, Amsterdam, 1985.
162. Chalmers, J. M., Mackenzie, M. W., Sharp, J. L. & Ibbet, R. N., *Anal. Chem.*, **59** (1987) 415.

APPENDIX: ANTIOXIDANTS AND LIGHT STABILISERS

I   HO—⟨ ⟩—R

Ia R = —CH_3
Ib R = —CH_2OH
Ic R = —$C(CH_3)_3$
Id R = —$CH_2COOC_{18}H_{37}$

Ie R = —NH—[triazine with SC₈H₁₇ groups]

II HO—[aryl]—R—[aryl]—OH

IIa R = —CH₂—
IIb R = —(CH₂)₂COO(CH₂)₆OOC(CH₂)₂—
IIc R = —(CH₂)₂CONH—NHOC(CH₂)₂—
IId R = —(CH₂)₂CONH(CH₂)₃HNOC(CH₂)₂—
IIe R = —(CH₂)₂CONH(CH₂)₆HNOC(CH₂)₂—
IIf R = —(CH₂)₂COO(CH₂)₂—S—(CH₂)₂OOC(CH₂)₂—

III HO—[aryl(CH₃)]—(CH₂)₂COO(CH₂CH₂O)₃OC(CH₂)₂—[aryl(CH₃)]—OH

IV [two aryl rings with R, OH, CH₃ substituents linked by X]

IVa X = —CH₂—   R = —C(CH₃)₃
IVb X = —CH₂CH(CH₃)CH₂— R = —CH₃

IVc X = —CH₂—   R = —[cyclohexyl with H, H₃C]

V HO—[aryl-CH₃]—(CH₂)₄—[aryl-CH₃]—OH

VI

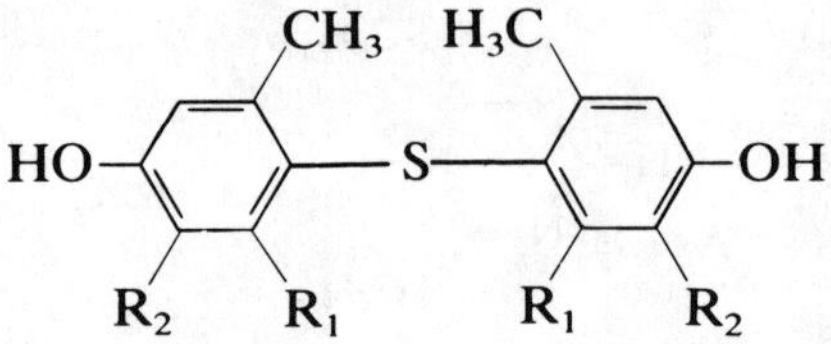

VIa $R_1 = -C(CH_3)_3$ $R_2 = -H$
VIb $R_1 = -H$ $R_2 = -C(CH_3)_3$

VII

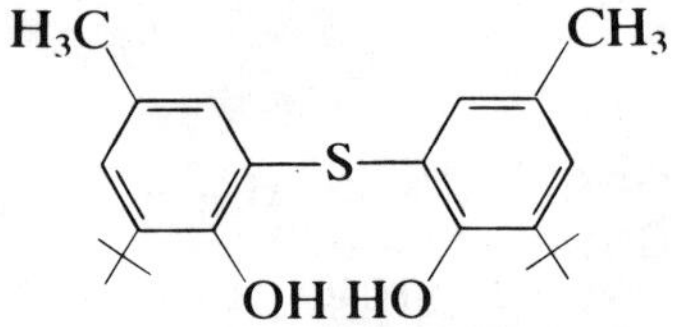

VIII R:

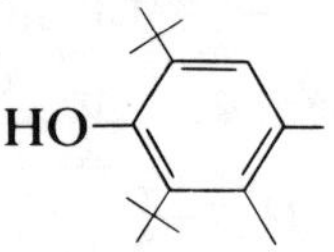

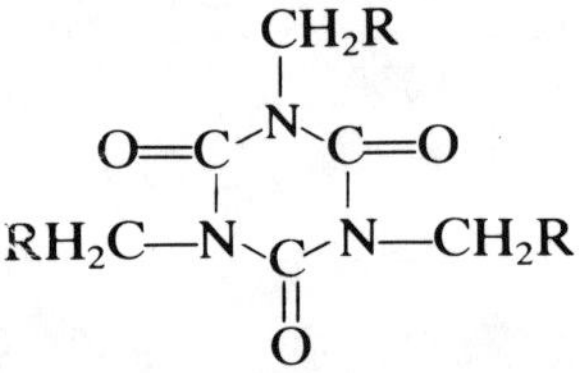

VIIIa

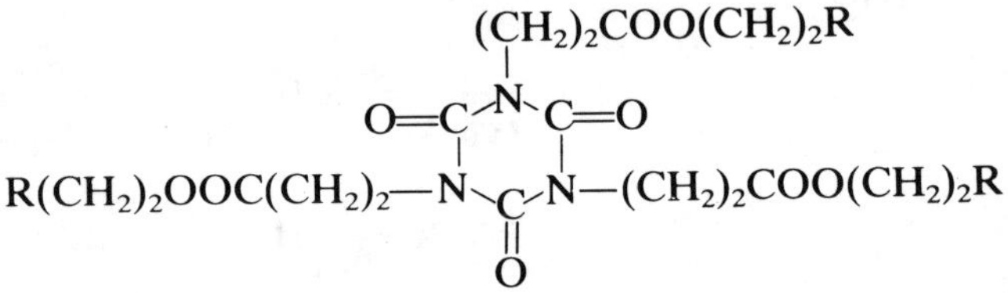

VIIIb

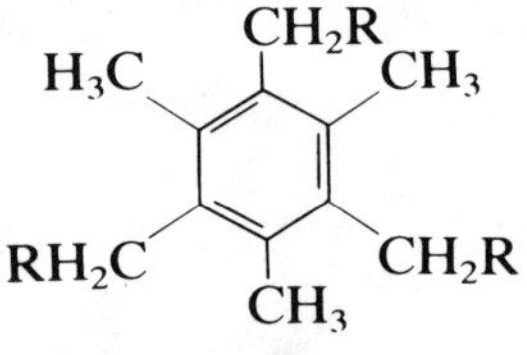

VIIId $[R(CH_2)_2COOCH_2]_3C-CH_2OH$
VIIIe $[R(CH_2)_2COOCH_2]_4$

VIIIc

IX [HO–C₆H₃(C(CH₃)₃)–]₂ CH—CH₂—CH—CH₃ with phenol (C(CH₃)₃, OH) substituent

X $H_{13}C_6OOC(CH_2)_2-C(CH_3)_2$—[C₆H₂(OH)₂]—$C(CH_3)_2-(CH_2)_2COOC_6H_{13}$

XI $S(CH_2CH_2COOC_nH_{2n+1})_2$ **XIa** $n = 12$ **XIb** $n = 14$ **XIc** $n = 16$

XII $[R-C_6H_4-O]_3P$

XIIa R = H
XIIb R = C_9H_{19}

XIII $[(C(CH_3)_3)_2-C_6H_3-O]_3P$

XIV $H_{37}C_{18}$—P(O—CH₂)₂C(CH₂—O)₂P—$C_{18}H_{37}$

	R_1	R_2
XVa	H	H
XVb	OH	CH_3
XVc	H	CH_3
XVd	H	nC_8H_{17}
XVe	H	$nC_{10}H_{21}$

XV (2-R₁-C₆H₄)—C(=O)—(2-HO-4-OR₂-C₆H₃)

XVI

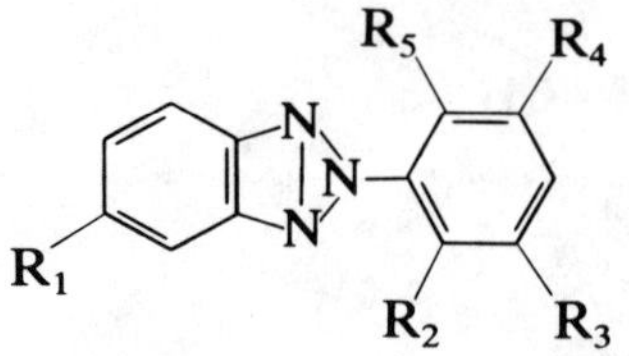

	R_1	R_2	R_3	R_4	R_5
XVIa	H	OH	H	CH_3	H
XVIb	H	OH	$C(CH_3)_3$	$C(CH_3)_3$	H
XVIc	H	H	$CH_2C(CH_3)_3$	$CH_2C(CH_3)_3$	OH
XVId	Cl	OH	$C(CH_3)_3$	$C(CH_3)_3$	H
XVIe	Cl	OH	CH_3	$C(CH_3)_3$	H
XVIf	H	OH	H	$nC_{18}H_{37}$	H
XVIg	H	OH	$(H_3C)_2CC_6H_5$	$(H_3C)_2CC_6H_5$	H

XVII

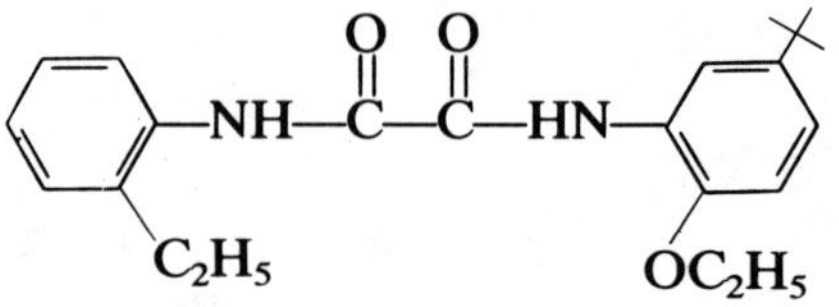

XVIII

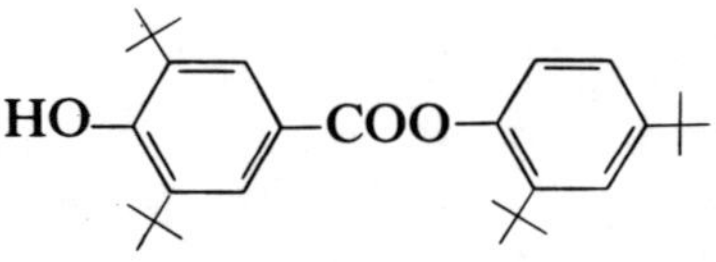

XIX   R = ... N—(H, CH_3, CH_2—)

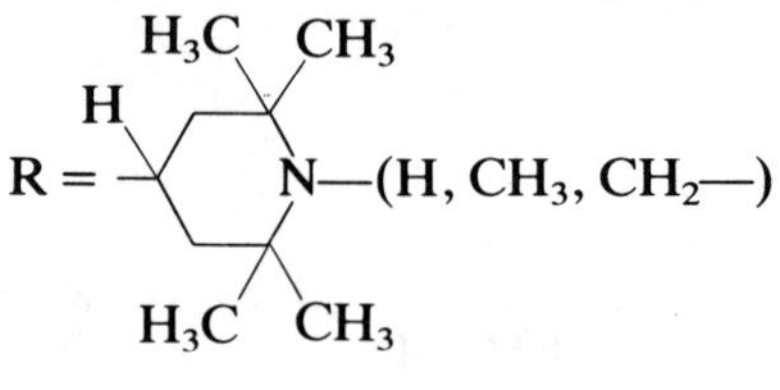

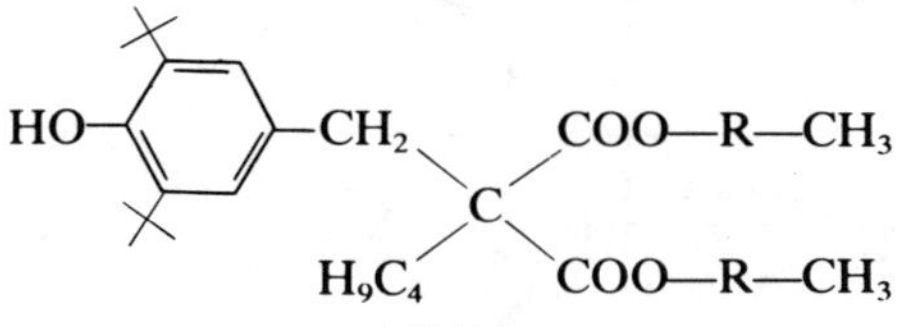

XIXa

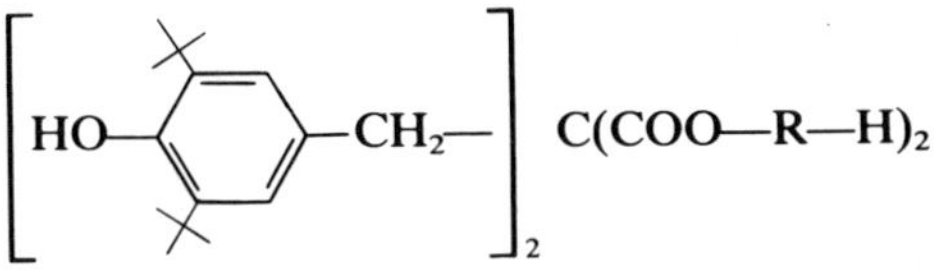

$$\left[HO-\underset{\times}{\overset{\times}{\bigcirc}}-CH_2- \right]_2 C(COO-R-H)_2$$

XIXb

$$H_3C-R-OOC-(CH_2)_8-COO-R-CH_3$$
XIXc

$$H-R-OOC-(CH_2)-COO-R-H$$
XIXd

$$H-\!\!\left[-O-R-(CH_2)_2-OOC-(CH_2)_2-CO-\right]_n\!OCH_3 \quad \textbf{XIXe}$$

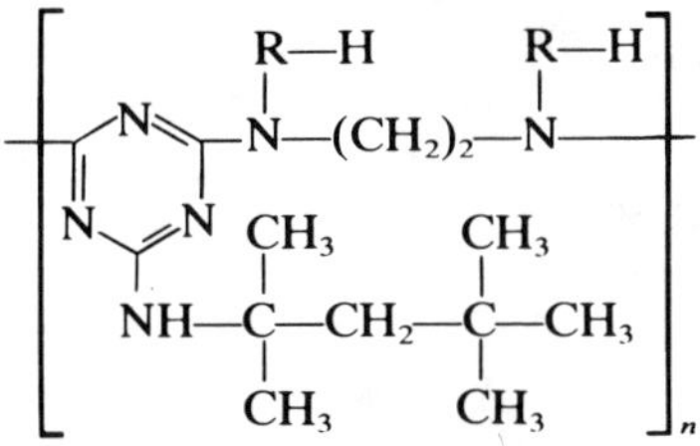

XIXf

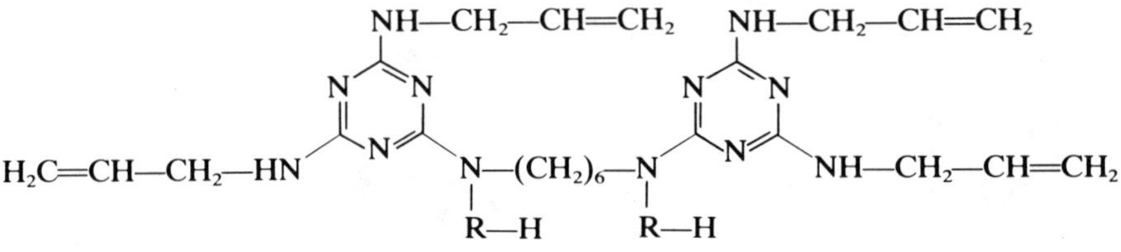

XIXg

XIXh

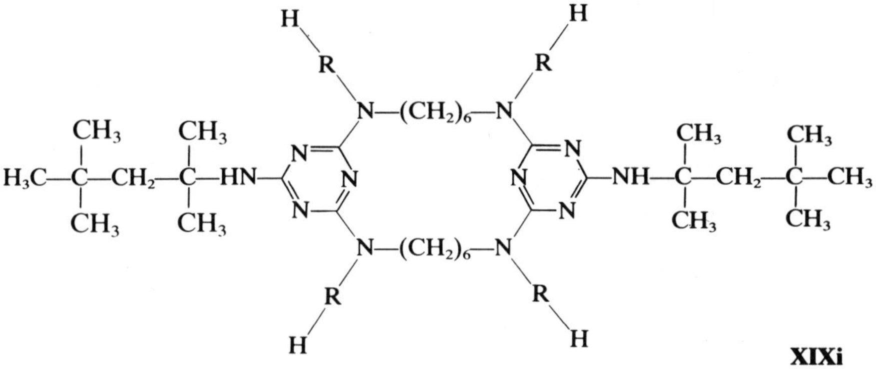

XIXi

XX

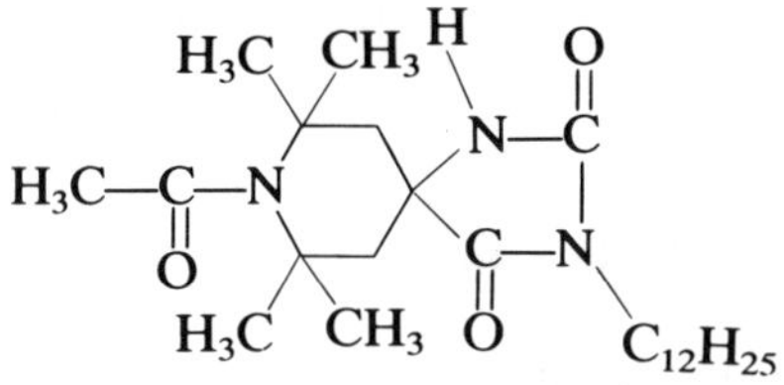

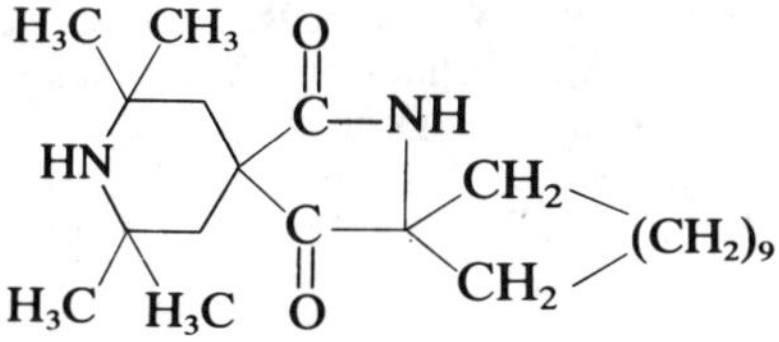

XXa

XXIa R = CH$_3$
XXIb R = CH$_2$CHC$_4$H$_9$
                    |
                   C$_2$H$_5$

XXI (C$_6$H$_5$)$_2$C=C—COOR
                      |
                      CN

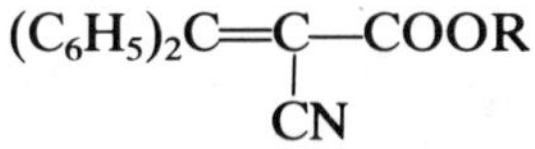

XXII

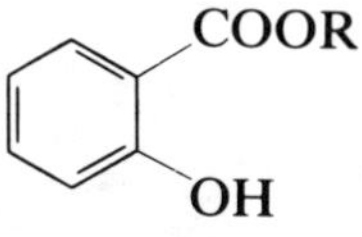

XXIIa R = CH$_3$
XXIIb R = C$_6$H$_5$

XXIII

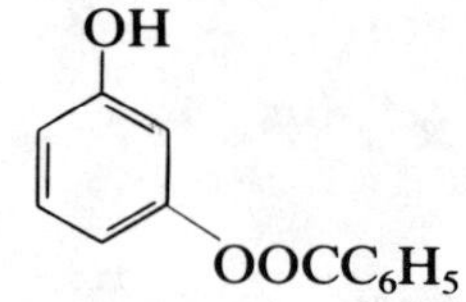

[structure: phenol with OOCC$_6$H$_5$ group]

XXIV

[structure: bis(3,5-di-tert-butyl-2-hydroxyphenyl sulfide) nickel complex with NH$_2$—C$_4$H$_9$]

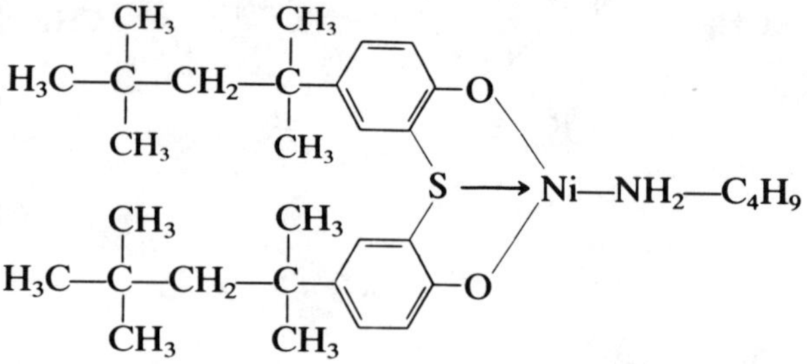

BIBLIOGRAPHY

Freitag, W., *J. Chromatogr.*, **450** (1988) 430.
Freitag, W., Wurster, R. & Mady, D., *J. Chromatogr.*, **450** (1988) 426.
Hirata, Y. & Okamoto, Y., *J. Microcol. Sep.*, **1** (1989) 46.
Rotchova, J. & Pospisil, J., *Plaste und Kautschuk*, **36** (1989) 289.
Bartle, K. D. & Kithinji, J. P., *Analyst*, **115** (1990) 125.
Arpino, P. J., Dilettato, D., Khoa Nguyen & Bruchet, A., *J. High Resolut. Chromatogr.*, **13** (1990) 5.

Index

Accelerated testing
  PPO blends, 156–8
  see also CRD...; HPUV tester
Acrylic monomers, determination of additives in, 273, 292
AgeRite Resin D, 271–2
AgeRite White, 271, 272
Agidol-123, 273
Alkoxyl radicals, reactions with phosphites, 49–54
Alkyl phosphites, antioxidant action of, 36
Alkylperoxyl radicals, reaction of dithiophosphates with, 65–7
Aluminium, greenhouse covers backed with, 178, 179
Analytical methods, 211–311
  purpose of, 217–18
Armostat additives, 230
Aromatic phosphites, higher-temperature antioxidant action of, 55
Arrhenius plot, photoyellowing of PPO blends, 154
Aryl phosphites
  autoxidation inhibited by, 39, 40, 43

Aryl phosphites—*contd.*
  chain-breaking antioxidant action of, 26
  high-temperature inhibition by, 56
  hydroperoxide-decomposing antioxidant action of, 27
  reaction with peroxyl radicals, 48
Aryl phosphonites, reaction with peroxyl radicals, 45, 48
Ascorbic acid (vitamin C), 275
Atlas HPUV tester, 136–7, 157
  see also HPUV tester
Atmul 83, 230
Attenuated total reflectance (ATR-IR) spectroscopy
  chromatography detection using, 224
  PPO photo-oxidation studied by, 141–3
Autoxidation
  chain length calculated, 18, 19
  change from non-steady to quasi-steady condition, 17
  high-temperature inhibition by phosphites, 54–7
  kinetics of inhibition, 16–20

Autoxidation—*contd.*
  low-temperature inhibition by
    phosphites, 37–45
  non-steady conditions, 16–17, 20
  quasi-steady conditions, 17–19
  radical chain mechanism of, 26
Aviation fuel, determination of
    antioxidants in, 273–4

Benzophenone
  polymer UV stability affected by,
    101
  polymers affected by, 120, 124
Benzophenone derivatives
  chromatographic separation of,
    249, 250
  extraction of, 231
  polymer UV stability affected by,
    161, 162, 175–8
  see also Chimassorb 81; Cyasorb
    UV-531; Uvinul M-410
Benzotriazole derivatives
  chromatographic separation of,
    249, 250, 264
  ketone photolysis in LDPE films
    affected by, 187–9
  polymer UV stability affected by,
    161
  synergism with HALS, 189–90
  see also Tinuvin 320; . . . 326;
    . . . 327; . . . 328; . . . P
BHT
  autoxidation inhibited by, 43, 44
  chemical name for, 214
  chromatographic separation of,
    246, 251, 253, 258, 261, 267,
    274, 283, 285, 292
  extraction of, 231, 232, 233, 234
  structure of, 210, 306
  volatility of, 238
4,4′-Bis(2,6-di-*tert*-butylphenol),
    chromatographic separation
    of, 283
Bonded-phase chromatography
    (BPC), 226–7
  functional groups used, 244

Butadiene–acrylonitrile (BAN)
  copolymers, determination
    of stabilisers in, 285–6
tert-Butylhydroquinone (TBHQ),
    structure of, 275
tert-Butyl-4-hydroxyanisole (BHA)
  chromatographic separation of,
    274–6, 285
  structure of, 275
tert-Butyloxyl radicals, reactions with
    phosphites, 49–54

CAO-5 antioxidant, chromatographic
    separation of, 291
Carbon blacks, irradiation-stability of
    polypropylene affected by,
    120–1
Carbon dioxide, supercritical, 294,
    295, 296
Carbonyl compounds
  formation during photo-oxidation
    of HDPE, 191, 192
  photolysis of, 172–3
  quenching of, 184–90
    benzotriazole UV absorbers
      used, 187–9
    hindered-amine light
      stabilis(z)ers used, 184–6
    synergism of benzotriazole and
      HALS UV absorbers,
      189–90
Chain-breaking antioxidants
  dithiophosphates as, 65, 97
  effectiveness of, 37
  phosphites as, 35–54
Chain-length calculation, inhibited
    autoxidation of
    hydrocarbons, 18, 19
Chimassorb 80, 210, 214
  chromatographic separation of,
    258, 283, 297
  extraction of, 232
Chimassorb 944, 209, 214
  chromatographic separation of,
    281–3
  extraction of, 235, 239

Chimassorb development products, 289
Chimassorb N705, 210, 214
Chromatographic analysis, 220–1
  steps in, 221
  techniques listed, 221
Column chromatography, 223–4
  amount of elution solvent required, 224
  sample required, 223
Competitors' products, analysis of, 218
CRD tester, 157
  PPO blends studied by, 158, 159, 160, 162, 164
Critical antioxidant concentration, definition of, 41
Crodinstab additives, 231
Cumene, oxidation inhibited by phosphites, 37–8, 40, 42–3, 44
Cumene hydroperoxide
  catalytic decomposition of
    by cyclic phosphites, 34
    by dithiophosphates, 67–8, 83, 89–91
  decomposition of
    by nickel dithiophosphates, 89–92
    by zinc dithiophosphates, 81–4
  ionic decomposition in presence of dithiophosphates, 68, 74, 76, 79, 89, 90
  radical side reactions in decomposition by phosphites, 35
  stoichiometric reaction of phosphites with, 28–30, 31
9-Cyanoanthracene, 145
Cyasorb UV-9, 214, 231, 233, 249, 260
Cyasorb UV-24, 214, 249, 260
Cyasorb UV-531, 99, 100, 101, 214
  chromatographic separation of, 246, 249, 253, 260, 261, 267
  extraction of, 230, 231, 232, 233
Cyasorb UV-1084, 101, 214, 249, 260, 283

Cyasorb UV-5441, 214, 267
Cyclic phosphites, catalytic decomposition of hydroperoxides by, 30–4

Dashboard films
  chromatographic separation of additives in, 264
  extraction of, 232
Decalin, oxidation in presence of
  iron dithiophosphate, 95
  thiophosphoryl disulphide, 68
  zinc dithiophosphates, 82
Dektac device, 164
Dialkyl arylphosphonites, reaction with peroxyl radicals, 45
Dialkyl dithiophosphoric acids, 61
  binding to polymers, 102–4
  as catalyst in decomposition of hydroperoxides, 74, 76
  preparation of, 63
Diaryl hydrogen phosphites, high-temperature inhibition by, 56
Diazobicyclo[2.2.2]octane (DABCO), PPO photo-oxidation affected by, 145, 146, 150
2,6-Di-*tert*-butyl-4-ethylphenol, chromatographic separation of, 292
2,6-Di-*tert*-butyl-4-methylphenol
  chromatographic separation of, 274
  inhibition by, 39, 43
  see also BHT
2,6-Di-*tert*-butyl-4-methylphenyl neopentylene phosphite, oxidation of cumene inhibited by, 37–8
2,6-Di-*tert*-butyl-4-methylphenyl *o*-phenylene phosphite
  decomposition of cumyl hydroperoxide by, 31
  oxidation of tetralin inhibited by, 38
Di-*tert*-butyl peroxalate (DTBPO), reaction of phosphites with, 50–1

9,10-Dicyanoanthracene, 145
Di-2-(ethylhexyl) phthalate, 233, 250
Dilauryl thiodipropionate (DLTDP),
    214
  chromatographic separation of,
    253, 258, 261, 262, 283, 297
  extraction of, 230, 231, 233
2,4-Dimethyl-6-*tert*-butylphenol,
  chromatographic separation
    of, 274
5,5-Dimethyl-1-pyrrolidine-*N*-oxide,
  superoxide trapped by, 145,
    146
Dimyristyl thiodipropionate
  (DMTDP), 214
Dioctyl phosphite, chromatographic
  separation of, 283
Distearyl thiodipropionate (DSTDP),
    214
  chromatographic separation of,
    253, 262, 267, 269, 282, 283,
    297
  extraction of, 230, 231, 233
Dithiophosphates
  antioxidant action of, 69–79
  as antioxidants for polymers,
    97–104
  chain-breaking action of, 65
  iron complex, 91, 95–7
  metal complexes
    antioxidant action of, 79–97
    general structure, 62
    reactions with hydroperoxyl
      radicals, 68–9
    reactions with peroxyl radicals,
      65–6
  nickel complexes, 62, 88–92, 93
  as peroxidolytic agents, 67–9
  as radical trapping agents, 65–7
  reactions with alkylperoxyl
    radicals, 65–6
  reactions and characteristics of,
    63–5
  synergistic effects, 62
  zinc complexes, 65–6, 79–88
Dithiophosphoric acids
  amides, 61–2
  ammonium salt, 102, 103

Dithiophosphoric acids—*contd.*
  binding to polymers, 102–4
  as catalyst in decomposition of
    hydroperoxides, 74, 76
  characteristics, 63, 65
  decalin oxidation affected by, 74
  kinetics of reaction with cumeme
    hydroperoxide, 75
  oxidation products on reaction with
    cumeme hydroperoxide, 75
  reactions, 64
Domains of realisation concept, 9–12
  boundary strips for, 10

Eastman RMB, 249
Electrochemical detectors, HPLC,
    263, 273, 276
Electron-beam irradiation, polymers
  affected by, 116
Electron scavenging, radiation-
  induced polymer
  degradation reduced by,
    118–19
Electron transfer mechanism, PPO
  resins, 144–7, 148–9
Electron transfer quenching
  mechanism, 146
Energy transfer techniques,
  radiation-induced polymer
  degradation reduced by, 119
Epoxidised soya-bean oil, 233, 250
Epoxy compounds, extraction of, 232
Erucamide
  chromatographic separation of,
    253, 261, 283, 297
  extraction of, 230, 231
ESR spectroscopy, radical formation
  in PPO studied by, 138, 139
Ester-type antioxidants: *see* DLTDP;
  DMTDP; DSTDP
Ethanox 330, 214
  chromatographic separation of,
    251, 253, 267, 289
  extraction of, 233
Ethanox 702, 214, 258, 285
Ethanox 754, 214, 253, 258

6-Ethoxy-1,2-dihydro-2,2,4-
  trimethylquinoline
  (ETMDQ), 273
Ethylene phosphites, decomposition
  of hydroperoxides by, 33
Ethylene–propylene rubber, gamma-
  irradiation effects, 115, 120
EVA copolymer, UV stability of,
  181, 183, 184
Extraction methods
  compatibility with chromatographic
    systems, 239–41
  liquid–liquid extraction, 233–4,
    238–9
  requirements, 228–9
  solid–liquid extraction, 229–33,
    235–8
  permeability of polyolefins to
    extraction solvent, 235–8
  solubility of additive in
    extraction solvents, 229, 235
  temperature/time effects, 238

Fats and oils, antioxidants in, 274–6,
  285
Field desorption mass spectroscopy
  (FDMS), 272
Fillers, UV stability of LDPE films
  affected by, 181
Flame ionisation detectors (FID),
  295–6
Fluorescence detector, HPLC, 263
Fluorescent lamp light, spectrum of,
  136, 137
Food antioxidants, chromatographic
  separation of, 263, 274–6,
  285, 292
Förster mechanism, 189
Fourier transform infrared (FT-IR)
  spectroscopic detector, 298
Free-radical mechanism, PPO resins,
  141–4

Gallates
  chromatographic separation of, 274
  structure of, 275

Gamma-irradiation, polymers
  affected by, 114, 115, 117
Gas chromatography (GC), 222
  compared with liquid
    chromatography, 222–3
  HALS reactions monitored by,
    202, 203
  high-resolution capillary, 287–8
  molecular weight limit for, 24
  techniques listed, 221
Gel permeation chromatography
  (GPC), 227, 277
  see also Size exclusion
    chromatography
General Electric CRD tester, 157
  see also CRD tester
Goodrite 3114, 214
  chromatographic separation of,
    251, 253, 254, 258, 269, 270
  extraction of, 230, 233
Goodrite 3125, 214
Gradient-elution separation, 227, 299
  NP-HPLC, 250–1
  RP-HPLC, 259–62
Greenhouse film covers
  aluminium backing for, 178, 179
  factors affecting lifetime of, 178
  UV stability of, 177–80
Group A phenols, 5, 9
  correlation equations for, 6, 7
  equations for domain boundary
    strips, 6, 7
  equations for oxidation rate, 13
  temperature limit of inhibiting
    action of, 15
Group B phenols, 5, 9
  correlation equations for, 6, 7
  equations for domain boundary
    strips, 6, 7
Group C phenols, 5, 9
  correlation equations for, 7
Guard columns, 236

HALS-1. See Tinuvin 770
HALS. See Hindered amine light
  stabilis(z)ers
Hatcol 200 plasticiser, analysis of, 285

High-density polyethylene (HDPE)
  extraction of, 231
  gamma-irradiation effects, 115
  photo-oxidation of, chemical
    changes, 190–202
  UV stabilisation of, 175–7, 184
High-energy irradiation
  polymers subjected to
    applications, 110
    inhibition of degradation, 118–30
    primary reactions, 110–12
    secondary reactions, 112–17
    types, 110
High-impact polystyrene (HIPS)
  blended with PPO resin, 136
  accelerated testing of, 156–8
  photostabilisation of, 158–65
  photoyellowing of, 138, 152–5
  photodegradation of, 137
High-performance liquid
    chromatography (HPLC),
    221, 225–7, 242–76
  compared with HSLC, 290, 292
  detection and identification used,
    262–8
  limits of detection, 268
  miscellaneous analysis of
    stabilisers, 268–76
  normal-phase, 227, 243–51
    gradient-elution separation,
      250–1
    isocratic separation, 245–50
  pressure in column, 226
  recent developments, 287–298,
    229–300
  reversed-phase, 227, 251–62
    gradient-elution separation,
      259–62
    isocratic separation, 252–9
    standard samples, 246, 253, 258,
      263
  types of, 226–7
High-performance thin-layer
    chromatography (HP-TLC),
    225
High-resolution capillary gas
    chromatography (HR-
    CGC), 287–98

High-resolution capillary supercritical
    fluid chromatography, 294–8
  solvents used, 295
High-speed liquid chromatography
    (HSLC), 288–92
  advantages of, 291
  applications in quality assurance,
    291
  compared with HPLC, 290, 292
Hindered-alkyl phosphites,
  autoxidation inhibited by,
    37–8
Hindered-amine light stabilis(z)ers
    (HALS)
  chromatographic separation of,
    247–8
  extraction of, 235
  ketone photolysis in LDPE films
    affected by, 185–6
  polyethylene affected by, 175–83,
    184
  PPO blends affected by, 159–61
  reactions during photo-oxidation of
    HDPE, 202–6
  kinetics, 205
  see also Chimassorb 944; Hostavin
    TMN-20/VPN-20; Tinuvin
    144; . . . 622; . . . 765;
    . . . 770
Hindered-aryl phenylphosphites,
  reactions with tert-butyloxyl
    radicals, 51, 53
Hindered-aryl phosphites
  autoxidation inhibited by, 39, 40,
    43
  chain-breaking antioxidant action
    of, 36
  hydrolysis of, 56
  reaction with peroxyl radicals, 48–9
Hindered cyclic arylene phosphites,
  reaction with peroxyl
    radicals, 49
Hindered-phenol antioxidants
  chromatographic separation of,
    264
  extraction of, 240
  polypropylene stained by
    irradiation of, 123

Hindered-phenol antioxidants—
  contd.
  see also . . . NKF; BHT; Ethanox
    702; . . . 754; Goodrite 3114;
    Irganox 245; . . . 1010;
    . . . 1024MD; . . . 1035;
    . . . 1076; Topanol CA;
    Vulkanox BKF
Hostavin TMN-20, 247, 248, 262
Hostavin VPN-20, 214, 230
HPUV tester, 136–7, 157
  PPO blends studied by, 138, 159,
    160, 162, 163, 164
Hydrocarbon mixtures, kinetics of
  oxidation, 21
Hydrocarbons
  inhibited autoxidation of
    induction period calculated, 18,
      19
    kinetics of, 16–20
  inhibited oxidation of
    by dithiophosphates, 61
    key reactions, 4–5, 7, 8
    mechanisms, 3–5
Hydroperoxide formation, inhibited
  autoxidation of
  hydrocarbons, 19–20
Hydroperoxides
  catalytic decomposition of, 68
    by phosphites, 30–4
  decomposition of
    by dithiocarbamates, 68
    by phosphites, 27–35
      catalytic decomposition
        reactions, 30–4
      higher-temperature reactions,
        54–7
      radical side reactions, 35
      stoichiometric reactions, 27–30
    by phosphonites, 30, 33
    by thiophosphoryl disulphide, 71
  photolysis of, 173–4
  stoichiometric reactions of
    with phosphites, 27–30
    with sulphur dioxide, 68
  two-stage decomposition process
    for, 70
Hydroquinone, determination of, 273

Hydroxybenzophenone derivatives
  polymer UV stability affected by,
    101, 161, 162
  PPO photoyellowing affected by,
    161, 162
  radicals scavenged by, 124, 126
Hydroxybenzotriazole derivatives,
  polymer UV stability
  affected by, 161
Hydroxylamines, radicals scavenged
  by, 124
2-Hydroxy-4-octyloxybenzophenone
  (HOBP)
  polymer UV stability affected by,
    101
  PPO photoyellowing affected by,
    161–2

Induction period
  definition of, 41
  inhibited autoxidation of
    hydrocarbons, 18, 19
  phosphites compared with BHT, 44
Infrared (IR) spectroscopy, 219
  detectors used in HPLC, 264, 266
  HALS reactions monitored by,
    202, 203, 204
  photo-oxidation of polyethylene
    monitored by, 190, 191
Inhibited oxidation
  hydrocarbons
    key reactions, 4–5, 7, 8
    mechanisms, 3–5
    parametric equations for rate of,
      12–13
Inhibiting effectiveness, factors
  affecting, 13
Inhibiting period, 13
Inhibition temperatures, theoretical
  approach to, 14–16
Ion-exchange chromatography (IEC),
  221, 227
Ion-pair chromatography (IPC), 221,
  227
Ion scavenging, radiation-induced
  polymer degradation
  reduced by, 118–19
Ion suppression technique, 259

Ionol, 231
Irgafos 166, 214
  chromatographic separation of, 258, 297, 298
  extraction of, 230, 231, 232
Irgafos TNPP, 214, 231, 233, 253
Irgafos TPP, 214, 283
Irganox 243, 214
  chromatographic separation of, 258, 283–5, 289, 297
  extraction of, 234
Irganox 259, 214
  chromatographic separation of, 264, 289, 291
  extraction of, 230, 232
Irganox 415, 215
Irganox 565, 215
Irganox 858, 283
Irganox 1010, 158, 215
  chromatographic separation of, 251, 253, 258, 261, 267, 281, 283, 289, 291, 292, 297
  extraction of, 230, 231, 232, 233, 235
Irganox 1019, 215, 289
Irganox 1024MD, 215, 258, 297
Irganox 1035, 215
  chromatographic separation of, 258, 282, 291, 297
  extraction of, 230, 231
Irganox 1076, 101, 210, 215
  chromatographic separation of, 251, 258, 261, 264, 267, 270–1, 281, 289, 292, 297
  extraction of, 230, 231, 232, 233
Irganox 1081, 215
Irganox 1098, 215, 289
Irganox 1330, 215
  chromatographic separation of, 297
  extraction of, 232, 239, 240
Irganox 3114, 215, 289, 297
Irganox 3125, 215, 289
Irganox PS-800, 215
Irganox PS-801, 215
Irganox PS-802, 215
Irgaphos development products, 289
Iron di-isobutyl dithiophosphate, 92, 95

Iron di-isobutyl dithiophosphate—*contd.*
  oxidation of cumene hydroperoxide affected by, 95–7
  oxidation of decalin affected by, 95
Iron dithiophosphate, 91, 95–7
Irradiation
  polymers affected by, 109–31
  see also Radiation...
Isocratic separation, 227, 299
  NP-HPLC, 245–50
  RP-HPLC, 252–59

Kemamide E, chromatographic separation of, 292
Ketone photolysis, 172–3
  effect of HALS, 186, 187
  quenching of, 184–90
Kharasch-type addition, 102
Kinetic topology
  inhibited oxidation of hydrocarbons, 7–12
  meaning of term, 2
Kumagawa extraction, 229, 230

Linear low-density polyethylene (LLDPE), UV stabilization of, 183, 184
Liquid chromatography (LC)
  advantages of, 222–3
  classical techniques, 223–5
  compared with gas chromatography, 222–3
  extraction methods used, 228–39
    compatibility with chromatographic system, 239–41
    liquid–liquid extraction, 233–4, 338–9
    solid–liquid extraction, 229–33, 235–38
  high-speed, 288–92
  microcolumn liquid chromatography, 292–4
  modern techniques, 225–8
  reason for using, 217–19

Liquid chromatography—contd.
  sample preparation for, 237–8,
    241–2
  sampling for, 228–41
  techniques listed, 221
Liquid–liquid chromatography
  (LLC), 221, 226
Liquid–liquid extraction procedures,
  233–4, 238–9
Liquid–solid chromatography (LSC),
  221, 226
Liquid–solid extraction procedures,
  229–3, 235–8
Low-density polyethylene (LDPE)
  extraction of, 231, 232
  gamma-irradiation effects, 115
  nickel dithiophosphates affecting
    photo-oxidative stability,
    100
  UV stabilization of, 177–82
    combined stabilis(z)ers used, 180
    comparison of various films, 184
    effect of fillers, 181
    film effects, 178–9
Lubricating oil, determination of
  additives in, 285

Mass spectrometric (MS) detectors,
  266–8, 273, 294
Melt-mixing procedure, 242
Metal dithiophosphates
  antioxidant action of, 78–97
  reaction with
    hydroperoxyl radicals, 66–7
    peroxyl radicals, 63–4
  UV stability of, 98–9
Methyl ether hydroquinone,
  determination of, 273
Methylene Blue, 145
4,4′-Methylenebis(2,6-di-*tert*-
  butylphenol), chromato-
  graphic separation of, 283
Microcolumn liquid chromatography,
  292–4
Mobilizer oil, radicals scavenging
  affected by, 125, 126, 127,
  128, 129

1-Monostearin, chromatographic
  separation of, 283

Natural rubber, binding of
  dithiophosphates to, 102,
  103
Nickel-based UV stabilis(z)ers
  EVA copolymers affected by, 181
  polyethylene affected by, 178
  see also Chimassorb N705
Nickel dibutyl dithiophosphate,
  antioxidant mechanisms of,
  92–3
Nickel dithiophosphates, 62, 88–92,
  93
  cumene hydroperoxide
    decomposed by, 89–91
  photoantioxidant efficiency of, 99
  tert-butyl hydroperoxide
    decomposed by, 92, 94
Nitroxyl radical, photolysis of, 205
Nonox WSP, 215
Nordihydroguaiaretic acid (NDGA)
  chromatographic separation of,
    274
  structure of, 275
Normal-phase (NP) HPLC, 227,
  243–51
  gradient-elution separation, 250–1
  isocratic separation, 245–50
  mobile phase used, 244
Norrish type I reaction, 172, 188
  quenching of, 188–9, 207
Norrish type II reaction, 172, 173
  quenching of, 189
  vinyl groups formed during, 196
Noryl resins, 136
  applications, 136
  photoyellowing of, 159
  see also PPO resins
Nuclear magnetic resonance (NMR)
  spectroscopy
  HPLC detection using, 272, 273
  PPO photo-oxidation studied by,
    141
  see also ^{31}P-NMR spectroscopy

Oils and fats, antioxidants in, 274–6, 285
Oleamide, chromatographic separation of, 253, 261, 283, 292
Optimisation of antioxidant action, theoretical approach to, 1–21
Optimum inhibitor, 13–14
  definition of, 19
Oxygen, irradiated polymers affected by, 113–14
Ozone, irradiated polymers affected by, 130

Paper chromatography, 221, 224
Parametric equations, rate of inhibited oxidation, 12–13
Pellicular (liquid chromatography) packings, 243
Peroxyl radicals
  generation of, 45
  reaction with
    dithiophosphates, 65–7
    hydroxylamine ethers, 198
    phosphites, 45–9
  scavenging of, 121
Phenolic antioxidants
  chromatographic separation of, 263, 264, 273
  classification of, 5, 9
  extraction of, 231, 240
  see also Hindered-phenol antioxidants
Phenol-inhibited oxidation
  classification of phenols, 5, 9
  correlation equations for activation energy of key reactions, 6
  correlation equations for rate constants of key reactions, 7
  domains of realisation, 9–12
  range of operating conditions for, 8
  temperature limits, 15
Phenothiazine, 120
Phenyl phosphites, autoxidation inhibited by, 39

o-Phenylene phosphites
  decomposition of hydroperoxides by, 31–3
  high-temperature inhibition by, 57
Phosphite esters. *See* Phosphites
Phosphites
  autoxidation inhibited by, 37–45
  chain-breaking antioxidant action of, 35–54
  higher-temperature antioxidant action of, 54–7
  hydroperoxide-decomposing antioxidant action of, 27–35
    catalytic decomposition reactions, 30–4
    radical side reactions, 35
    stoichiometric reactions, 27–30
  induction period for autoxidation, concentration effects, 41, 44
  reactions with
    alkoxyl radicals, 49–54
    alkylperoxyl radicals, 45–9
    tert-butyloxyl radicals, 49, 52, 53–4
  stoichiometric factor for, 41
Phosphonite esters. *See* Phosphonites
Phosphonites
  hydroperoxide-decomposing antioxidant action of, 30, 33
  reactions with
    alkoxyl radicals, 53
    alkylperoxyl radicals, 45, 46–7
Phosphorus-based antioxidants, 23–104
Photo-oxidation
  polyethylene, mechanisms, 172–4
  PPO resins
    electron transfer mechanism, 144–7, 148–9
    free-radical mechanism considered, 141–4
    mechanisms, 138–51
Photostabilization
  PPO blends, 158–65
    hindered amines used, 159–61
    UV screening agents used, 161–5

Photoyellowing
  PPO blends, 138
  light intensity effects, 155–6
  temperature effects, 154
  wavelength effects, 152–4
Pigments, polyethylene UV stability affected by, 177
Piperidines, polypropylene stabilized by, 122, 123–4
Planar chromatographic techniques, 224–5
  see also Paper. . . ; Thin-layer chromatography
Plasticization effects
  irradiated polymers, 126–7
  PPO blends, 161, 164
^{31}P-NMR spectroscopy
  nickel dithiophosphate studied by, 92, 94
  thionophosphoric acids studied by, 76, 77
  thiophosphoryl disulphide studied by, 71, 72, 97
  zinc dithiophosphates studied by, 82, 83–6
Polphrino, chromatographic separation of, 289
Polyamide, gamma-irradiation effects, 115
Polyani–Semenov equations, 6
Poly(2,6-dimethyl-1,4-phenylene oxide). See PPO resins
Polyester, electron-beam irradiation effects, 116
Poly(ether imide), electron-beam irradiation effects, 116
Poly(ethyl ethyl ketone), electron-beam irradiation effects, 116
Polyethylene
  determination of additives in, 282
  extraction of additives, 230–3
  gamma-irradiation effects, 115, 116, 117
  oxidation inhibited by phosphites, 55
  photo-oxidation of, chemical changes, 190–202

Polyethylene—contd.
  UV stabilization of, 175–83
  HALS used, 176–83, 190–206
  see also High-density. . . ; Low-density. . .
Poly(ethylene terephthalate), gamma-irradiation effects, 115
Polyimide, electron-beam irradiation effects, 116
Polymer hydroperoxides, decomposition by phosphites, 27, 30
Polymers
  additives in, radiation effects on, 130
  diffusion of reagents in, 21
  dithiophosphates as antioxidants for, 97–104
  irradiation effects, 109–31
  oxidation of, rate constants for, 21
  phosphite-inhibited oxidations of, 45
  radiation-induced reactions in
    additives affected, 130
    effect of oxygen, 113–14
    electron/ion scavenging to prevent, 118–19
    energy transfer techniques to prevent, 119
    inhibition of degradation, 118–30
    new stabilizer system to prevent, 127–9
    ozone effects, 130
    physical properties affected, 114–17
    plasticization effects, 126–7
    primary reactions, 110–12
    radical scavenging to prevent, 120–6
    secondary reactions, 112–17
    stabilization mechanisms summarized, 131
    solubility of additives in, 101–3
Polyolefins
  extraction of additives, 231, 235–8

Polyolefins—*contd.*
  radiation stabilis(z)ers, mechanisms summarized, 131
Polyoxymethylene, gamma-irradiation effects, 115
Poly(1,4-phenylene isophthalamide), electron-beam irradiation effects, 116
Poly(phenylene oxide), electron-beam irradiation effects, 116
Polypropylene
  chromatographic analysis of antioxidants in, 280
  effects of stabilis(z)ers on post-irradiation oxidation, 122, 123, 125
  extraction of additives, 230, 232–3, 235
  gamma-irradiation effects, 114, 120, 122, 123–9
  oxidation inhibited by phosphites, 55
  photo-oxidative stability of, 98, 99, 100
Polystyrene, electron-beam irradiation effects, 116
Polysulphone, electron-beam irradiation effects, 116
Polytetrafluoroethylene
  gamma-irradiation effects, 115
  irradiation-stability affected by carbon blacks, 120–1
Polyurethane
  determination of stabilisers in, 284–5
  extraction of, 234
Poly(vinyl chloride)
  chromatographic analysis of additives in, 280, 285
  extraction of additives, 231, 233, 234
  gamma-irradiation effects, 115
PPO resins
  absorption spectrum of, 143
  blended with HIPS, 136
  accelerated testing of, 156–8
  photostabilisation of, 158–65
  photoyellowing of, 138, 152–6

PPO resins—*contd.*
  direct photolytic cleavage mechanism for, 140, 143–4
  mechanism for photo-oxidation, 138–51
  electron transfer mechanism, 144–7
  free-radical mechanism discounted, 141–4
  stabilization implications, 147–51
  photodegradation of, 137–8
  photostabilization of, 158–65
  hindered amines used, 159–61
  UV screeners used, 161–5
  photoyellowing of, 138
  environmental factors, 152–6
  light intensity effects, 155–6
  light wavelength effects, 152–4
  temperature effects, 154

Quality control analysis, 218

Radiation-induced polymer reactions
  dose–rate effects, 115–16
  inhibition of degradation, 118–30
  primary reactions, 110–12
  secondary reactions, 112–17
  timescales involved, 111
Radical scavenging, radiation-induced polymer degradation reduced by, 120–6
Range-of-operating-conditions concept, 8
Reactive antioxidants, 102–4
Refractive index (RI) detectors, 262, 263, 285
Resorcinol monobenzoate (RMB), 214, 249, 260
Reversed-phase (RP) HPLC, 227, 251–62
  gradient-elution separation, 259–62
  isocratic separation, 254–61, 252–9
Rose Bengal, 145
Rubber. *See* Natural rubber

INDEX

Salol (stabiliser), 215, 249, 260
Sample preparation, liquid
 chromatography, 241–2
Sampling, liquid chromatography,
 228–41
Sandovur EPU, 215, 232
Santonox, 215, 230, 273, 283
Santonox R, 215
 chromatographic separation of,
  251, 258, 264, 265, 282, 291
 extraction of, 231, 233
Santowhite Powder, 215, 233, 267
Sicostab AO-3, 274
Size exclusion chromatography
  (SEC), 221, 227–8, 277–86
 analysis of additive extracts,
  278–83
 applications, 278–300
 direct determination of stablisers in
  polymers, 283–6
 packings used, 277
Solid–liquid extraction procedures,
  229–33, 235–8
 boiling solvents used, 232–3, 238
 factors affecting efficiency, 237
 permeability of polyolefins to
  extraction solvent, 235–8
 solubility of additive in extraction
  solvents, 229, 235
 solvents below boiling point used,
  233
 Soxhlet and related extractors
  used, 229, 230–2
Soxhlet extraction, 229, 230–2
Spectrodensiometric, technique, 225
Spectroscopic analysis, 219–20
 direct methods, 219–20
 indirect methods, 220
Stearamide
 chromatographic separation of,
  253, 261, 283
 extraction of, 231
Sterically hindered. . . . *See*
 Hindered. . .
Stoichiometric inhibition factor
 definition of, 41
 dithiophosphates, 67
 phosphites compared with BHT, 44

Styragel (chromatography) columns,
 279, 280, 281, 286
Sulphur dioxide, reactions with
 hydroperoxides, 68
Sunkem MS, 215, 249, 260
Sunlight, wavelength distribution of,
 136, 137
Supercritical fluid chromatography
  (SFC), 221, 294
 advantages of, 294–5
 high-resolution capillary, 294–8
Supercritical solvents, 295
Superoxide
 evidence for, 145, 148
 formation in PPO resins, 144
Synergism, HALS/benzotriazole-type
 UV absorbers, 189–90

TBH, decomposition by nickel
 dithiophosphates, 92, 94
TBHQ. *See tert*-Butylhydroquinone
Tetralin, oxidation inhibited by
 phosphites, 38–9, 42–3, 44,
 57
Thin-layer chromatography (TLC),
 221, 224–5
 applications, 225
 detection in HPLC, 266
Thionophosphoric acids, 73, 76, 77
 antioxidant action of, 76–7
Thiophosphoryl disulphides, 69–73
 applications, 61
 binding to polymers, 103
 decalin oxidation affected by, 70
 oxidation in presence of
  hydroperoxides, 78
 polymer UV stability affected by,
  100, 101
 structure of, 62
Tinuvin 120, 215, 246, 258, 289
Tinuvin 144, 215
 chromatographic separation of,
  247, 289, 297, 298
 extraction of, 233
Tinuvin 292, 216, 297
Tinuvin 315, 216

Tinuvin 320, 216, 231, 233, 249
  chromatographic separation of,
    249, 258, 260, 289, 297
  extraction of, 231, 233
Tinuvin 326, 210, 216
  chromatographic separation of,
    246, 249, 258, 260, 289, 297
  extraction of, 232
Tinuvin 327, 216
  chromatographic separation of,
    246, 249, 258, 260, 289
  extraction of, 232
Tinuvin 328, 210, 216, 234, 258,
    283–5, 297
Tinuvin 440, 216, 297
Tinuvin 622, 209, 216
Tinuvin 765, 216
Tinuvin 770
  chemical name for, 216
  chromatographic separation of,
    247, 248, 262, 297
  extraction of, 230, 232
  polymer UV stability affected by,
    101
  PPO photo-oxidation affected by,
    145, 146, 150, 159–63
  reactions during photo-oxidation of
    HDPE, 202–7
  structure of, 209
Tinuvin 900, 216, 289
Tinuvin P, 215
  chromatographic separation of,
    246, 248, 258, 260, 292,
    297
  extraction of, 231, 232
Titanium dioxide, polyethylene UV
    stability affected by, 175–7
α-Tocopherol, 274, 275
Topanol CA, 216
  chromatographic separation of,
    251, 253, 258, 261, 267, 283
  extraction of, 230, 231, 232, 233
Topanol OC, 216, 297
Trialkyl phosphites, reaction with
    peroxyl radicals, 45
Tricresyl phosphate, chromatographic
    separation of, 285

2,4,5-Trihydroxybutyrophenone
    (THBP)
  chromatographic separation of, 274
  structure of, 275
1,2,4-Trimethoxybenzene, PPO
    photo-oxidation affected by,
    145, 146,150
2,2,4-Trimethyl-1,2-dihydroquinoline
    (TMDQ), 271
Triphenyl phosphites, reaction with
    peroxyl radicals, 48
Tris(*tert*-butyl-4-methylphenyl)
    phosphite, high-temperature
    inhibition by, 57
Tris(ethyl hexyl) trimellitate,
    extraction of, 234
Tris(nonyl phenyl) phosphite, 233,
    250, 292

Ultraviolet (UV) detectors, 262–4,
    265, 273, 276, 285
Ultraviolet (UV) screening agents,
    PPO bends stabilized by,
    161–5
Ultraviolet (UV) spectroscopy, 219
Ultraviolet (UV) stability
  metal dithiophosphates, 98–9
  polyethylene, 175–83
Ultraviolet (UV) stabilization
  mechanisms of, 183–206
  polyethylene stabilized with
    HALS, 190–206
  quenching of carbonyl compounds
    considered, 184–90
Uvinul 400, 216, 249, 260
Uvinul M-410, 216, 249, 260
Uvinul N-35, 216, 260
Uvinul N-539, 216
  chromatographic separation of,
    249, 260
  extraction of, 231, 233

Vacuum sublimation, 238

INDEX 329

Vinyl groups
 formation during
  photolysis of carbonyl
   compounds, 172, 196
  photolysis of hydroperoxides,
   174, 197
  photo-oxidation of HDPE, 191,
   192, 193–5
 mechanical properties of HDPE
  related to, 197, 201
trans-Vinylidene groups
 formation during
  catalytic reactions of nitroxyl
   radicals, 199
  photolysis of hydroperoxides,
   173, 197
  photo-oxidation of HDPE, 191,
   193, 195–6, 198
 mechanical properties of HDPE
  related to 203, 198
Vitamin C, 275
Vitamin E, 275
Vulkanox BKF, 216, 258
Vulkanox NKF, 216, 258, 264, 265

Wax removal, 235–7

Weston 618, 216
 chromatographic separation of,
  253, 269
 extraction of, 230, 233

Xenon arc exposure tests,
 disadvantage of, 164–5
Xenotest, polyethylene films exposed
 in, 188, 189

Yellowing, irradiated polypropylene,
 123
Yellowness Index (YI), PPO blends,
 138, 152, 153, 158, 159, 160,
 163

Zinc dithiophosphates, 78–87
 decomposition of cumene
  hydroperoxide by, 78–87
 effect of ligand structure, 63
 reaction with alkylperoxyl radicals,
  65–6

A113 0988011 7

SCI QD 381.9 .D47 M43 1990

Mechanisms of polymer
 degradation and